Energy: Exploration and Conversion

Energy: Exploration and Conversion

Edited by **David McCartney**

Syrawood Publishing House
New York

Published by Syrawood Publishing House,
750 Third Avenue, 9th Floor,
New York, NY 10017, USA
www.syrawoodpublishinghouse.com

Energy: Exploration and Conversion
Edited by David McCartney

International Standard Book Number: 978-1-68286-282-7 (Hardback)

Printed in China.

Contents

Preface

The demand for energy is constantly on the rise. This has led to advanced research in this field. The aim of this book is to present researches that have facilitated energy exploration. This book covers in detail some existent theories and innovative concepts revolving around energy exploration and conversion, such as energy storage, power supply, distributed generation, power transmission, energy generation, etc. With state-of-the-art inputs by acclaimed experts of this field, this book targets students and professionals alike.

Significant researches are present in this book. Intensive efforts have been employed by authors to make this book an outstanding discourse. This book contains the enlightening chapters which have been written on the basis of significant researches done by the experts.

Finally, I would also like to thank all the members involved in this book for being a team and meeting all the deadlines for the submission of their respective works. I would also like to thank my friends and family for being supportive in my efforts.

Editor

1

Application of Satellite-Based Spectrally-Resolved Solar Radiation Data to PV Performance Studies

Ana Maria Gracia Amillo [1], Thomas Huld [1,*], Paraskevi Vourlioti [1], Richard Müller [2] and Matthew Norton [1]

[1] European Commission, Joint Research Centre, Via Fermi 2749, Ispra 21027, Italy;
E-Mails: ana-maria.gracia-amillo@jrc.ec.europa.eu (A.M.G.A.); voula.vrlt@gmail.com (P.V.); matthew.norton@jrc.ec.europa.eu (M.N.)

[2] Deutscher Wetterdienst, Frankfurter Straße 135, Offenbach 63067, Germany;
E-Mail: Richard.Mueller@dwd.de

* Author to whom correspondence should be addressed; E-Mail: thomas.huld@jrc.ec.europa.eu;

Academic Editor: Jean-Michel Nunzi

Abstract: In recent years, satellite-based solar radiation data resolved in spectral bands have become available. This has for the first time made it possible to produce maps of the geographical variation in the solar spectrum. It also makes it possible to estimate the influence of these variations on the performance of photovoltaic (PV) modules. Here, we present a study showing the magnitude of the spectral influence on PV performance over Europe and Africa. The method has been validated using measurements of a CdTe module in Ispra, Italy, showing that the method predicts the spectral influence to within ±2% on a monthly basis and 0.1% over a 19-month period. Application of the method to measured spectral responses of crystalline silicon, CdTe and single-junction amorphous silicon (a-Si) modules shows that the spectral effect is smallest over desert areas for all module types, higher in temperate Europe and highest in tropical Africa, where CdTe modules would be expected to yield +6% and single- junction a-Si modules up to +10% more energy due to spectral effects. In contrast, the effect for crystalline silicon modules is less than ±1% in nearly all of Africa and Southern Europe, rising to +1% or +2% in Northern Europe.

Keywords: photovoltaic performance; spectral response; solar spectrum

---

1. Introduction

It is well known that the energy conversion efficiency of photovoltaic (PV) cells depends on the wavelength of the incoming light [1–4]. This so-called spectral response of the cell depends on the PV technology used. Since the spectrum of sunlight varies over time, the power output of PV cells and modules will depend not only on the total in-plane irradiance, but also on the instantaneous spectrum of the sunlight. As a result, the different spectral responses of PV technologies will cause PV modules to have different instantaneous power output, even if they have the same nominal power [5–8]. If the solar spectrum differs significantly between geographical locations, this may lead to a difference in the overall energy output from PV installations. Knowing the effect of the spectral variations at any point would therefore be useful for PV system planners and investors in order to choose the PV modules best suited for a given location.

Measuring the spectrum of sunlight is much more difficult and expensive than measuring the broadband radiation using pyranometers. Mainly for this reason, the number of sites where the spectrum has been measured consistently is relatively small. This, in turn, means that assessing the influence of the spectral variations on PV energy output at a given site is very difficult.

Even when the task of collecting spectrally resolved data is discounted, the analysis of the collected information still poses difficulties. Adequately characterizing a solar spectrum typically involves obtaining measurements at a resolution of 2 nm over the wavelength range of 300–1700 nm. Instead of one data point per measurement, as would be obtained with a pyranometer, several hundred data points are acquired. Since the variation of any one of these data points alone is not enough to influence the measurable output of a photovoltaic system, it becomes necessary to find some average value or values that will indicate a general shift in the distribution of light over the spectrum.

One such metric that has gained popularity within the photovoltaic community is the "average photon energy" (APE) proposed by Jardine *et al.* in [9]. Similar to an average photon wavelength [10], it represents the average energy of all of the photons impinging upon a target surface and is typically expressed in units of electron-volts. Unlike a spectral mismatch value or the "Z parameter" [11], it is a technology-independent parameter that describes only the spectral quality of solar irradiation. Some authors have argued that an APE value will describe a single spectral distribution of irradiance to within a level of certainty sufficient for PV analyses [12]. Other researchers have presented evidence that this is not the case [8]. With the present results, we hope to add to this discussion.

Recently, methods have been developed to estimate the spectral content of sunlight using data from geostationary satellites [13,14]. These data have uniform coverage over large geographical areas; for instance, the Meteosat class of satellites cover Europe, Africa and nearly all of Asia. The spatial resolution is generally a few kilometers. The satellite data record covers up to 30 years in some regions, and the time resolution is typically hourly or better.

It should therefore be possible to use such satellite retrieved data together with information about the spectral response of PV modules in order to estimate the overall influence of spectral variations on the PV energy output. However, this will of course only be possible if the spectrally-resolved radiation data are sufficiently accurate, so as to give meaningful estimates of the spectral variations in the PV output.

The aim of this paper is to investigate the possibility of using the satellite-based spectral irradiance data to estimate the effect of spectral variations on PV energy output and to quantify the magnitude of this effect over large geographical areas. The structure of the paper is as follows: Section 2 briefly describes the satellite-based spectral irradiance data. We then present the mathematical models used to estimate the PV performance in Section 3. Section 4 contains the validation of the spectrally-resolved irradiance data, both for global horizontal and direct normal irradiance. This section also describes the validation of the model for calculating the influence of spectral variations, using outdoor measured PV performance data. The application of the models to calculate the spectral effect over large geographical regions is then given in Section 5.

2. Satellite-Based Estimates of Spectrally-Resolved Solar Radiation Data

2.1. Retrieval of Spectrally-Resolved Solar Radiation Data from Satellite Images

The SPECMAGIC algorithm calculates spectrally-resolved solar irradiance using the effective cloud albedo [15] retrieved from geostationary meteorological satellites combined with data on the atmospheric content of aerosols, water vapor and ozone. Aerosol information is taken from the Monitoring Atmospheric Composition & Climate (MACC) project. The MACC aerosol data result from a data assimilation system for global reactive gases, aerosols and greenhouse gases. It consists of a forward model for aerosol composition and dynamics [16] and the data assimilation procedure described in detail in [17]. The data are generated by the European Centre for Medium Range Weather Forecast on a Gaussian T159, which is transformed to a $0.5 \times 0.5°$ regular latitude longitude grid. The water vapor profile results from the reanalysis of the European Centre for Medium-Range Weather Forecasts [18,19]. Monthly values have been used at a $0.25 \times 0.25°$ grid. For ozone, climatological standard profiles provided within the radiative transfer model "libRadtran" are used [20].

The algorithm, which is described in Müller *et al.* [13], calculates both global horizontal (GHI) and direct normal irradiance (DNI) in 30 spectral bands in the range from 300 nm–4,600 nm. The spatial resolution of the radiation data is three arc-minutes. Irradiance has been calculated using the Meteosat satellites covering Europe, Africa and nearly all of Asia [21]. The spectral bands used in the algorithm are identical to the 32 bands used by Kato *et al.* [22], except for the first two bands in the ultraviolet. The lower and upper limits of the 32 Kato bands are provided in Table A1 of the Appendix.

The uncertainty of the retrieval algorithm is typically below 1%–2%, which has been estimated by comparison with radiative transfer model based methods; please see [13] for further details. This uncertainty is significantly lower than the experimental error of about 5% of spectro-radiometric measurements [23], but does not include the errors induced by the used atmospheric input, which is likely the main error source for the estimation of solar surface irradiance. The accuracy of the retrieved spectrally-resolved irradiance depends on the specific sites and spectral region and is further discussed in Section 4.1 of this manuscript, as well as in [13]. Typically, the error is close to the measurement error of the ground-based spectro-radiometric measurements.

2.2. Calculation of Inclined-Plane Radiation

Given the global horizontal and direct normal (or horizontal) irradiance, the direct irradiance can be calculated on any arbitrary plane [24]. There exists a number of models for estimating the diffuse irradiance on an inclined plane [25,26]. In this work, we use a modified version of the model developed by Muneer [25]. The modification was necessary because the model uses the clearness index k_c (Equation (1)) as an input parameter:

$$k_c = \frac{G}{G_{ext}} \tag{1}$$

where G and G_{ext} are, respectively, the horizontal irradiance at the Earth's surface and at the top of the atmosphere. When using this model for spectrally-resolved irradiance, it becomes necessary to substitute both G and G_{ext} by the corresponding irradiances in each spectral band in order to calculate k_c for that band.

3. Models and Data for PV Performance Studies

The power output of a PV module or array depends on a number of different environmental influences. In this study, we will consider some of these in order to estimate how the output power varies with:

- Increased reflectivity at a shallow angle of incidence;
- The effect of the spectrum of the sunlight;
- The effects of module temperature and irradiance.

Some other effects will not be considered here, such as aging of the modules, local shadows or deposition of dust, dirt or snow on the modules. For some of these effects, the geospatial variation is poorly known, while other effects, such as shadowing, are likely to be extremely local and not suited for studies over large geographical regions.

3.1. Influence of Angle of Incidence on the Transmitted Irradiance

When light strikes the surface of a PV module at a sharp angle of incidence (light rays are nearly parallel to the module surface), there is increased reflection of the light, and the output is correspondingly lower. In the calculations described here, we use the model of Martin and Ruiz [27] to account for this effect. This model assumes that the angular dependence of reflectivity is the same for all wavelengths.

3.2. Effects of Changes in the Light Spectrum

The efficiency with which the energy of the photons in sunlight is converted into electrical energy in the PV cell depends on the photons' wavelength. Some photons have insufficient energy to excite an electron in the PV material, while others have excess energy that cannot be transferred to the electrons. The sensitivity of a given PV cell to different light wavelength is called the spectral response or spectral response curve, denoted $S_r(\lambda)$. The spectral response is a function of wavelength and at each wavelength is proportional to the conversion efficiency of the cell/module when illuminated with monochromatic light of that wavelength. This is not the case for multijunction PV cells, so this description does not apply to these PV technologies.

Some spectral response curves are shown in Figure 1. These spectral response curves were measured at the ESTI Laboratory of the European Commission Joint Research Centre. The methods used to measure the spectral response curves are described in [4,28].

Figure 1. Plots of the spectral response for five different module types: amorphous silicon (single junction), CdTe, $CuInSe_2$ (CIS), crystalline Si and Sanyo HIT (hybrid c-Si/a-Si). The curves are all normalized so that the integral equals one.

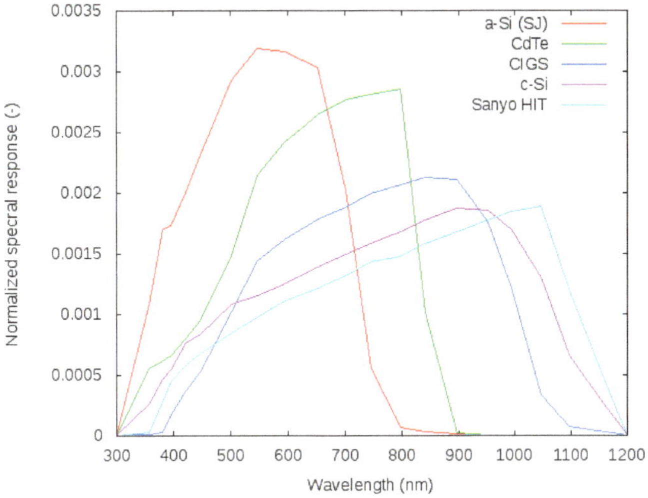

The short-circuit current I_{sc} is very nearly proportional to the number of electrons that are excited in the PV material per second. As shown in Equation (2), for a given wavelength λ, the contribution to I_{sc} is proportional to the spectral response multiplied by the intensity of the light at that wavelength. I_{sc} can then be found by integrating this over the wavelength range where $S_r(\lambda) > 0$:

$$I_{sc} = k \int S_r(\lambda)R(\lambda)d\lambda \tag{2}$$

Here, $R(\lambda)$ is the spectral light intensity and k a proportionality constant specific to the given cell/module and the measured spectral response curve.

The power of a PV module is not proportional to I_{sc}, since it also depends on the module voltage. However, for small changes in the power, the variation in the voltage is small, and the power will be approximately proportional to I_{sc}:

$$\Delta P \propto \Delta I_{sc}$$

This would be the case when the spectrum changes, but the total irradiance stays the same.

Following [6,8], the effect of spectral changes can then be estimated as the spectral mismatch, MM (Equation (3)):

$$MM = \frac{\int S_r(\lambda)R(\lambda)d\lambda}{\int S_r(\lambda)R_{STC}(\lambda)d\lambda} \cdot \frac{G_{STC}}{G_{POA}} = \frac{I_{sc}}{I_{sc,STC}} \tag{3}$$

Here, R_{STC} is the spectrum defined in the standard test conditions (STC), G_{STC} is the total broadband irradiance of the STC spectrum, while G_{POA} is the broadband irradiance in the plane of array (found by integrating $R(\lambda)$). $I_{sc,STC}$ is the short-circuit current under the STC spectrum and irradiance.

The mismatch factor MM can now be used to define an effective irradiance, G_{eff}, from the received in-plane of array irradiance G_{POA} (Equation (4)), which can be used to estimate the output power of the PV module.

$$G_{eff} = MM \cdot G_{POA} \tag{4}$$

3.3. Model for PV Performance

It is well known that the power output of a PV module depends on the received irradiance in a nonlinear way and also on the temperature of the module. The importance of these effects varies between PV technologies.

There exist a number of different models to account for these effects in the literature [29–32]. The model used in this work is described in Huld *et al.* [32]. According to this model, the instantaneous power can be described, as shown in Equation (5), as a function of the effective in-plane of array irradiance G_{eff} and the module's temperature T_{mod}:

$$P(G_{eff}, T_{\text{mod}}) = P_{\text{nom}} \frac{G_{eff}}{G_{STC}} \text{PR}_m(G'_{eff}, T') \tag{5}$$

where G_{STC}=1,000 W/m^2 is the STC irradiance and G'_{eff} and T' are irradiance and temperature values normalized to STC values according to Equation (6):

$$G'_{eff} \equiv G_{eff}/G_{STC}, \qquad T' \equiv T_{\text{mod}} - 25°\text{C} \tag{6}$$

PR_m is the module performance ratio; its dependence on G'_{POA} and T' is shown in Equation (7):

$$\text{PR}_m = 1 + k_1 \ln(G'_{eff}) + k_2 \left(\ln(G'_{eff})\right)^2 + T' \left[k_3 + k_4 \ln(G'_{eff}) + k_5 \left(\ln(G'_{ef})\right)^2\right] + k_6 T'^2 \tag{7}$$

The coefficients k_1 to k_6 must be found by fitting the model to experimental data measured on the module or module type of interest.

PR_m depends on the instantaneous irradiance and module temperature. Under steady-state (or slowly changing) conditions, the T_{mod} can be estimated from the ambient (or air) temperature T_{amb}, wind speed W and the in-plane of array irradiance in the following way (Equation (8)) [33,34]:

$$T_{\text{mod}} = T_{\text{amb}} + \frac{G_{POA}}{u_0 + u_1 W} \tag{8}$$

The coefficients u_0 and u_1 describe how much the PV module is heated by the solar radiation and cooled by the wind. They depend on the way the PV module is mounted (free-standing rack, ventilated or unventilated roof mounting, *etc.*) and also slightly on the module type.

In the applications of this model, in the following, we have used values for u_0 and u_1 taken from Köhl *et al.* [34].

3.4. Sources of PV Module Data

The models for PV power estimation given in Sections 3.1–3.3 contain coefficients that must be determined from measurements performed on the PV modules. For the angle-of-incidence (AOI) effect, we have used coefficients taken from the literature [27].

Data for the module temperature model have been taken from [34]. The coefficients for a free-standing rack are given in Table 1.

Table 1. Values of coefficients u_0 and u_1 used in Equation (8) for different photovoltaic (PV) technologies.

Module Type	u_0 (W/(°C.m^2))	u_1 (W.s/(°C.m^3))
c-Si	26.9	6.20
CdTe	23.4	5.44
a-Si	25.7	6.29

For the PV power model in Equation (7), coefficients for typical crystalline silicon modules can be found in [32]. The coefficients for CdTe and a-Si are based on measurements performed by the JRC ESTI laboratory and fitted using the methods described in [32].

4. Validation

Some validation results for the SPECMAGIC spectrally-resolved irradiance data have been reported in [13] for two sites in Europe, for global spectral irradiance on an inclined plane. Here, we will add results for one site for global horizontal spectral irradiance and one site for spectral direct normal irradiance (DNI). The DNI spectrum is important in this context, because it is needed to calculate the inclined-plane spectral irradiance.

The validation of the spectrally-resolved irradiance values derived from the SPECMAGIC algorithm is performed using measurements of global horizontal and direct (beam) normal spectral irradiance registered at two different locations, Ispra (Italy) and Nicosia (Cyprus), respectively. A detailed description of the two measured databases, including the procedure followed to register and record the spectral irradiance measurements, the instruments used, the wavelength range covered and the temporal resolution of the recordings, can be found in [35] for the global irradiance database and in [36] for the direct irradiance database.

The spectroradiometer system installed in Cyprus has been calibrated *in situ* by performing a calibration transfer from a reference set of spectroradiometers. This reference set was, in turn, calibrated at the European Solar Test Installation laboratories against an FEL (filament type) standard lamp calibrated at the National Physics Laboratory (NPL). The average uncertainty budget for the calibration of the outdoor spectroradiometers has been estimated using the known calibration drift of these devices, the noise levels present, the sensor linearity and the uncertainties involved in the outdoor transfer. This has resulted in an estimated expanded uncertainty of calibration of $\pm 5.4\%$. A full breakdown of this estimate has been described in [36]. The measurement uncertainty is more difficult to estimate due to the large number of variables that can affect each measurement, but accounting for the same uncertainty components again results in a first estimate of expanded uncertainty of ±6.69%.

However, a brief description of the measured databases and the period of time used for this particular analysis is presented in Table 2.

Table 2. Description of the measured spectral irradiance datasets used. GHI, global horizontal; DNI, direct normal irradiance.

Location	**Ispra (Italy) (45.81° N, 8.62° E)**	**Nicosia (Cyprus) (35.15° N, 33.42° E)**
Variable ($W/m^2/nm$)	Spectral GHI	Spectral DNI
Period of time analyzed	2009: January–December 2010: January–June	2013: August–September
Wavelength range (nm)	400–2,500	300–1,700
Wavelength step (nm)	2	2
Temporal resolution (minutes)	30	5
Initial number of measurements	10,411	7,091

While the measured data contain the spectral irradiance registered every few nm, the output of the satellite algorithm is the spectral irradiance already integrated into 30 bands ("Kato bands") of different wavelength widths in the range from 300 nm–4600 nm. Another difference between the measured databases and the estimated one is the time frequency of the data. While measurements are performed every few minutes, the frequency of the estimated values depends on the availability of the satellite images. For this validation, there are two estimated values available per hour for Nicosia and one per hour for Ispra. These differences between the measured and estimated data make it necessary to perform some modifications to the original measured values before comparing measured and estimated irradiance values in the validation. The procedure applied goes as follows:

1. For every satellite retrieved time point, the closest measurement in time to the moment when the satellite image was taken is selected from the measurement time series. Due to the frequency of the measured data, a 5-min window centered at the moment of the satellite data was applied for the Cyprus database and a 20-min window was used for the Ispra database.
2. The measured spectral irradiance values are integrated into bands according to the limits of the Kato bands available in the estimated database. However, due to the wavelength range of the equipment used to measure the spectral irradiance, not all of the 30 bands available in the estimated database can be considered.
3. It was decided to remove all of the moments when the Sun's elevation angle is below 5° .

Once the measured data have been treated according to the previously described procedure, measured and estimated irradiance values in every Kato band can be compared in order to validate the performance of the SPECMAGIC algorithm. The output of SPECMAGIC consist of irradiance in each of the Kato bands. However, since the widths of these bands vary strongly, the derived irradiance values are not very informative. Instead, we will show the results of the validation with these two sets of values:

- spectral irradiance in every band calculated by dividing the irradiance values in each Kato band by the width of the band in nm, R_λ (W/m^2/nm);
- normalized spectral irradiance R_n (1/nm), obtained by dividing the irradiance value in every Kato band by the sum of the irradiance values of all the Kato bands considered. In this way, the broadband irradiance, after the normalization, is equal to 1 W/m^2 for both the measurements and the estimates. These relative irradiance values per Kato band make it possible to see if some parts of the estimated spectrum have an excess or deficit relative to the measured averaged spectrum.

In the next two sections, the results of the validation of the estimated global horizontal and direct normal irradiance are presented. Two statistics are applied, general mean bias deviation $gMBD$ and relative general mean bias deviation $rgMBD$. The general mean bias deviation ($gMBD$, W/m^2) is calculated as the difference between the total irradiance estimated by the SPECMAGIC algorithm for all of the considered Kato bands (K) and the total measured irradiance, according to Equation (9).

$$gMBD = \sum_{i=K_1}^{K_2} G_i^{sat} - \sum_{i=K_1}^{K_2} G_i^{meas} \tag{9}$$

where G_i^{sat} and G_i^{meas} are, respectively, the average irradiance in the Kato band i from the satellite database and the measured one (W/m^2). K_1 and K_2 are the first and last of the Kato bands available in the considered measured datasets of global horizontal and direct normal irradiance.

The relative general mean bias deviation ($rgMBD$) is the result of dividing $gMBD$ by the sum of the average measured irradiance in every Kato band considered, following Equation (10).

$$rgMBD = \frac{gMBD}{\sum_{i=K_1}^{K_2} G_i^{meas}} \tag{10}$$

In addition to these, the mean bias deviation in absolute and relative terms is calculated for every Kato band separately. Other statistics have been calculated, as well, like the root mean square deviation ($RMSD$) or the mean absolute bias (MAB), but are provided as extra information in the Supplementary Material [37].

4.1. Global Horizontal Radiation

From the initial database for Ispra with measurements at more than 10,000 time points, after applying the process described above to select the closest moments to the available satellite data for this location, the validation of the SPECMAGIC results is performed using 3252 time points. Due to the wavelength range measured of global horizontal irradiance, it is only possible to analyze the Kato bands from 7–27, which cover the wavelength range between 407.5 nm–2153.5 nm. The $gMBD$ of 17.58 W/m^2, 6.12% in $rgMBD$, shows that the algorithm tends to overestimate the solar resource in this location. This same trend was observed when the same algorithm was applied to estimate the broadband global horizontal irradiance in the same location [21]. In that other validation, for which data registered in 2005 were used, the resulting errors of 13.85 W/m^2 $gMBD$ or 9.02% in $rgMBD$ are similar to the deviations shown in the present paper. It should be noted that in the analysis performed in [21], the Ispra location showed the highest annual bias out of 16 stations.

Table 3 contains the results obtained for every Kato band considered. The average measured global horizontal spectral irradiance value (R_λ) along with the bias in absolute MBD ($W/m^2/nm$) and relative terms $rMBD$ (%) are shown. The same statistics are shown for the validation performed on the normalized irradiance values.

Table 3. Average global horizontal measured spectral irradiance and bias values between satellite and measured databases calculated for the spectral irradiance and normalized values. MBD, mean absolute bias; $rMBD$, relative MBD.

	Kato Band		**Spectral Irradiance, R_λ**			**Normalized Irradiance, R_n**	
	Lower Limit (nm)	**Upper Limit (nm)**	**Average Meas ($W/m^2/nm$)**	**MBD ($W/m^2/nm$) 10^{-2}**	**$rMBD$ (%)**	**MBD (1/nm) 10^{-5}**	**$rMBD$ (%)**
7	407.5	452.0	0.42	4.5	10.9	6.6	4.5
8	452.0	517.7	0.50	4.1	8.1	3.3	1.9
9	517.7	540.0	0.49	3.8	7.7	2.6	1.5
10	540.0	549.5	0.49	4.0	8.2	3.3	1.9
11	549.5	566.6	0.48	4.1	8.6	3.9	2.3
12	566.6	605.0	0.46	5.0	10.8	7.0	4.4
13	605.0	625.0	0.45	4.5	10.1	5.8	3.7
14	625.0	666.7	0.43	3.8	8.6	3.6	2.4
15	666.7	684.2	0.44	2.6	5.9	−0.3	−0.2
16	684.2	704.4	0.39	1.8	4.6	−1.9	−1.4
17	704.4	742.6	0.37	1.9	5.2	−1.1	−0.9
18	742.6	791.5	0.34	1.8	5.2	−1.1	−0.9
19	791.5	844.5	0.31	1.5	4.9	−1.2	−1.2
20	844.5	889.0	0.30	1.4	4.8	−1.3	−1.3
21	889.0	974.9	0.17	1.3	7.9	1.0	1.7
22	974.9	1,045.7	0.22	1.1	4.8	−0.9	−1.2
23	1,045.7	1,194.2	0.13	0.4	2.7	−1.4	−3.2
24	1,194.2	1,515.9	0.064	0.2	3.2	−0.6	−2.7
25	1,515.9	1,613.5	0.074	0.5	6.9	0.2	0.8
26	1,613.5	1,964.8	0.029	−0.2	−7.3	−1.3	−12.7
27	1,964.8	2,153.5	0.020	−0.03	−1.7	−0.5	−7.3

Considering the spectral irradiance values in each Kato band, the bias remains below 11% for every band. The highest deviations are observed below 667 nm (Kato band 14). Above that wavelength, the deviations decrease to the point of becoming negative for the last two bands (wavelengths above 1613 nm), even though the width of the bands is considerable bigger than for the first bands.

Using the normalized values, the share of the irradiance contained in every band can be analyzed. As shown in the last two columns in Table 3, the satellite's estimate results in a distribution of the spectral irradiance that gives more weight to the visible range than there is in the measured spectrum. Conversely, for wavelengths above 700 nm, the estimated values represent a lower share of the broadband irradiance. This is not the case for Kato Bands 21 and 25, which cover a wavelength range affected by the water vapor absorption, that show a positive bias. This could mean that the SPECMAGIC algorithm underestimates this effect for the Ispra site or that the input data related to this effect are not accurate enough for this particular location. However, it should be noted that apart from the two last bands, the relative bias is less than 5% in all considered Kato bands.

Figures 2 and 3 show the satellite and measured spectral and normalized irradiance values per Kato band, along with the corresponding bias (MBD). The satellite and measured values for the different Kato bands are assigned to the mid-wavelength of every band (X axis). For the spectral irradiance (Figure 2), there is a positive bias over nearly the entire spectrum, consistent with the positive bias observed in the broadband irradiance analysis. In contrast, the normalized spectral irradiance shows much better agreement between measured and estimated spectrum, with a positive bias of 2%–5% around 500–800 nm and a slight negative bias beyond 1200 nm.

Figure 2. Average satellite retrieved and measured global horizontal spectral irradiance per Kato band. Bias per Kato band.

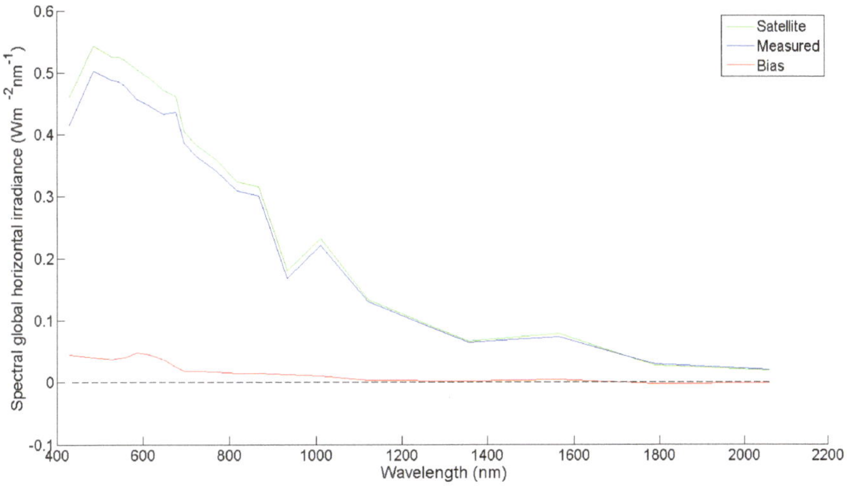

Figure 3. Average satellite retrieved and measured normalized global horizontal irradiance per Kato band. Bias per Kato band.

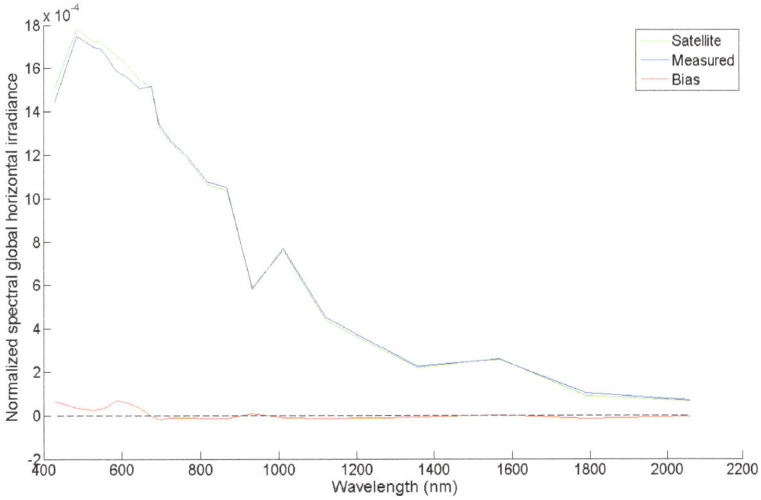

Table 4. Average direct normal measured spectral irradiance and bias values between satellite and measured databases calculated for the spectral irradiance and normalized values.

	Kato Band		Spectral Irradiance, $R_{DNI,\lambda}$			Normalized Irradiance, $R_{DNI,n}$	
	Lower Limit (nm)	Upper Limit (nm)	Average Meas (W/m^2/nm)	MBD (W/m^2/nm) 10^{-2}	$rMBD$ (%)	MBD (1/nm) 10^{-5}	$rMBD$ (%)
4	306.8	327.8	0.06	1.9	32.7	2.7	24.6
5	327.8	362.5	0.17	1.7	10.4	1.1	3.6
6	362.5	407.5	0.34	−1.4	−4.1	−6.3	−9.9
7	407.5	452.0	0.66	−3.7	−5.7	−14.0	−11.4
8	452.0	517.7	0.88	−4.2	−4.8	−17.5	−10.6
9	517.7	540.0	0.93	−4.2	−4.5	−18.0	−10.3
10	540.0	549.5	0.93	−2.2	−2.4	−14.5	−8.3
11	549.5	566.6	0.94	−1.9	−2.1	−14.1	−8.0
12	566.6	605.0	0.91	0.3	0.4	−9.9	−5.8
13	605.0	625.0	0.92	0.2	0.2	−10.0	−5.9
14	625.0	666.7	0.87	1.7	2.0	−6.9	−4.2
15	666.7	684.2	0.85	4.3	5.0	−2.2	−1.4
16	684.2	704.4	0.76	3.0	4.0	−3.3	−2.3
17	704.4	742.6	0.69	3.1	4.5	−2.4	−1.9
18	742.6	791.5	0.65	6.3	9.8	3.7	3.1
19	791.5	844.5	0.58	5.0	8.5	2.0	1.9
20	844.5	889.0	0.56	8.9	15.9	9.2	8.8
21	889.0	974.9	0.27	6.2	22.5	7.8	15.1
22	974.9	1,045.7	0.42	7.9	19.0	9.2	11.8
23	1,045.7	1,194.2	0.23	3.9	16.5	4.2	9.4
24	1,194.2	1,515.9	0.13	2.0	15.8	2.1	8.8
25	1,515.9	1,613.5	0.16	3.9	25.1	5.1	17.5

Similarly to the validation performed on the broadband irradiance [21], data corresponding to clear sky conditions and overcast situations have been analyzed separately also for the spectrally-resolved irradiance. The results are presented in the Supplementary Material [37].

4.2. Direct Normal Radiation

From the initial database for Nicosia with more than 7,000 time points, once the measured values are integrated into Kato bands and the closest moments to the available satellite data have been selected, the validation of the algorithm is performed using 790 time points. Due to the wavelength range measured of direct normal irradiance, it is only possible to analyze the Kato bands from 4–25, which cover the wavelength range from 306.8–1613.5 nm.

The general bias, $gMBD$, of 34.6 W/m^2 is higher than the one obtained in the validation of the global horizontal irradiance. However, due to the higher direct irradiance values at this site, the relative bias of 6.49% $rgMBD$ is similar in both datasets. The positive bias in the broadband irradiance is also observed over most of the wavelength range in the spectral direct irradiance. These results agree with those obtained in the first validation of the SPECMAGIC algorithm, being a bit higher than those reported in [13]. The accuracy observed on that occasion was in the range of the uncertainty of surface measurements, which is about 5%, with the exception of the part of the spectrum above 1200 nm (Kato Band 24 and above), where higher deviations were observed.

Table 4 contains the results obtained for every considered Kato band in the estimation of the direct normal irradiance ($R_{DNI,\lambda}$). The average measured spectral irradiance value, along with the bias in absolute MBD (W/m^2/nm) and relative terms $rMBD$ (%) are shown. The same statistics are shown for the validation performed on the normalized irradiance values.

The results obtained for the spectral direct irradiance show a tendency to underestimate the part of the spectrum below 566 nm with the exception of the first two available bands and to overestimate the rest of the wavelength range measured. The bias for the various Kato bands are generally higher than the ones obtained for the global horizontal irradiance. However, the relative mean bias deviation values are similar in both datasets.

The same trend to underestimate in the lowest part of the spectrum and overestimate in higher wavelength ranges is also observed when considering the normalized values. This is the opposite of what was observed for the global horizontal spectral irradiance.

Figures 4 and 5 show, respectively, the satellite and measured spectral and normalized irradiance values per Kato band, along with the corresponding bias.

In both figures, a significant overestimation is noticeable at wavelengths above 1200 nm, as was observed also by Müller *et al.* [13], in the first validation of the SPECMAGIC algorithm.

Due to the small number of months analyzed and the fact that direct normal irradiance is very low in the case of overcast sky conditions, the validation of the algorithm has not been performed selecting clear and overcast sky conditions, as has been done for the global horizontal irradiance.

Figure 4. Average satellite retrieved and measured spectral direct normal irradiance per Kato band. Bias per Kato band.

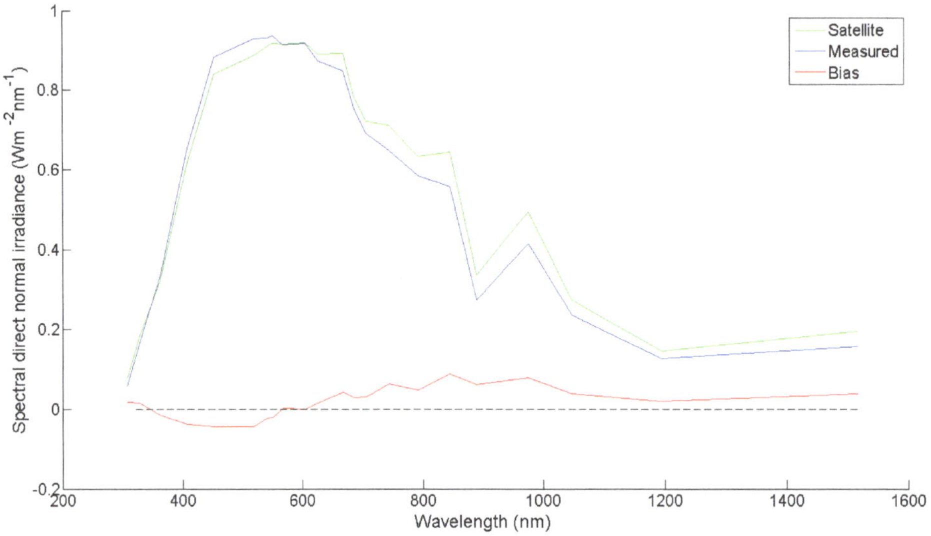

Figure 5. Average satellite retrieved and measured normalized direct normal irradiance per Kato band. Bias per Kato band.

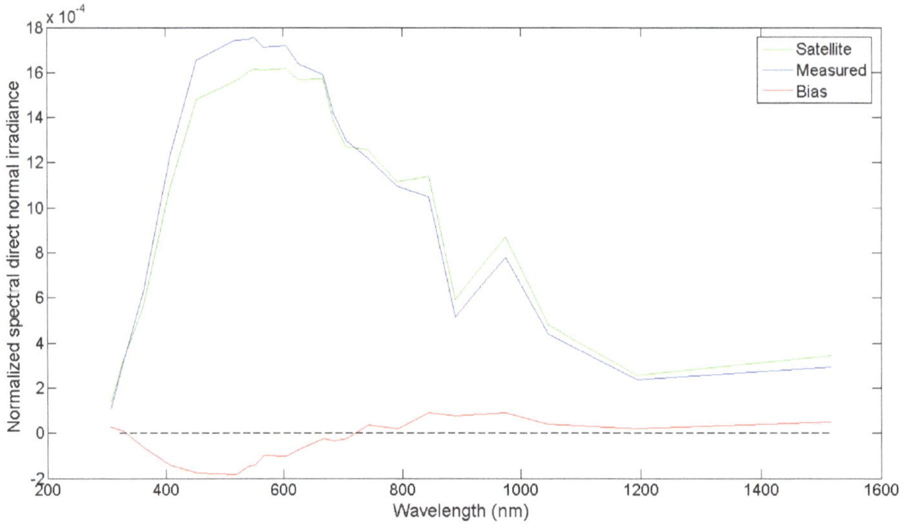

4.3. Average Photon Energy Calculation

The average photon energy can be used to describe the spectral irradiance distribution at a certain location to within a level of certainty, as shown by Minemoto *et al.* [12,35]. Despite the applications of the APE values and due to the lack of spectrally-resolved irradiance data, in this section, we have calculated the APE values considering the output of the SPECMAGIC algorithm. In order to validate the suitability of considering those satellite-based values for the APE calculation, we have calculated the APE with three different datasets:

- global horizontal spectral irradiance measured every 2 nm ($meas$);
- global horizontal spectral irradiance measured every 2 nm, but integrated into Kato bands ($measKB$);
- satellite-derived global horizontal spectral irradiance, which is already integrated into Kato bands ($satKB$).

By comparing the first two, we can analyze whether the integration of the spectral irradiance into Kato bands affects the APE values that would result from considering the real measurements recorded every few nm. Comparing the APE values resulting from the last two datasets will help with analyzing the suitability of the satellite data for the estimation of the APE values at a certain location. The data used for this analysis, both measured and satellite-based values, correspond to Ispra (Italy) and were used in previous sections. A total number of 3252 data points have been considered, as explained in Section 4.1. For the estimation of the APE values, only the wavelength range between 400 nm and 1050 nm is used, which covers the Kato bands from 7–22, inclusive. A full description of the calculation procedure for the APE can be found in Norton *et al.* [35].

The APE values for the three datasets are shown in Table 5, while the average monthly APE values for the three datasets are shown in Figure 6.

Table 5. Average photon energy (APE) values calculated for the spectral global horizontal irradiance values measured and estimated for Ispra, using data from January 2009 to June 2010.

Dataset	APE Value
$Meas$	1.8468
$MeasKB$	1.8409
$SatKB$	1.8477

Figure 6. Monthly APE values calculated for the spectral global horizontal irradiance values measured and estimated for Ispra.

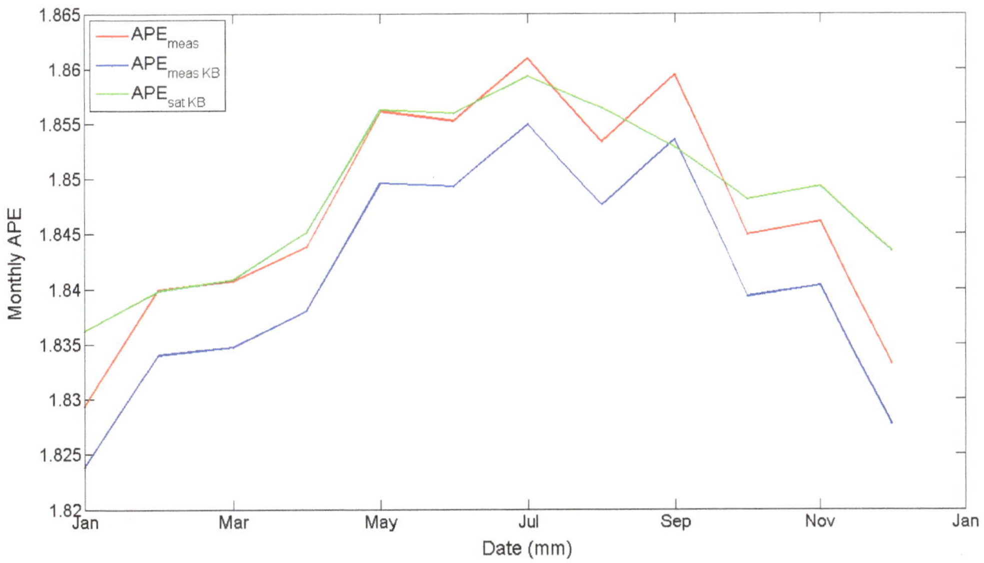

Calculating the APE values using the measured data integrated into Kato bands ($MeasKB$) reduced the APE values by about 0.32% with regard to the values obtained when considering the measurements recorded every 2 nm ($Meas$). When APE values are calculated using the satellite retrieved values, APE tends to be higher than those obtained from the measured data, especially for the winter months (from October–February). However, considering the yearly average values, both APE values from $MeasKB$ and $SatKB$ are very similar, with a bias of 0.37%. This is consistent with the results shown in Figure 3 that showed a slight positive bias in the blue end of the spectrum, leading to a higher APE. In view of these results, we can conclude that the SPECMAGIC output values could well be applied to estimate the APE values at any location.

4.4. Validation of the Model for PV Performance under Varying Spectrum

The ability of the algorithm to retrieve spectral irradiance values from satellite images has been validated in the previous sections. In this section, these satellite values will be applied to validate the model used to estimate the changes in the performance of PV modules under different solar spectra. For this, outdoor measurement data from a CdTe module will be used.

The PV module considered was measured outdoors during the period of time between May 2010, and November, 2011, in the test installation at the Joint Research Centre (JRC) ESTI Laboratory located in Ispra, Italy. The measurement system for long-term outdoor measurements of PV modules is described in [38] (see also [4]). The PV modules are mounted on a metal rack facing south at an inclination angle of 45° from the horizontal plane. The module temperatures, T_{mod}, are measured by PT100 sensors on the back of the modules. The in-plane of array irradiance, G_{POA}, is measured using a calibrated silicon reference cell mounted in the same plane as the modules. Measurements are made approximately every four minutes during daytime, provided the in-plane irradiance exceeds the threshold of 50 W/m^2.

To validate the spectral performance model, the effective long-term mismatch factor for the CdTe module, MM_{CdTe}, is calculated using both satellite-derived irradiance data and real measurements of the module short-circuit current. First, the spectral mismatch factor for I_{sc}, $MM_{Isc,n}$, is calculated by inserting the module short-circuit current data for each measurement moment n into Equation (3). The measured module short-circuit current, I_{sc}, at each moment must be corrected to a temperature of 25°C. This is done according to Equation (11), where α is the temperature coefficient of the module, and T_{mod} (°C) is the module temperature. α was found from indoor measurements to have the value of $\alpha = -0.00055\ ^\circ\mathrm{C}^{-1}$.

$$I_{sc\_25} = \frac{I_{sc}}{1 - (\alpha \cdot (T_{mod} - 25^\circ\mathrm{C}))} \tag{11}$$

The overall spectral mismatch factor $\langle MM_{\mathrm{meas}\_Isc} \rangle$ can then be calculated in the following way (similar to [6,8]). In the absence of spectral effects, the short-circuit current would be proportional to the measured irradiance: $I_{sc} = (G_{POA}/G_{\mathrm{STC}}) \cdot I_{sc\_STC}$, where I_{sc_STC} is the short circuit current under STC conditions and G_{STC} is the STC irradiance of 1000 W/m^2. The instantaneous spectral mismatch factor is then calculated as the deviation of the measured I_{sc} from this "ideal" value:

$$MM_{Isc,n} = I_{sc} \cdot G_{\mathrm{STC}} / (G_{POA} \cdot I_{sc\_STC})$$

I_{sc_STC} was found from the outdoor measurements using the (temperature-corrected) I_{sc} values during times when $G_{POA} \simeq 1000\ \mathrm{W/m^2}$ and the air mass AM $\simeq 1.5$.

The overall measured spectral mismatch for the time period of the measurements is then calculated using Equation (12):

$$\langle MM_{\mathrm{meas}\_Isc} \rangle = \frac{\sum_{n=1}^{N} MM_{Isc,n} \cdot I_{sc\_25,n}}{\sum_{n=1}^{N} I_{sc\_25,n}} \tag{12}$$

To estimate the overall mismatch factor for CdTe using the satellite data, $\langle MM_{\mathrm{CdTe}\_sat} \rangle$, the instantaneous spectral correction factor $MM_{\mathrm{CdTe}\_sat,n}$ for each measurement moment n is first calculated by inserting the satellite spectral irradiance data $R_{SAT}(\lambda)$ and the CdTe module spectral response, $S_{r,CdTe}(\lambda)$, into Equation (3). The mismatch factor for the entire measurement period is then calculated by summing these values weighted by the irradiance intensity at which they occur, as described in Equation (13):

$$\langle MM_{\mathrm{CdTe}\_sat} \rangle = \frac{\sum_{n=1}^{N} MM_{\mathrm{CdTe}\_sat,n} \cdot G_{POA,n}}{\sum_{n=1}^{N} G_{POA,n}} \tag{13}$$

here, $G_{POA,n}$ is the plane-of-array irradiance, in this case measured using a calibrated silicon reference cell. However, this approach introduces an additional calculation step, because this reference cell has its own spectral mismatch value. To correct for this, it is necessary to divide the satellite mismatch value for the CdTe module, $\langle MM_{\mathrm{CdTe}\_sat} \rangle$, by the mismatch of the silicon reference cell, $\langle MM_{\mathrm{Si}\_sat} \rangle$, as shown in Equation (14).

$$\langle MM_{\mathrm{pred}\_sat} \rangle = \frac{\langle MM_{\mathrm{CdTe}\_sat} \rangle}{\langle MM_{\mathrm{Si}\_sat} \rangle} \tag{14}$$

Here, $\langle MM_{\mathrm{pred}\_sat} \rangle$ is the predicted overall mismatch between the CdTe module and the irradiance measured by the Si reference cell. $\langle MM_{\mathrm{Si}\_sat} \rangle$ is calculated in the same manner as for $\langle MM_{\mathrm{CdTe}\_sat} \rangle$, but inserting the satellite irradiance data $R_{SAT}(\lambda)$ and the Silicon reference cell spectral response, $S_{r,Si}(\lambda)$, into Equation (3), to calculate for every moment n for the spectral mismatch factor $MM_{\mathrm{Si}\_sat,n}$.

For the period of time the CdTe module was measured outdoors, there are approximately 35,000 satellite data points and 100,000 measurements from the outdoor field. In order to perform the validation of the model in the most accurate way and considering the frequency of the measurements from the outdoor field, only those data from the satellite and the outdoor field datasets, whose time difference was below five minutes, were selected. In the end, almost 10,000 data points are used for this analysis.

For a given moment n, three different spectral correction factors are calculated ($MM_{Isc,n}$, $MM_{\mathrm{CdTe}\_sat,n}$ and $MM_{\mathrm{Si}\_sat,n}$). From these values, the overall mismatch factors are calculated ($\langle MM_{\mathrm{meas}\_Isc} \rangle$, $\langle MM_{\mathrm{CdTe}\_sat} \rangle$ and $\langle MM_{\mathrm{Si}\_sat} \rangle$; with these last two values, $\langle MM_{\mathrm{pred}\_sat} \rangle$ is calculated). The results of these calculations are shown in Table 6.

Both mismatch values ($\langle MM_{\mathrm{meas}\_Isc} \rangle$ and $\langle MM_{\mathrm{pred}\_sat} \rangle$) show that, for the location analyzed, the effective irradiance, G_{eff}, received by the CdTe module would be slightly higher than the in-plane of array irradiance registered in the measured database, G_{POA}. This would improve the estimated performance of the module for this location with regard to the estimations obtained if the spectral effect were not taken into account. Considering the mismatch factor resulting from the measurements as the real value for the module, $\langle MM_{\text{meas\_}Isc}\rangle$, using the satellite data to estimate the effective irradiance would imply an overestimation of 0.09%. This result shows that using the satellite information provides accurate results, at least for the location analyzed. However, it would be necessary to consider other PV technologies, as well as data from other locations.

Table 6. Mismatch factors obtained for the CdTe module using measured and estimated input data. Weighted average for the measurement period May, 2010–November, 2011.

Mismatch Factor	Value (%)
Predicted CdTe mismatch $\langle MM_{\text{CdTe\_}sat}\rangle$	1.00
Predicted Si mismatch $\langle MM_{\text{Si\_}sat}\rangle$	0.61
Predicted mismatch between CdTe and Si sensor $\langle MM_{\text{pred\_}sat}\rangle$	0.39
Measured mismatch between CdTe and Si sensor $\langle MM_{\text{meas\_}Isc}\rangle$	0.30

The comparison of the mismatch factors has also been made month by month. The numbers are provided in Supplementary Material [37]. Figure 7 shows the monthly variation in the mismatch factors. It can be seen that the annual variability is higher for the ground-based measurement values than for the satellite-based values.

Figure 7. Monthly values of the CdTe module mismatch factor calculated from outdoor measurements and satellite data.

The spectral mismatch shows a clear variation over the seasons, with generally higher values in summer and lower values in winter. This pattern is found both in the ground-based data ($\langle MM_{\text{meas\_}Isc} \rangle$) and in the satellite estimates values ($\langle MM_{\text{pred\_}sat} \rangle$). In most cases, both values tend to agree, in the sense that both are above or below one simultaneously. As was shown above, when considering the whole time period, the satellite estimate of the mismatch factor is only slightly higher than that obtained from the I_{sc} measurements. However, when looking at monthly values, there are larger deviations. Considering the common available months in the two years (from May–November), the satellite-based estimates for a given month are more similar between the years than the ground-based calculation. This may indicate a problem either with the measurements of the irradiance or in the measurement of I_{sc}. If this is the case, the main conclusion must be that the difference between the satellite-based and ground-based calculation is well within the uncertainty of the ground-based measurements. However, this comparison does show that the satellite-based estimate can resolve the variations in the mismatch over the seasons, which is an indication that the method can work over a range of different conditions (different weather types and varying Sun elevation angles). Considering the satellite data results in a general overestimation of the mismatch factor ($\langle MM_{pred\_sat} \rangle > \langle MM_{meas\_Isc} \rangle$). However, there are some months, around the winter period, where an underestimation is observed. The bias varies between -2.13% and 1.66%, although for half of the months, the bias remains below 1%.

5. Geospatial Mapping of PV Performance Variations due to Solar Spectrum Variations

5.1. Comparison with Results from the Literature

The validation presented in the previous section suffers from the fact that actual measured data for PV modules were available for only one single site. It would therefore be useful to compare the results we have obtained with those published by other groups. It will not be a true validation, since there are a number of differences in the approaches, but it will still be useful to see how well these studies compare.

Two recent studies have reported values for the spectral mismatch of different module types at various locations: Alonso-Abella *et al.* [6] and Dirnberger *et al.* [8]. In the first study, four different sites in Germany, Spain and Algeria were used, while the second study reports results for Freiburg, Germany.

There are a number of differences in approaches, which means that the comparison will not be exact. The two studies considered an inclination angle of 30°, while our results are for 20° inclination. Especially for a-Si, the spectral response varies between manufacturers, and the one used in [8] seems to have a slightly broader spectral response curve. The time period is different, and the results in [8] show some variation between years. Finally, these two studies considered only the influence of the spectrum on I_{sc}, and the mismatch is essentially the deviation in monthly or annual average I_{sc}. In contrast, our study uses the spectrally-corrected irradiance as input to a model for the module power. If there is a correlation between module temperature and the spectral effect (and this can be seen in the seasonal variation in Figure 7), the overall mismatch in the energy produced will be different from that considering only I_{sc}.

Both studies have presented results as monthly average mismatch for the different PV technologies and locations. Figure 8 below shows similar results for four different locations: Freiburg [8] and Madrid, Jaen and Tamanrasset [6].

Figure 8. Monthly average spectral mismatch $\langle MM \rangle$ for four different sites for three different PV technologies, for modules mounted south-facing at a 20° inclination. Results are for the year 2005.

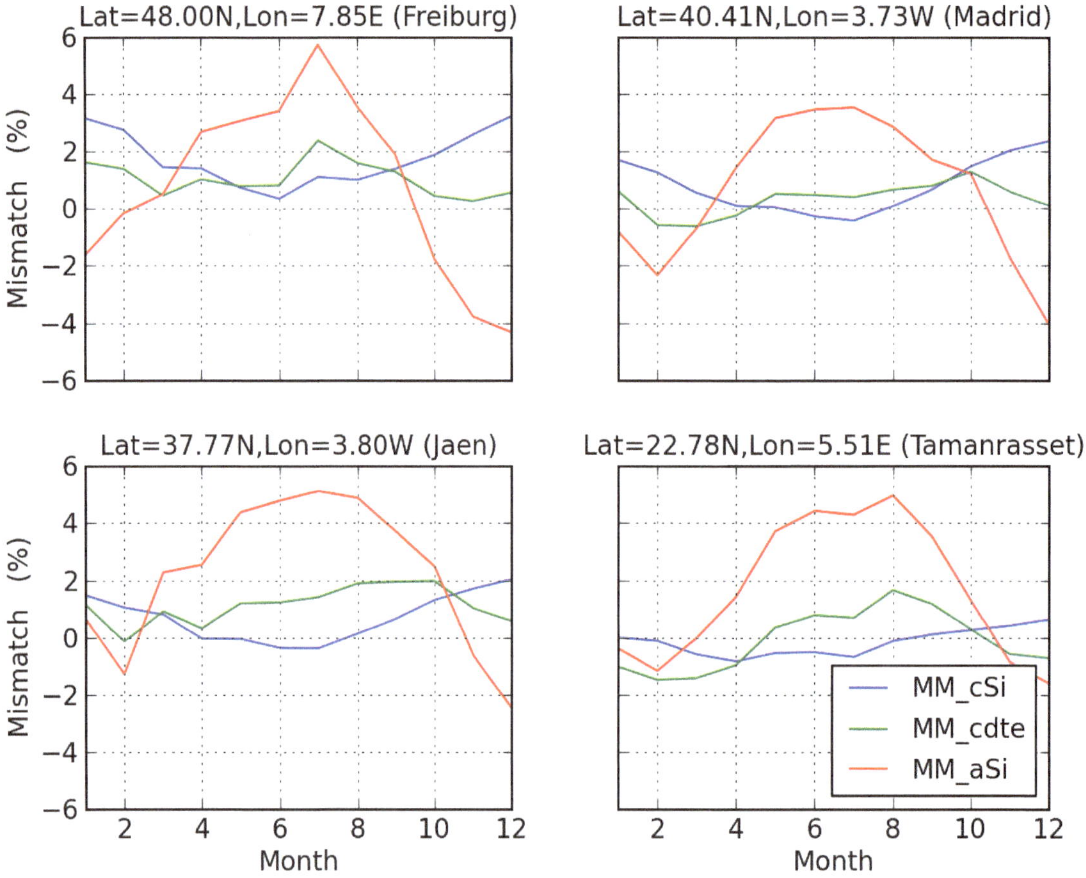

A comparison with the results of the previous studies shows that there are strong similarities: the seasonal variation of a-Si is stronger than for the other two module types; peak positive mismatch for a-Si is in summer for all four sites, while the a-Si mismatch is slightly negative in winter. The spectral effect is smaller for CdTe and c-Si, and there is less seasonal variation.

Some differences are also evident: our results show a somewhat stronger seasonal effect for c-Si than the other studies, with the lowest values in summer. The overall effect for c-Si is stronger than the values found in [6]. For CdTe, the seasonal variation is less clear in our results, and the overall effect is somewhat lower than the one found in [8].

Table 7 shows annual average mismatch for the three PV technologies for five locations, comparing our results with those of [6,8]. In the table, also Stuttgart is shown [6]. Our results differ from these authors by up to $\pm 2\%$. Given the differences in approaches, this is not surprising. An interesting case is the values from Freiburg and Stuttgart, which are geographically close. Here, our results show very similar values, while the results of [8] (Freiburg) and [6] (Stuttgart) vary more between each other and in opposite directions from our results ([8] shows higher values of $\langle MM \rangle$, while [6] has lower values).

Table 7. Annual average mismatch $\langle MM \rangle$ (in percent) for four different locations and three different PV technologies. The comparison is between the results obtained in this work compared with values found in recent studies.

	This Work			**Recent Studies** [6,8]		
Location	**c-Si**	**CdTe**	**a-Si**	**c-Si**	**CdTe**	**a-Si**
Freiburg	+1.3	+1.1	+2.0	+1.4	+2.4	+3.4
Stuttgart	+1.3	+1.1	+1.8	−0.6	−0.5	−0.4
Madrid	+0.6	+0.4	+1.3	+0.6	−1.2	−0.5
Jaen	+0.5	+0.6	+2.8	−0.2	−0.4	0.4
Tamanrasset	−0.2	−0.1	+1.7	−0.8	+0.1	+2.2

5.2. Mapping the Influence of the Spectrum on PV Performance

Using the data and models described in Sections 2 and 3, it is now possible to estimate the performance of different PV module types over large geographical regions. The calculations are performed to obtain estimates of the PV power output at each hour of a year, using the following steps:

1. From the global horizontal and direct horizontal irradiance, calculate the inclined-plane direct and diffuse irradiance in each spectral band (Section 2.2).
2. Correct the in-plane of array irradiance for AOI effects (Section 3.1).
3. From the corrected direct and diffuse irradiance, use Equation (3) to calculate the effective irradiance.
4. Calculate the module temperature from the AOI-corrected irradiance, ambient temperature and wind speed using Equation (8).
5. Use Equations (5)–(7) to calculate the instantaneous module power.

In order to estimate the overall effect of spectral variations, we have also performed these calculations leaving out Step 3 in the list above, so the power will be estimated without taking into account the solar spectrum.

The calculation steps in the above list have been performed using one year of irradiance data derived from the Meteosat-7 satellite at 0° longitude. The coverage area of this satellite is Europe, Africa and parts of western Asia. Temperature and wind data were taken from the European Centre for Medium-Range Weather Forecast (ECMWF) ERA-Interim reanalysis, with a spatial resolution of 0.75° latitude/longitude and a time resolution of 3 h [39]. The study year is 2005. For this calculation, the modules were assumed to be mounted on a fixed open-rack facing the Equator at an inclination angle of 20° from the horizontal plane.

The resulting maps show the estimated total energy production by a $1\text{k}W_\text{p}$ array of modules of the given module type. The annual energy production $E_{\text{tot},spec}$ is calculated following Equation (15):

$$E_{\text{tot},spec} = \sum_{t_h=1}^{8760} P\left(G_{eff}(t_h), T_{\text{mod}}(t_h)\right) \tag{15}$$

where t_h is the h-th hour in the year.

An example of these maps of annual energy PV output is shown in Figure 9, where the PV output estimation for a CdTe module is calculated for the year 2005 considering the effects of irradiance and temperature on PV performance, as well as the spectral effects.

Figure 9. Total yearly energy output (kWh) from a $1kW_p$ array of CdTe modules, mounted at an Equator-facing plane with an inclination of 20°, for the year 2005.

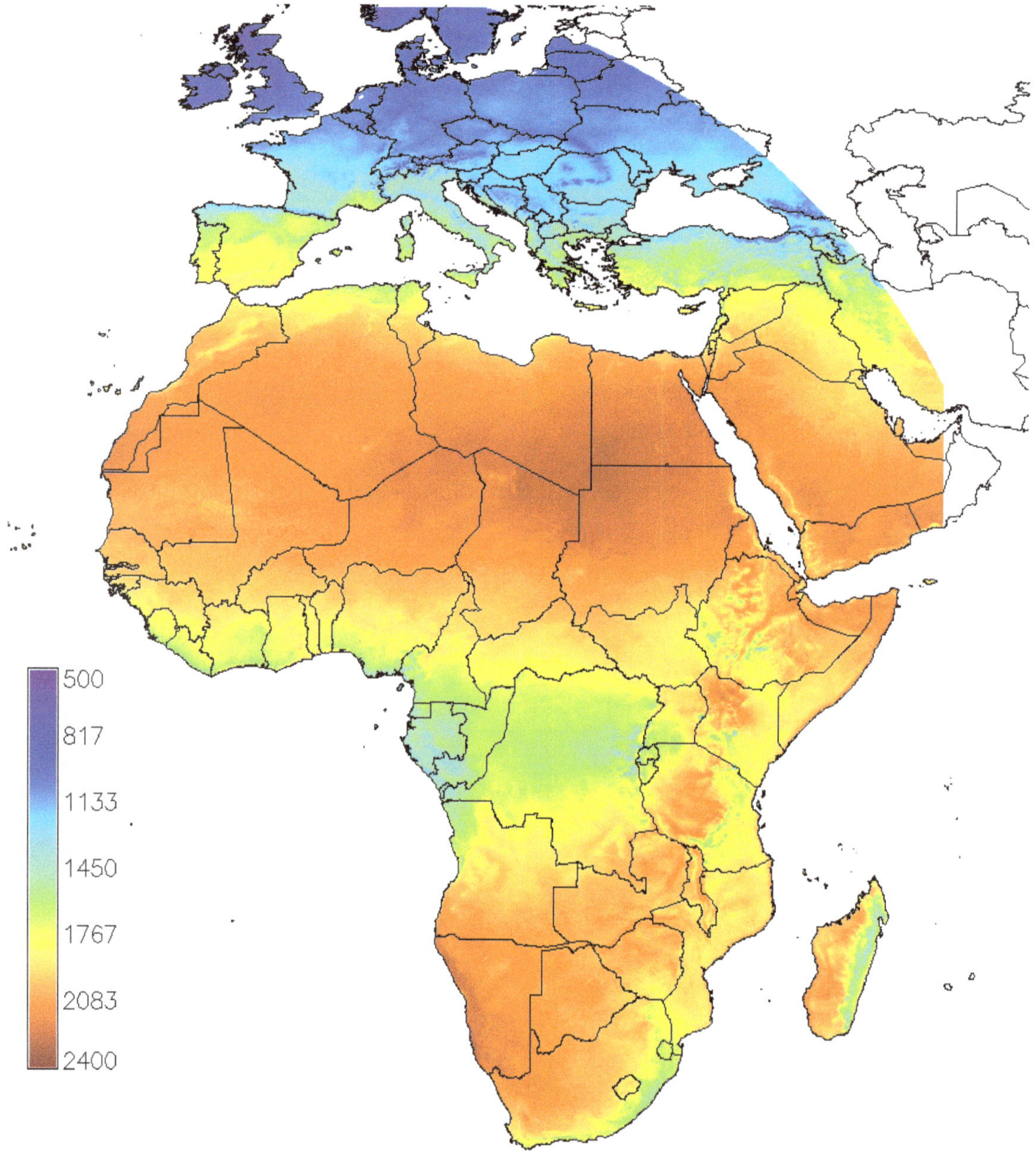

A similar map was calculated for the total energy E_{tot}, but leaving out the spectral correction step (Step 3 in the above list). From these two energy output estimates, we can then calculate a measure of the annual average spectral mismatch, $\langle MM \rangle$ (Equation (16)):

$$\langle MM \rangle = \frac{E_{tot,spec} - E_{tot}}{E_{tot}} \tag{16}$$

A map of $\langle MM \rangle$ for CdTe modules is shown in Figure 10. From the figure, we see that in large regions, the spectral effect is very small. This is the case in desert areas (north Africa, Arabia and Namibia) and in Southern Europe (Turkey, southern Spain). In most of Europe, the effect is a little larger, at around 0.01–0.02 (corresponding to 1%–2% extra energy output), increasing as we move north. The largest effect is seen in humid tropical regions, such as equatorial West Africa, where the effect can be up to about 0.05 (5% extra energy). The region with a negative spectral effect at the easternmost edge of the image is probably an artifact caused by the increased uncertainty in the solar radiation estimates at the edge of the satellite image.

Figure 10. Effect of spectral variations on PV output from CdTe modules mounted Equator-facing at an inclination of 20°. Annual average spectral mismatch for the year 2005.

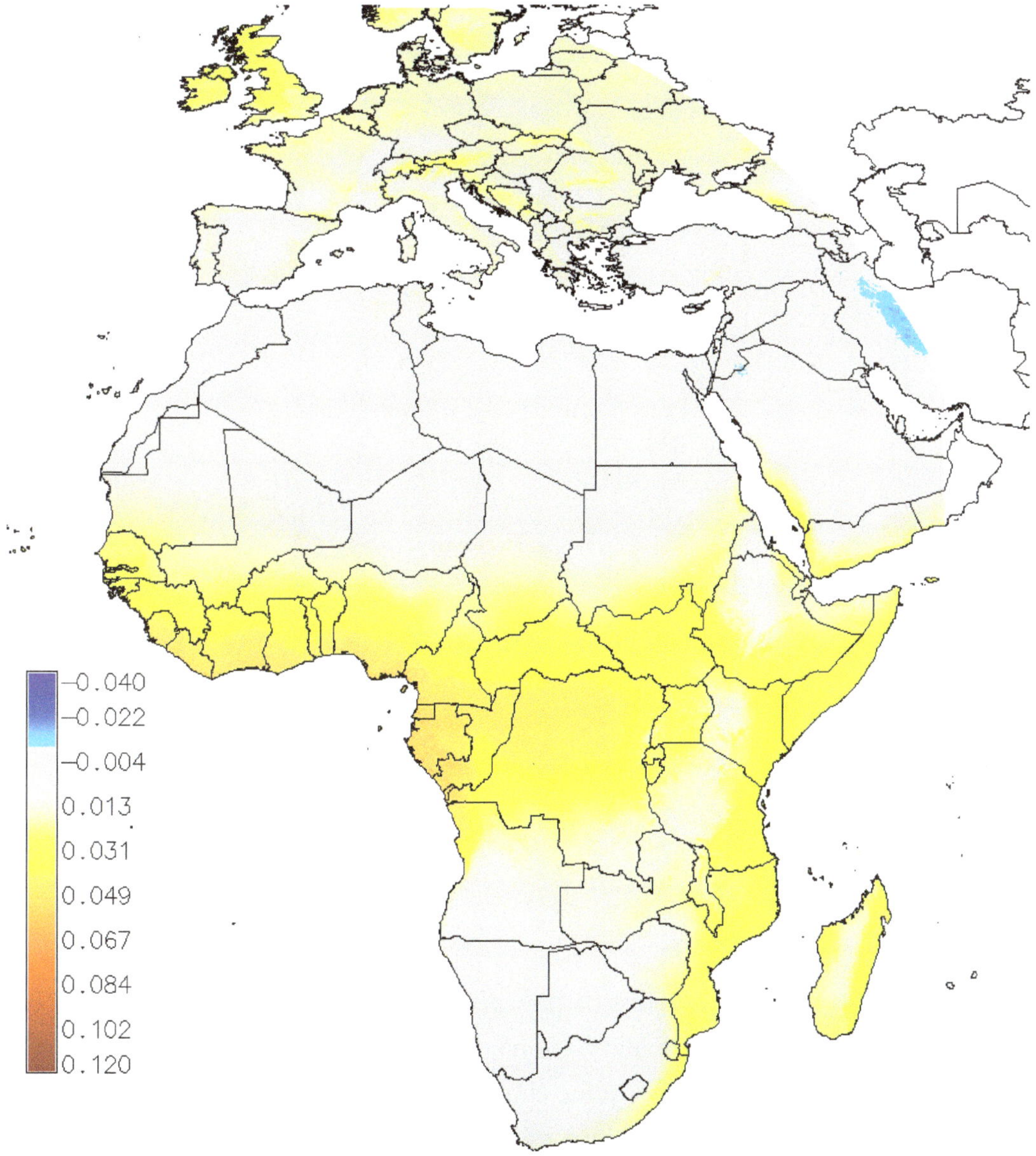

The same calculation was carried out also for crystalline silicon modules and for a single-junction amorphous silicon module (see the spectral response curves in Figure 1). The results are shown in

Figures 11 and 12 respectively, showing again the spectral effect as calculated in Equation (16). The color scale is the same in Figures 10–12 to make the comparison easier.

Figure 11. Effect of spectral variations on PV output from crystalline silicon modules (inclination 20°); annual average spectral mismatch for the year 2005.

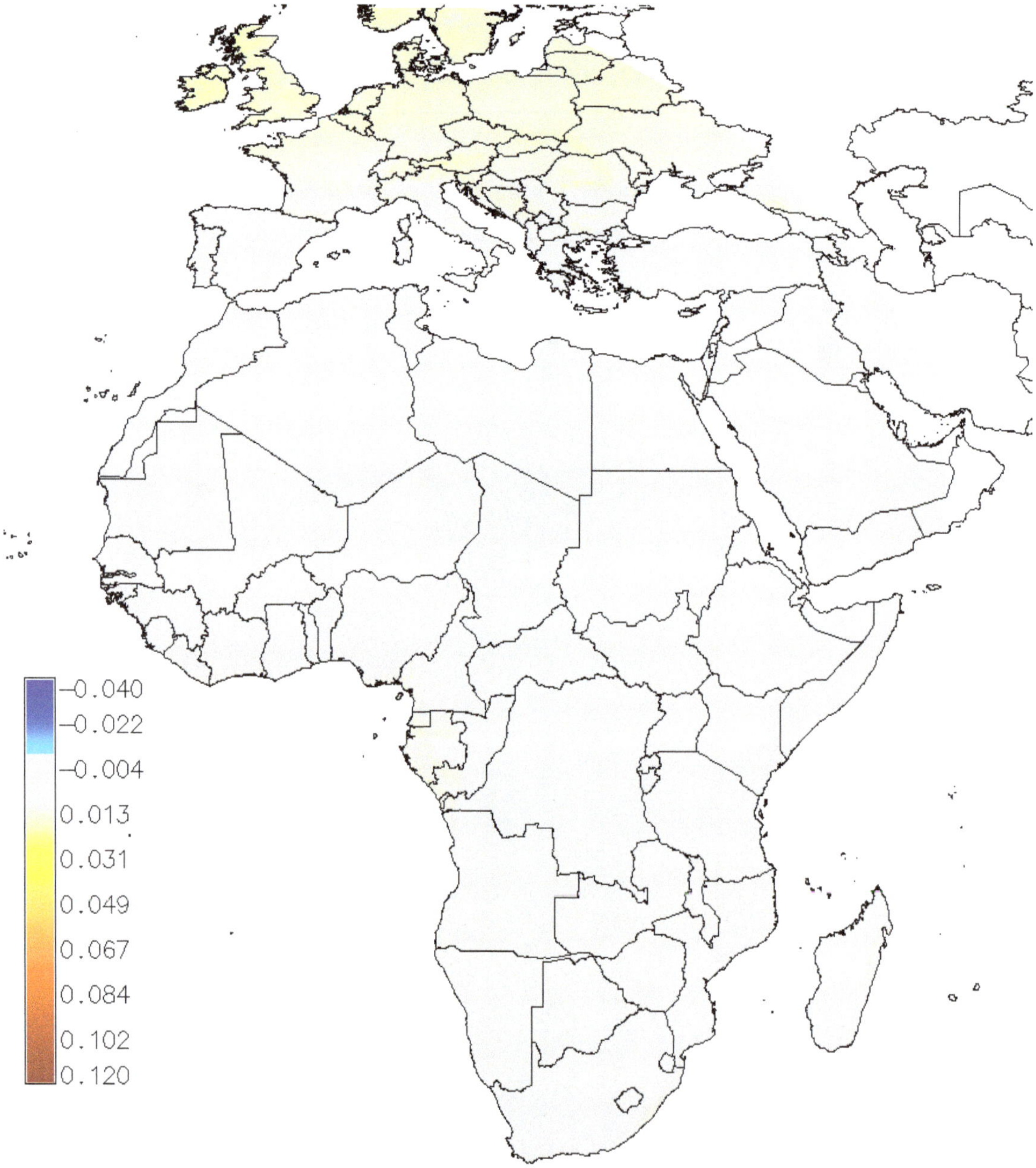

It is clear from Figure 11 that the influence of the spectral variations is much less pronounced for crystalline silicon modules. In all of Africa and in Southern Europe, the absolute value of the effect is less than 1%. In Central Europe, this increases somewhat to about +2%, and the strongest effect is found in Northern Europe, but the effect is still limited to about +3%.

In contrast, the effect of the spectral variations is much more pronounced for amorphous silicon (Figure 12). The effect is nearly always positive, again with the lowest values found in desert areas (from 0 to +2%). In Europe, the effect is somewhat stronger, at +3% or +4%. As for CdTe, the strongest effect is found in the tropical areas of Africa, where the effect may reach as high as +10%, meaning that the actual energy output of the module may be up to 10% higher than would be predicted without taking into account the spectral effects.

Figure 12. Effect of spectral variations on PV output from a single-junction amorphous module; annual average spectral mismatch for the year 2005.

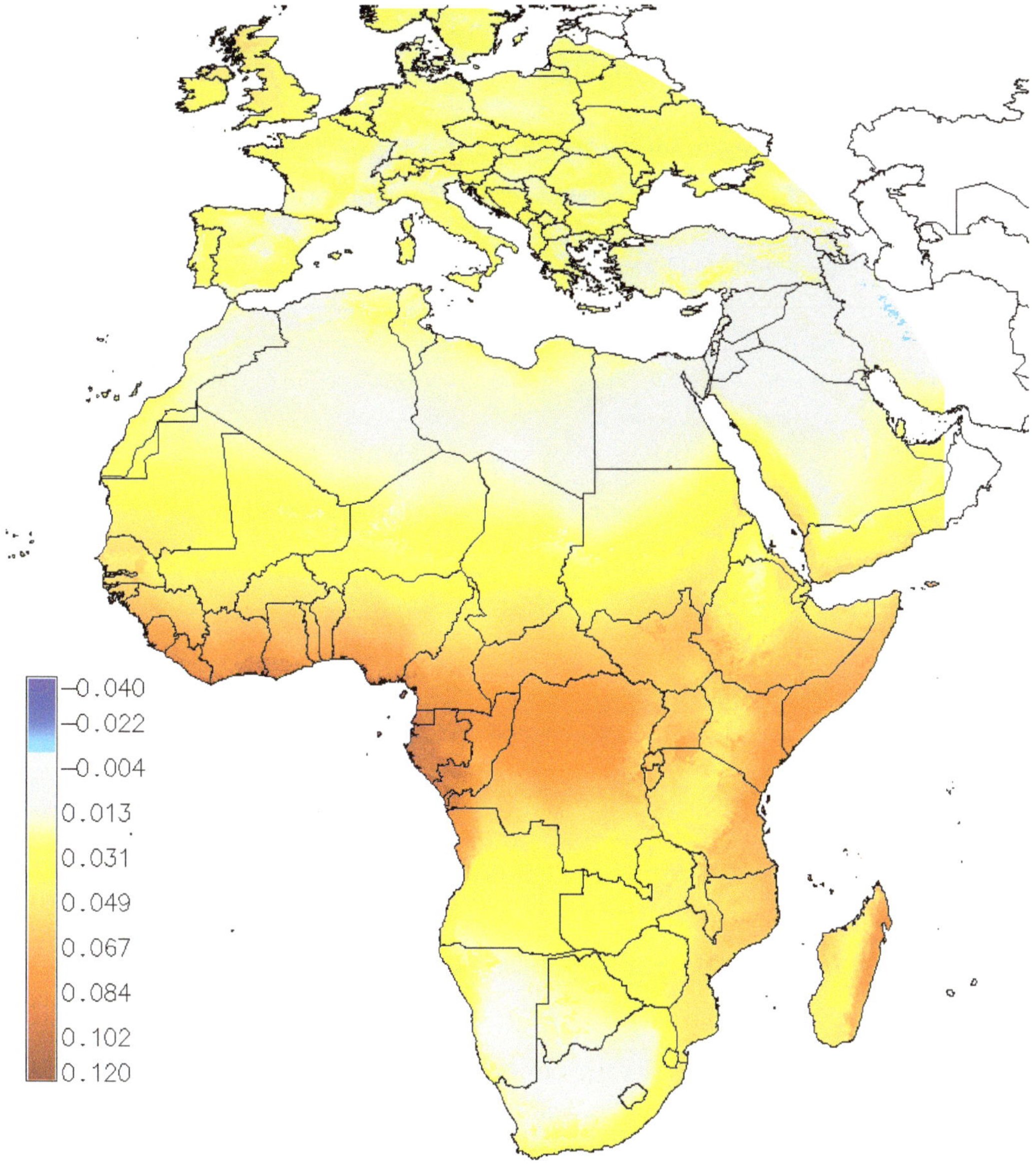

The overall pattern of the effects is that the effect is stronger in areas with a high level of diffuse irradiance, such as northern Europe and tropical West Africa. The effect is also stronger for PV technologies with more narrow spectral response curves, with the strongest effect seen for amorphous silicon that is only sensitive to radiation with a wavelength less than 750 nm.

5.3. Spectral Effect Dependence on Inclination Angle

The results shown in the previous section have all been calculated using a single inclination angle for the PV modules (20° from the horizontal plane, Equator-facing). If the inclination angle is changed, the amount of in-plane irradiance changes, but also the ratio of direct to diffuse irradiance is affected. In order to study the effect of changing inclination angle, we have performed the calculation of MM for a number of different inclination angles for four different locations. These locations represent a humid equatorial climate, Köppen–Geiger (K-G) classification “Aw” (1°48’N, 15°20’E), a desert climate, K-G= “BWh” (24°55’N, 46°25’E), a Mediterranean climate, K-G = “Csa” (44°5’N, 5°4’E), and a maritime Northern European climate, K-G = “Cfb” (51°97’N, 4°56’E). The inclination angle was varied from 0° to 60°, south-facing. The calculation was made for one year (2011) for three different module types. The resulting values of MM are shown in Figure 13.

Figure 13. Effect of spectral variations on PV output for three different module types as a function of inclination angle; annual average spectral mismatch for the year 2011.

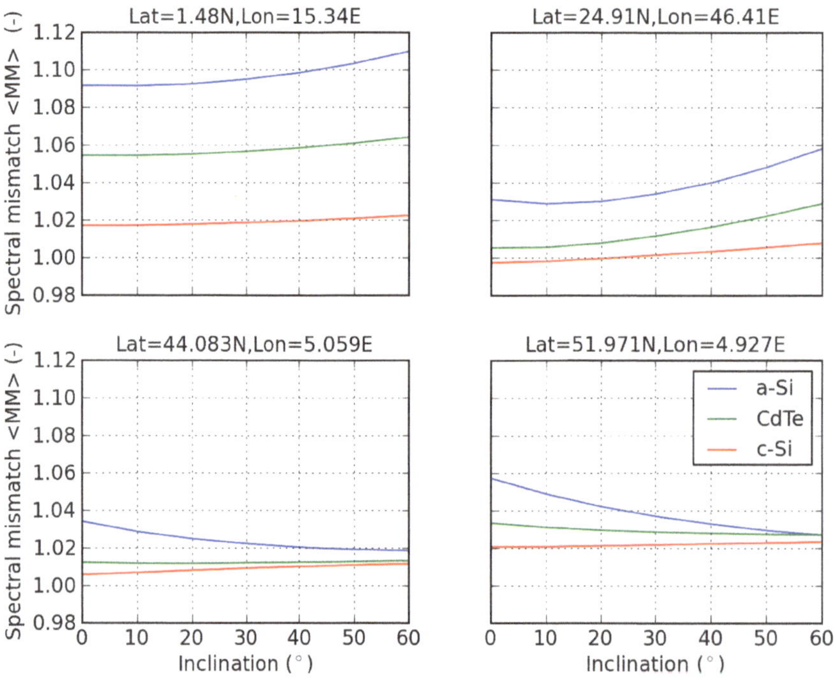

The spectral performance of c-Si modules varies only slightly with inclination angle. The effect is somewhat more pronounced for CdTe and even more for single-junction a-Si. Furthermore, at latitudes close to the Equator, the tendency is towards an increased effect at steep inclinations, while at more northerly locations, the variation is in the opposite direction. This may be due to the changing ratio of direct to diffuse light. Near the Equator, the amount of direct sunlight decreases with increasing inclination, while at higher latitudes, the amount of direct light increases (until the inclination angle equals the latitude) while the diffuse light decreases.

5.4. Mapping Average Photon Energy

The average photon energy (APE) is one way to reduce the complexity of variations in the spectrum of sunlight to a single parameter. As the name implies, this number is simply the average energy of the photons arriving in a given plane located at the Earth’s surface within a given time period. The averaging may be performed for all of the wavelengths in the sunlight or may be restricted to a certain band.

In the study presented here, APE has not been used to estimate PV performance, as the full spectrum information is available from the satellite-based estimates. However, it could still be useful to calculate APE in order to compare these results with other studies.

The annual average APE was calculated for irradiance in the same plane (20° Equator-facing) as the results shown in the previous sections. For the present study, we chose the interval 300–1050 nm for the calculations. In each spectral band, the photon energy is assumed to be equal to that of the central wavelength in that band.

The annual average of APE can then be found as an irradiance-weighted average of the hourly APE maps, according to Equation (17):

$$\langle APE \rangle = \frac{\sum_{t_h=1}^{8760} APE_{t_h} \cdot G_{POA,t_h}}{\sum_{t_h=1}^{8760} G_{POA,t_h}} \tag{17}$$

The resulting map of APE for the year 2005 is shown in Figure 14.

Figure 14. Average photon energy for the year 2005, in electron-volts. For comparison, APE for the standard test conditions (STC) spectrum in this wavelength range is 1.841 eV.

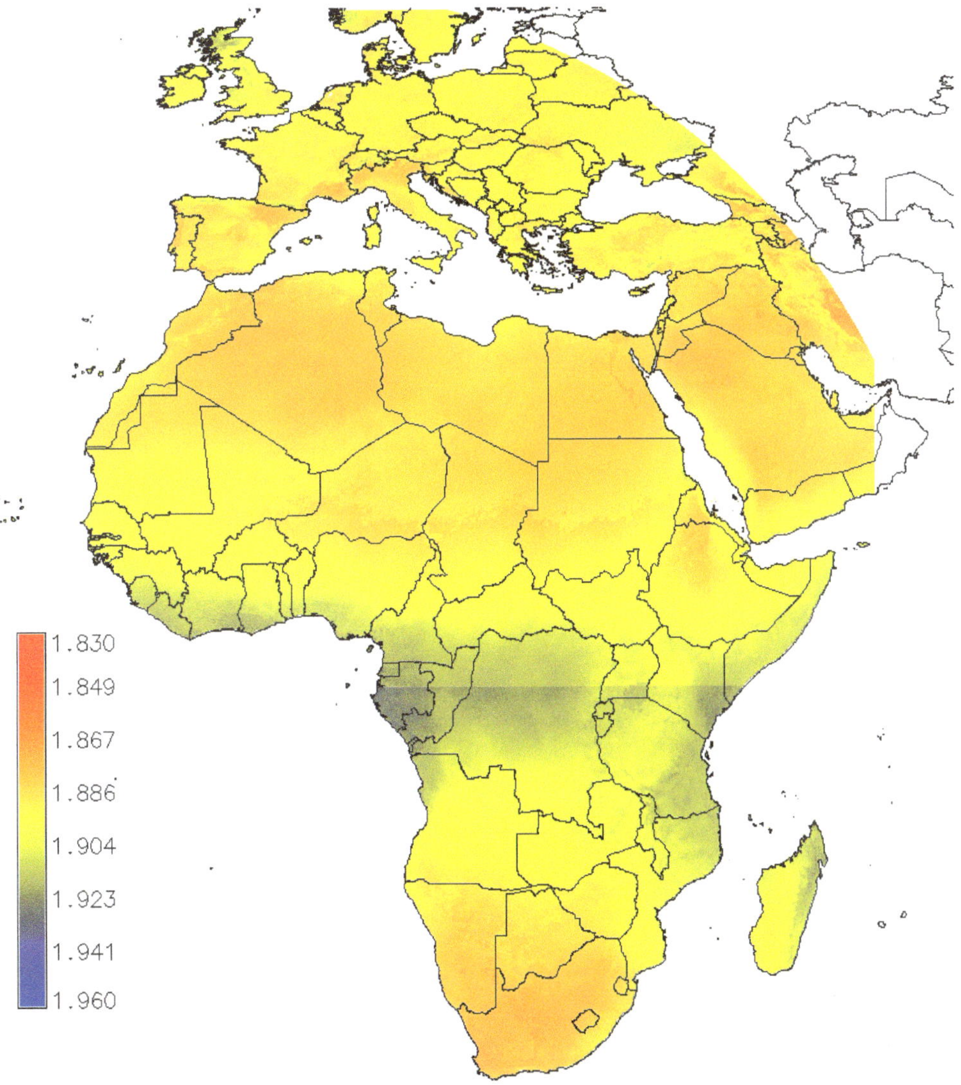

It is clear from the map that the overall pattern of variation in APE is similar to the pattern of PV performance variations shown in Figures 10–12, especially for the PV technologies with the most restricted spectral range. However, these maps will not by themselves show how good is the correlation between APE and the mismatch factor or whether APE can be used to predict MM with reasonable accuracy. A detailed discussion of this is outside the scope of this paper, but a few examples will be shown to give an idea of the issue. Figure 15a shows the instantaneous values of MM for a-Si for a location in Gabon. Figure 15b shows $\langle MM \rangle$ values for CdTe for 1,000 random locations in the study area.

Figure 15. (**a**) Instantaneous MM for a-Si as a function of APE for every hour during 2005 at a location in Gabon (1°30'S, 10°15'E). Values are shown if the global irradiance $G_{POA} > 200 \mathrm{W/m^2}$. (**b**) Annual average $\langle MM \rangle$ for CdTe for 1,000 random locations in the study area.

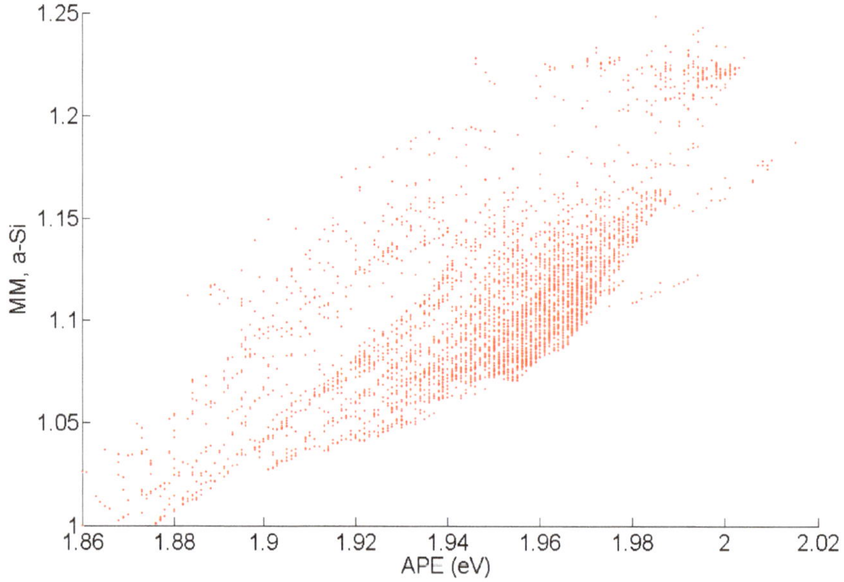

a)

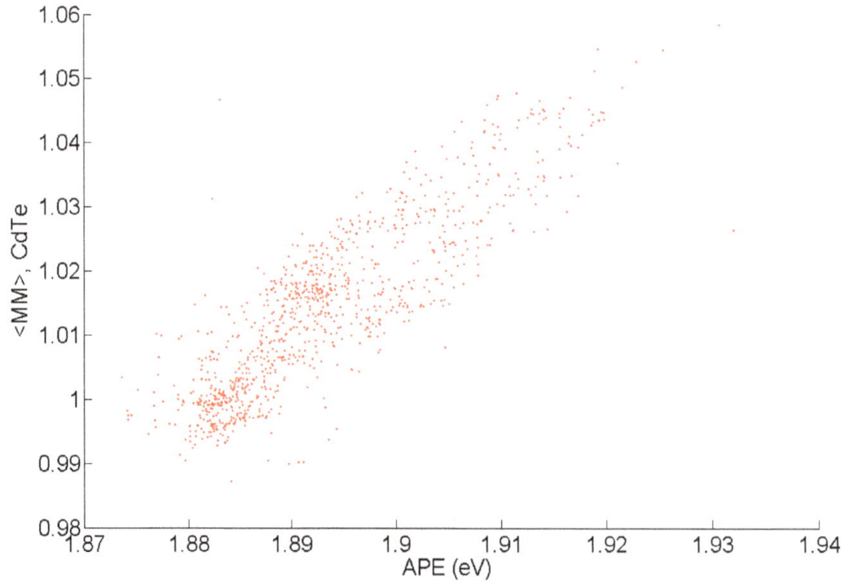

b)

While it is clear that there is indeed some relation between APE and MM, it is also clear from the figures that there is quite a lot of variation in both MM and $\langle MM \rangle$ for any given value of APE. Thus, MM for a-Si at this location may vary by almost $\pm 10\%$ for the same APE. This range is almost twice as high as the range of values of $\langle MM \rangle$ for a-Si in the whole study area. For the $\langle MM \rangle$ of CdTe, the values at different locations with the same average APE may vary by up to three percentage points, which is about half the total range of $\langle MM \rangle$ for CdTe.

These results show that at least in these cases, APE would not be a good indicator of the spectral mismatch of PV systems. In addition, there is the problem that APE must be estimated in some way for all locations where solar spectral measurements are not available, which would introduce a further uncertainty in the estimate of the mismatch. For these reasons, we would suggest that it is preferable to use spectrally-resolved radiation data from satellites to estimate the spectral mismatch for PV studies, if these data are available.

6. Conclusions

We have presented a method for calculating the effect of time-varying sunlight spectrum on the performance of PV modules using spectrally-resolved irradiance data estimated from satellite data. The satellite data necessary for the method may be calculated from the Meteosat satellites covering Eurasia and Africa.

Validation of the satellite-based spectral data performed in previous papers and in this study shows that the agreement between estimated and measured spectral data is good, both for the spectrum of global horizontal irradiance and direct normal irradiance. The method to estimate the spectral effect on PV modules was also checked against measured data from a CdTe module and found to give good agreement.

Applying spectral response data for crystalline silicon, CdTe and single-junction amorphous silicon modules to the spectral satellite data, we were able to construct maps of the annual average spectral effect on these module types over Europe and Africa. Generally, the effect is small for all three module types in desert areas, indicating that the average spectrum here is quite close to the standard spectrum. A small positive effect of +2 to +4% is seen in most of Central and Northern Europe for CdTe and a-Si modules. The strongest effect is seen in tropical Africa, where CdTe modules may produce up to about 6% more energy due to spectral effects, while for a-Si modules, the effect may reach +10%. In contrast, in this area, c-Si modules show little overall effect. The magnitude of this effect is large enough that it should be taken into account when estimating PV energy production in these areas.

For the time being, the study has been performed only for single-junction PV technologies, such as crystalline silicon, CdTe or single-junction a-Si modules. Multijunction PV technologies have a more complicated behavior under varying spectra due to the varying spectral sensitivity of the different layers. This study also considered only fixed-mounted flat-plate PV modules. Concentrating PV modules use only the direct normal irradiance, which will have a different spectrum, and, also, often use multijunction cells. We plan to use the methods developed for this study to investigate the spectral effects for these PV technologies in the near future.

Acknowledgments

The authors gratefully acknowledge the spectral irradiance data provided by the FOSS Research Centre for Sustainable Energy, Department of Electrical and Computer Engineering, University of Cyprus.

Part of the work presented here was carried out within the European Metrology Research Programme project "ENG55": "Towards an energy-based parameter for photovoltaic classification". The EMRP is jointly funded by the EMRP participating countries within the European Association of National Metrology Institutes and the European Union.

Author Contributions

Ana Maria Gracia Amillo performed most of the validation work described in the paper and wrote most of the section on the validation and the Supplementary Material.

Thomas Huld worked on the calculation of spectral irradiance from satellites using the SPECMAGIC code, wrote the code for calculating the PV spectral influence over large geographical regions and wrote most of the sections about methods and results.

Paraskevi Vourlioti worked on the validation of DNI and wrote the relevant part of the section on validation.

Richard Müller is the main developer of the SPECMAGIC code and performed the calculations of satellite-based spectral irradiance for the validation section.

Matthew Norton worked on the measurements of DNI spectra and part of the comparison of APE between measurements and satellite data.

Appendix

Table A1. Lower and upper limits in nm of the 32 Kato bands.

Kato Band	Lower Limit (nm)	Upper Limit (nm)
1	240.1	272.5
2	272.5	283.4
3	283.4	306.8
4	306.8	327.8
5	327.8	362.5
6	362.5	407.5
7	407.5	452.0
8	452.0	517.7
9	517.7	540.0
10	540.0	549.5
11	549.5	566.6
12	566.6	605.0
13	605.0	625.0
14	625.0	666.7

Table A1. *Cont.*

Kato Band	Lower Limit (nm)	Upper Limit (nm)
15	666.7	684.2
16	684.2	704.4
17	704.4	742.6
18	742.6	791.5
19	791.5	844.5
20	844.5	889.0
21	889.0	974.9
22	974.9	1,045.7
23	1,045.7	1,194.2
24	1,194.2	1,515.9
25	1,515.9	1,613.5
26	1,613.5	1,964.8
27	1,964.8	2,153.5
28	2,153.5	2,275.2
29	2,275.2	3,001.9
30	3,001.9	3,635.4
31	3,635.4	3,991.0
32	3,991.0	4,605.7

Conflicts of Interest

The authors declare no conflicts of interest.

References

1. Gottschalg, R.; Infield, D.G.; Kearny, M.J. Experimental study of variations of the solar spectrum of relevance to thin film solar cells. *Solar Energy Mater. Solar Cells* **2003**, *79*, 527–537.
2. Gottschalg, R.; Betts, T.R.; Williams, S.R.; Sauter, D.; Infield, D.G.; Kearny, M.J. A critical appraisal of the factors affecting energy production from amorphous silicon photovoltaic arrays in a maritime climate. *Solar Energy* **2004**, *77*, 909–916.
3. Minemoto, T.; Nagae, S.; Takakura, H. Impact of spectral irradiance distribution and temperature on the outdoor performance of amorphous Si photovoltaic modules. *Solar Energy Mater. Solar Cells* **2007**, *91*, 919–923.
4. Nikolaeva-Dimitrova, M.; Kenny, R.; Dunlop, E.; Pravettoni, M. Seasonal variations on energy yield of a-Si, hybrid, and crystalline Si PV modules. *Prog. Photovolt. Res. Appl.* **2010**, *18*, 311–320.
5. Gottschalg, R.; Betts, T.R.; Eeles, A.; Williams, R.R.; Zhu, J. Influences on the energy delivery of thin film photovoltaic modules. *Solar Energy Mater. Solar Cells* **2013**, *119*, 169–180.

6. Alonso-Abella, M.; Chenlo, F.; Nofuentes, G.; Torres-Ramirez, M. Analysis of spectral effects on the energy yield of different PV (photovoltaic) technologies: The case of four specific sites. *Energy* **2014**, *67*, 435–443.
7. Nofuentes, G.; Garcia-Domingo, B.; Munoz, J.V.; Chenlo, F. Analysis of the dependence of the spectral factor of some PV technologies on the solar spectrum distribution. *Appl. Energy* **2014**, *113*, 302–309.
8. Dirnberger, D.; Blackburn, G.; Müller, B.; Reise, C. On the impact of solar spectral irradiance on the yield of different PV technologies. *Solar Energy Mater. Solar Cells* **2015**, *132*, 431–442.
9. Jardine, C.; Betts, T.; Gottschlg, R.; Infield, D.G.; Lane, K. Influence of spectral effects on the performance of multijunction amorphous silicon cells. In Proceedings of the 17th European Photovoltaic and Solar Energy Conference, Munich, Germany 22–26 October 2001; Volume 44.
10. Belluardo, G.; Wagner, J.; Tetzlaff, A.; Moser, D. Evaluation of spectral effect on module performance using modeled average wavelength. In Proceedings of the 28th European Photovoltaic Solar Energy Conference and Exhibition, Paris, France, 30 September 2013.
11. Peharz, G.; Siefer, G.; Bett, A.W. A simple method for quantifying spectral impacts on multi-junction solar cells. *Solar Energy* **2009**, *83*, 1588–1598.
12. Minemoto, T.; Nakada, Y.; Takahashi, H.; Takakura, H. Uniqueness verification of solar spectrum index of average photon energy for evaluating outdoor performance of photovoltaic modules. *Solar Energy* **2009**, *83*, 1294–1299.
13. Müller, R.; Behrendt, T.; Hammer, A.; Kemper, A. A new algorithm for the satellite-based retrieval of solar surface irradiance in spectral bands. *Remote Sens.* **2012**, *4*, 622–647.
14. Behrendt, T.; Kuehnert, J.; Hammer, A.; Lorenz, E.; Betcke, J.; Heinemann, D. Solar spectral irradiance derived from satellite data: A tool to improe thin film PV performance estimations? *Solar Energy* **2013**, *98*, 100–110.
15. Mueller, R.; Trentmann, J.; Träger-Chatterjee, C.; Posselt, R.; Stöckli, R. The role of the effective cloud Albedo for climate monitoring and analysis. *Remote Sens.* **2011**, *3*, 2305–2320.
16. Morcrette, J.; Boucher, O.; Jones, L.; Salmond, D.; Bechtold, P.; Beljaars, A.; Benedetti, A.; Bonet, A.; Kaiser, J.; Razinger, M.; *et al.* Aerosol analysis and forecast in the European Centre for Medium-Range Weather Forecasts Integrated Forecast System: Forward modeling. *J. Geophys. Res.* **2009**, *114*, doi:10.1029/2008JD011235.
17. Benedetti, A.; Morcrette, J.; Boucher, O.; Dethof, A.; Engelen, R.; Fisher, M.; Flentje, H.; Huneeus, N.; Jones, L.; Kaiser, J.; *et al.* Aerosol analysis and forecast in the European Centre for Medium-Range Weather Forecasts Integrated Forecast System: 2. Data assimilation. *J. Geophys. Res.* **2009**, *114*, doi:10.1029/2008JD011115.
18. Dee, D.; Uppala, S. Variational bias correction of satellite radiance data in the ERA-interim reanalyis.. *Q. J. R. Meteorol. Soc.* **2009**, *135*, 1830–1841.
19. Betts, A.; Ball, J.; Bosilovich, M.; Viterbo, P.; Zhang, Y.; Rossow, W. Intercomparison of water and energy budgets for five Mississippi subbasins between ECMWF re-analysis (ERA-40) and NASA Data Assimilation Office fvGCM for 1990–1999. *J. Geophys. Res.* **2003**, *108*, doi: 10.1029/2002JD003127.

20. Mayer, B.; Kylling, A. Technical note: The libRadtran software package for radiative transfer calculations-Description and examples of use. *Atmos. Chem. Phys.* **2005**, *135*, 1855–1877.
21. Gracia Amillo, A.; Huld, T.; Müller, R. A New Database of Global and Direct Solar Radiation Using the Eastern Meteosat Satellite, Models and Validation. *Remote Sens.* **2014**, *6*, 8165–8189.
22. Kato, S.; Ackerman, T.; Mather, J.; E.E.C. The k-distribution method and correlated-k approximation for a shortwave radiative transfer model. *J. Quant. Spectrosc. Radiat. Transf.* **1999**, *62*, 109–121.
23. Gueymard, C. Prediction and validation of cloudless shortwave solar spectra incident on horizontal, tilted, or tracking surfaces. *Solar Energy* **2008**, *82*, 260–271.
24. Šúri, M.; Hofierka, J. A New GIS-based Solar Radiation Model and Its Application to Photovoltaic Assessments. *Trans. GIS* **2004**, *8*, 175–190.
25. Muneer, T. Solar radiation model for Europe. *Build. Serv. Eng. Res. Technol.* **1990**, *4*, 153–163.
26. Perez, R.; Seals, R.; Ineichen, P.; Stewart, R.; Menicucci, D. A new simplified version of the Perez diffuse irradiance model for tilted surfaces. *Solar Energy* **1987**, *39*, 221–231.
27. Martin, N.; Ruiz, J. Calculation of the PV modules angular losses under field conditions by means of an analytical model. *Solar Energy Mater. Solar Cells* **2001**, *70*, 25–38.
28. Bahrs, U.; Zaaiman, W.; Mertens, M.; Ossenbrink, H. Automatic large area spectral response facility. In Proceedings of the 14th European Photovoltaic SolarEnergy Conference, Barcelona, Spain, 30 June–4 July 1997; pp. 2296–2298.
29. King, D.; Boyson, W.; Kratochvil, J. *Photovoltaic Array Performance Model*; Technical Report SAND2004-3535; Sandia National Laboratories: Albuquerque, NM, USA, 2004.
30. Friesen, G.; Chianese, D.; Pola, I.; Realini, A.; A., B. Energy rating measurements and predictions at ISAAC. In Proceedings of the 22nd European Photovoltaic SolarEnergy Conference, Milan, Italy, 3–7 September 2007; pp. 2754–2757.
31. Skoplaki, E.; Palyvos, J. On the temperature dependence of photovoltaic module electrical performance: A review of efficiency/power correlations. *Solar Energy* **2009**, *83*, 614–624.
32. Huld, T.A.; Friesen, G.; Skoczek, A.; Kenny, R.A.; Sample, T.; Field, M.; Dunlop, E.D. A power-rating model for crystalline silicon PV modules. *Solar Energy Mater. Solar Cells* **2011**, *95*, 3359–3369.
33. Faiman, D. Assessing the outdoor operating temperature of photovoltaic modules. *Prog. Photovolt. Res. Appl.* **2008**, *16*, 307–315.
34. Koehl, M.; Heck, M.; Wiesmeier, S.; Wirth, J. Modeling of the nominal operating cell temperature based on outdoor weathering. *Solar Energy Mater. Solar Cells* **2011**, *95*, 1638–1646.
35. Norton, M.; Gracia Amillo, A.; Galleano, R. Comparison of Solar Spectral Irradiance Measurements using the Average Photon Energy Parameter. Available online: http://re.jrc.ec.europa.eu/pvgis/spectralpaper/comparison-solar- spectral.pdf (accessed on 17 April 2015).
36. Norton, M.; Paraskeva, V.; Galleano, R.; Makrides, G.; Kenny, R.; Georghiou, G. High quality measurements of the solar spectrum for simulation of multi-junction photovoltaic cell yields. In Proceedings of the 29th European Photovoltaic Solar Energy Conference and Exhibition, Amsterdam, The Netherlands, 22–26 September 2014; pp. 2002–2007.

37. Supplementary material to the present paper. Available online: http://re.jrc.ec.europa.eu/pvgis/spectralpaper/supplementary.docx (accessed on 17 April 2015).
38. Kenny, R.; Dunlop, E.; Ossenbrink, H.; Müllejans, H. A practical method for the energy rating of c-Si photovoltaic modules based on standard tests. *Prog. Photovolt. Res. Appl.* **2006**, *14*, 155–166.
39. Dee, D.P.; Uppala, S.M.; Simmons, A.J.; Berrisford, P.; Poli, P.; Kobayashi, S.; Andrae, U.; Balmaseda, M.A.; Balsamo, G.; Bauer, P.; *et al.* The ERA-Interim reanalysis: Configuration and performance of the data assimilation system. *Q. J. R. Meteorol. Soc.* **2011**, *137*, 553–597.

2

Estimation of State of Charge for Two Types of Lithium-Ion Batteries by Nonlinear Predictive Filter for Electric Vehicles

Yin Hua [1], Min Xu [1,*], Mian Li [1,2], Chengbin Ma [2] and Chen Zhao [2]

[1] National Engineering Laboratory for the Automotive Electronic Control Technology, Shanghai Jiao Tong University, 800 Dongchuan Rd., Shanghai 200240, China; E-Mail: hua.yin1986@163.com

[2] University of Michigan-Shanghai Jiao Tong University Joint Institute, Shanghai Jiao Tong University, 800 Dongchuan Rd., Shanghai 200240, China; E-Mails: mianli@sjtu.edu.cn (M.L.); chbma@sjtu.edu.cn (C.M.); zhaochen815@aliyun.com (Z.C.)

* Author to whom correspondence should be addressed; E-Mail: EngineLabXu@126.com or mxu@sjtu.edu.cn

Academic Editors: Paul Stewart and Chris Bingham

Abstract: Estimation of state of charge (SOC) is of great importance for lithium-ion (Li-ion) batteries used in electric vehicles. This paper presents a state of charge estimation method using nonlinear predictive filter (NPF) and evaluates the proposed method on the lithium-ion batteries with different chemistries. Contrary to most conventional filters which usually assume a zero mean white Gaussian process noise, the advantage of NPF is that the process noise in NPF is treated as an unknown model error and determined as a part of the solution without any prior assumption, and it can take any statistical distribution form, which improves the estimation accuracy. In consideration of the model accuracy and computational complexity, a first-order equivalent circuit model is applied to characterize the battery behavior. The experimental test is conducted on the $LiCoO_2$ and $LiFePO_4$ battery cells to validate the proposed method. The results show that the NPF method is able to accurately estimate the battery SOC and has good robust performance to the different initial states for both cells. Furthermore, the comparison study between NPF and well-established extended Kalman filter for battery SOC estimation indicates that the proposed NPF method has better estimation accuracy and converges faster.

Keywords: state of charge; lithium-ion battery; electric vehicles; nonlinear predictive filter

---

1. Introduction

Global warming, the petroleum crisis, and legislation pushing for higher fuel economy and lower emissions, are leading to the development of electric vehicles (EVs) [1,2]. As the key component of any electric vehicle, the energy storage system attracts more and more attention. A variety of electrochemical energy storage devices are currently used in EV applications, such as lithium-ion (Li-ion) battery, nickel metal hydride (NiMH) battery, lead acid (LA) battery, and ultracapacitor (UC). Among them, Li-ion batteries are viewed as the most promising energy storage units for EVs, for its high energy density, high power density, low self-discharging rate, and long lifespan [3,4].

However, strict requirements should be satisfied when using Li-ion batteries, and a battery management system (BMS) is required to provide the functions of monitoring, estimation, and protection to ensure the safe operations of Li-ion batteries. The state of charge (SOC), acting the similar role as the fuel meter for the internal combustion engine system, is the most important factor for batteries which should be accurately estimated by the BMS. The battery SOC indicates the residual capacity of the battery system and has significant importance in predicting the remaining driving range of EVs. Besides, accurate SOC estimation can also prevent the batteries from over-charging and over-discharging conditions and thus can extend the battery cycle life [5]. However, since the battery SOC cannot be directly measured and it is affected by many factors, such as current, temperature and battery age, estimation of the battery SOC is still a challenging problem that needs to be solved.

A number of SOC estimation methods have been proposed. Each method has its own advantages and limitations. Generally, these methods can be mainly classified into two kinds: (1) direct measurement based estimation and (2) model based estimation. The first kind of method directly uses the measurements from battery system to calculate the SOC, such as current integration method [6], open circuit voltage (OCV) based method [7]. The current integration method is easy to implement with low computation, but it suffers from the low estimation accuracy due to the accumulative errors caused by current sensor noises. In addition, it is also difficult to obtain the initial SOC when using the current integration method. Therefore, the open circuit voltage method is usually used complementarily with the current integration method to recalibrate the SOC and to provide the initial SOC. However, a long rest time of the tested battery is required to reach the open circuit voltage, which is usually unrealistic for real world applications.

In the second kind of method, the battery model is utilized when estimating the battery SOC. One of the model based methods for SOC estimation is based on the black-box battery models, such as neural networks (NN) [8], fuzzy logic (FL) [9], and support vector machine (SVM) [10]. Eddahech *et al.* [8] developed a recurrent neural network as a SOC predictor that takes into account operational conditions, the results show that the predictor allows very precise SOC estimation. Salkind *et al.* [9] utilized the fuzzy logic to estimate the battery SOC by using the training datasets obtained by impedance spectroscopy and coulomb counting techniques. Anton *et al.* introduced a support vector machine based SOC estimator for a high-capacity lithium iron manganese phosphate ($LiFeMnPO_4$) battery cell, using cell current, cell voltage, and cell temperature as independent variables. The results show that the SVM SOC estimator maintains a high level of accuracy. According to the literatures, the black-box model based methods can be quite accurate if sufficient experimental data is used to train the model. However, their performance highly depends on the quantity and quality of the training data set, a large amount of offline battery tests are necessary to obtain a good model which can be very time-consuming. Optimum state filtering method is another kind of model based method for battery SOC estimation. This method usually performs SOC estimation based on an equivalent circuit battery model [11,12]. Many different state filtering methods have been investigated, such as extended Kalman filter (EKF) [13–15], sigma point Kalman filter (SPKF) [16–18], adaptive extended Kalman filter (AEKF) [19], adaptive unscented Kalman filter (AUKF) [20], particle filter (PF) [21] and others [22–26]. Plett [13–17] established the EKF and UKF based SOC estimation methods using different orders of equivalent circuit battery models for simultaneous state and parameters estimation of LiPB packs. Both of them obtained very good results, and the methods were robust to different initial states. Han *et al.* adopt the AEKF method for SOC estimation by adaptively updating the process and measurement noise covariance which improved the estimation accuracy. Similarly, Hu *et al.* applied the AUKF for SOC estimation. All of the Kalman filter (KF) based methods achieve very good estimation performance because of online state error correction capability. However, there are some shortcomings for KF based methods. For instance, the statistic distribution of the process noise is assumed to be the zero mean white Gaussian process and prior knowledge of the noise covariance should be known before estimating the battery SOC. In fact, it is difficult to obtain the accurate information of the process noises for real world applications, and the filter performance will decrease or even diverge with inaccurate noise information. Besides, the assumption of the zero-mean Gaussian process noises usually cannot be met in practice which can decrease the estimation accuracy [27]. PF is another optimum state filtering methodology for SOC estimation [21]. It is able to represent any probability density function for the state by using Monte Carlo sampling methods which improves the estimation accuracy. However, the computational effort is high for PF due to the large amount of particles which makes it difficult to apply for real-world applications.

In this paper, the nonlinear predictive filter (NPF) is proposed to estimate the SOC of Li-ion batteries and the proposed method is evaluated on the battery cells with different chemistries. The NPF method was firstly proposed by Crassidis *et al.*, for spacecraft attitude estimation and obtained satisfactory results [28]. It is a nonlinear optimum state estimation method implemented on continuous-discrete time systems, which is particularly suitable for battery systems [29]. In the NPF method, the nonlinear dynamic system is treated as a preliminary model with a to-be-determined model error part, where the preliminary model describes the system dynamic and the model error mainly represents the system process noise. The significant advantage of NPF method is that the process noise is treated as an unknown model error determined as a part of the solution, and it is able to represent any distribution form. Compared to the aforementioned direct measurement based estimation method, the NPF method can provide more accurate results with better robust performance of initial values due to its capability of correcting the state error online. In comparison with black-box model based method, the large amount of training data sets is not required for NPF method which saves the computational effort. The KF based method usually assumes a zero mean white Gaussian process noise and requires the prior knowledge. In contrary, the process noise in NPF can be any type without any restricting pre-assumptions. Therefore, the NPF based SOC estimation method has better estimation performance and is more practical for real-world applications. Compared to the PF method, the NPF method has lower computation, since a large number of particles needs to be used and numerous matrix operations are required for PF method which greatly increase the requirements for hardware system. Additionally, the aforementioned literatures mainly perform their methods on one type of batteries; the robustness performance of different battery types is not discussed. In this study, the NPF method is performed on $LiCoO_2$ (LCO) and $LiFePO_4$ (LFP) battery cells, the experimental results show that the NPF based method can accurately estimate the battery SOC with good robustness to different initial values. Meanwhile, the estimation result of LFP battery suffers from a lower accuracy than that of LCO battery due to the characteristics of the flat open circuit voltage. Furthermore, the comparison study between NPF and EKF with the same experimental conditions indicates that the proposed NPF method has better estimation accuracy and faster convergence rate.

The rest of the paper is organized as follows. In Section 2, a first-order equivalent circuit battery model is introduced. The description for SOC estimation of the Li-ion battery cell using NPF is presented in Section 3. In Section 4, the estimation results are analyzed and discussed to verify the proposed method. Finally, the conclusion is provided in Section 5.

2. Battery Model

2.1. Model Structure

In order to apply NPF for SOC estimation, a suitable battery model is required to characterize the electrochemical properties of Li-ion batteries, including: ohmic resistance, charge transfer and diffusion. Different types of battery models have been proposed in the literature. Among them, the equivalent circuit models (ECMs) are the most commonly used ones for battery state estimation. The ECMs capture the battery input-output dynamics through electrical circuit elements, such as resistor, capacitor and voltage source, and can be easily used for model based estimation. Hu *et al.* introduced a comprehensive study for different types of ECMs, and their results indicate that the first-order ECM achieves an excellent compromise between accuracy and complexity [30]. Therefore, as shown in Figure 1, a first-order ECM composed of an open circuit voltage (OCV) source, a resistor, and an RC network, is used in this study. The resistor represents the electrical resistance of battery components with the accumulation and dissipation of charge in the electrical double-layer. The RC network describes the charge transfer and diffusion effect, and the voltage source indicates the battery's open circuit voltage which is a function of the battery SOC.

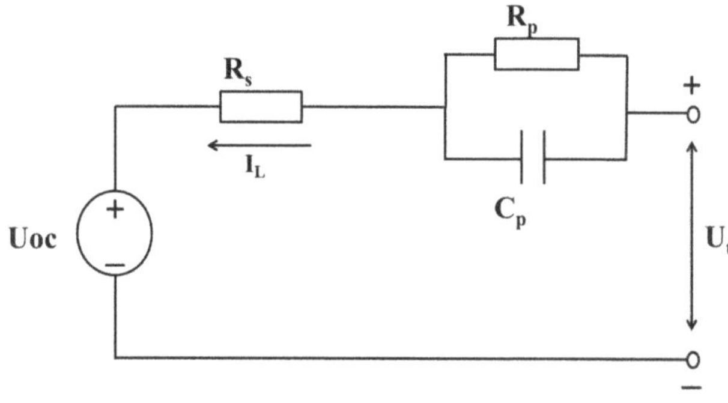

Figure 1. Schematic diagram of the first-order battery model.

The SOC definition for the Li-ion battery is formulated as:

$$SOC(t) = SOC(0) + \int_0^t \frac{\eta I_L(t)dt}{C_n} \tag{1}$$

where $SOC(0)$ is the initial SOC value, $SOC(t)$ is the battery SOC at time t, η is the Coulombic efficiency (in this paper η is assumed as 1), I_L is the input current (positive for charge, negative for discharge) and C_n is the nominal capacity.

Equations (2) and (3) describe the electrical behavior of the Li-ion battery:

$$\dot{U}_p(t) = -(R_p C_p)^{-1} U_p(t) + C_p^{-1} I_L(t) \tag{2}$$

$$U_t(t) = U_{oc}(SOC(t)) + U_p + I_L(t) R_s \tag{3}$$

where U_p is the polarization voltage, R_p and C_p are the polarization resistance and capacitance, respectively, U_t is the battery terminal voltage, and R_s is the ohmic resistance.

A spline function is employed to describe the relationship between the battery open circuit voltage U_{oc} and battery SOC, given by Equation (4):

$$U_{oc}(t) = Spline(SOC(t)) \tag{4}$$

2.2. Model Parameter Identification

2.2.1. Experimental Setup

In this work, two different types of battery cells, $LiCoO_2$ (LCO) and $LiFePO_4$ (LFP), are tested. The standard specifications of the tested cells are listed in Tables 1 and 2. The schematic diagram of the battery test bench is shown in Figure 2. It consists of a Takasago ZX-800LA electric power, a Kikusui PLZ150U electric load, a NI cDAQ-9174 data acquisition system, a host PC, and a thermal chamber. The ZX-800LA electric power can charge the battery cell with the maximum current of 80A at the maximum voltage of 84V, while the PLZ150U electric load is able to provide the maximum discharge current of 30A with the maximum voltage of 150V. The electric power and load are remotely controlled by the host PC to determine the charge/discharge power load of the tested battery cell. The data acquisition system with a sampling rate of 10 Hz is used to capture the current, voltage, and temperature of the tested battery and to transfer the obtained data to the host PC.

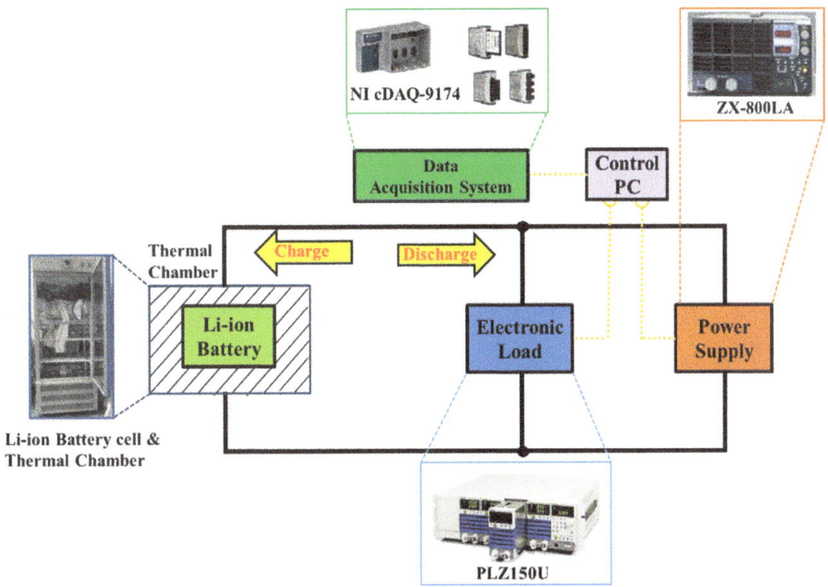

Figure 2. Schematic of battery test bench.

Table 1. Battery cell (LCO) specification.

Item	Specification
Cell Dimensions (mm)	Ø 18 × 69
Cell Weight (g)	48.2
Cell Capacity (nominal, Ah)	2.6
Cell Voltage (nominal, V)	3.7
Gravimetric Energy Density (nominal, Wh/kg)	180
Volumetric Energy Density (nominal, Wh/L)	464
Operating Temperature	−20 °C to 60 °C

Table 2. Battery cell (LFP) specification.

Item	Specification
Cell Dimensions (mm)	Ø 32 × 113
Cell Weight (g)	205
Cell Capacity (nominal, Ah)	4.5
Cell Voltage (nominal, V)	3.3
Gravimetric Energy Density (nominal, Wh/kg)	71
Volumetric Energy Density (nominal, Wh/L)	161
Operating Temperature	−30 °C to 55 °C

2.2.2. Parameter Identification

For the first-order battery model, the values of the model parameters as well as the SOC-OCV relationship need to be identified. A series of battery tests, including: capacity test, pulse current test, and open circuit voltage test, are conducted to extract these parameters. The content of these tests are described as follows:

(1) Capacity test: The capacity test discharges the battery cell from the fully charged state (upper-limit voltage) to the fully discharged state (lower-limit voltage) with 0.5 C rate, and the cell capacity is referred as the total Ampere-hours drained out of the battery during the test. The cut-off voltages used during the test for LCO battery are $V_{\max} = 4.2$ V, $V_{\min} = 2.8$ V, and the cut-off voltages for LFP battery are $V_{\max} = 3.6$ V, $V_{\min} = 2$ V. The experimental results of the capacities for the tested LCO and LFP cells are 2.62 Ah and 4.29 Ah, respectively.

(2) Pulse current test: To identify the values of the electrical circuit elements in the first-order ECM, a pulse current test is conducted on the battery cells at 10% SOC intervals starting from 0.9 to 0.3. During the test, the environment temperature is controlled at 25 °C. The detailed test procedure can be found in [31]. In this study, the time period between two current pluses, when no current is applied, is used for parameters identification. The current and voltage profiles during this time period are shown in Figure 3. The ohmic resistance R_s can be expressed as:

$$R_s = \frac{U_s}{I_L} \tag{5}$$

where U_s is the instantaneous voltage response within one second, and I_L is the current before the rest time period. The polarization voltage U_t at time t can be expressed by:

$$U_t = U_p(1 - e^{-\frac{t}{R_p C_p}}),\ U_p = R_p I_L \quad (6)$$

where U_p is the maximum polarization voltage during the rest time period. The polarization resistance and capacitance can be identified by minimizing the difference between the model output and voltage measurement using the nonlinear least square method. The identification results of R_s, R_p, C_p for LCO and LFP cells are shown in Figure 4. It can be seen that the parameters vary at different battery SOC points. However, to reduce the complexity of the battery model, R_s, R_p, C_p at different SOC points are averaged to obtain the final model parameters and the results are shown in Table 3.

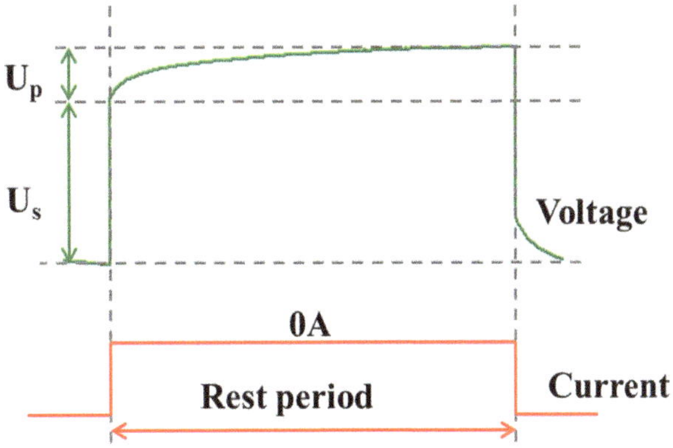

Figure 3. Voltage profile during the rest time period.

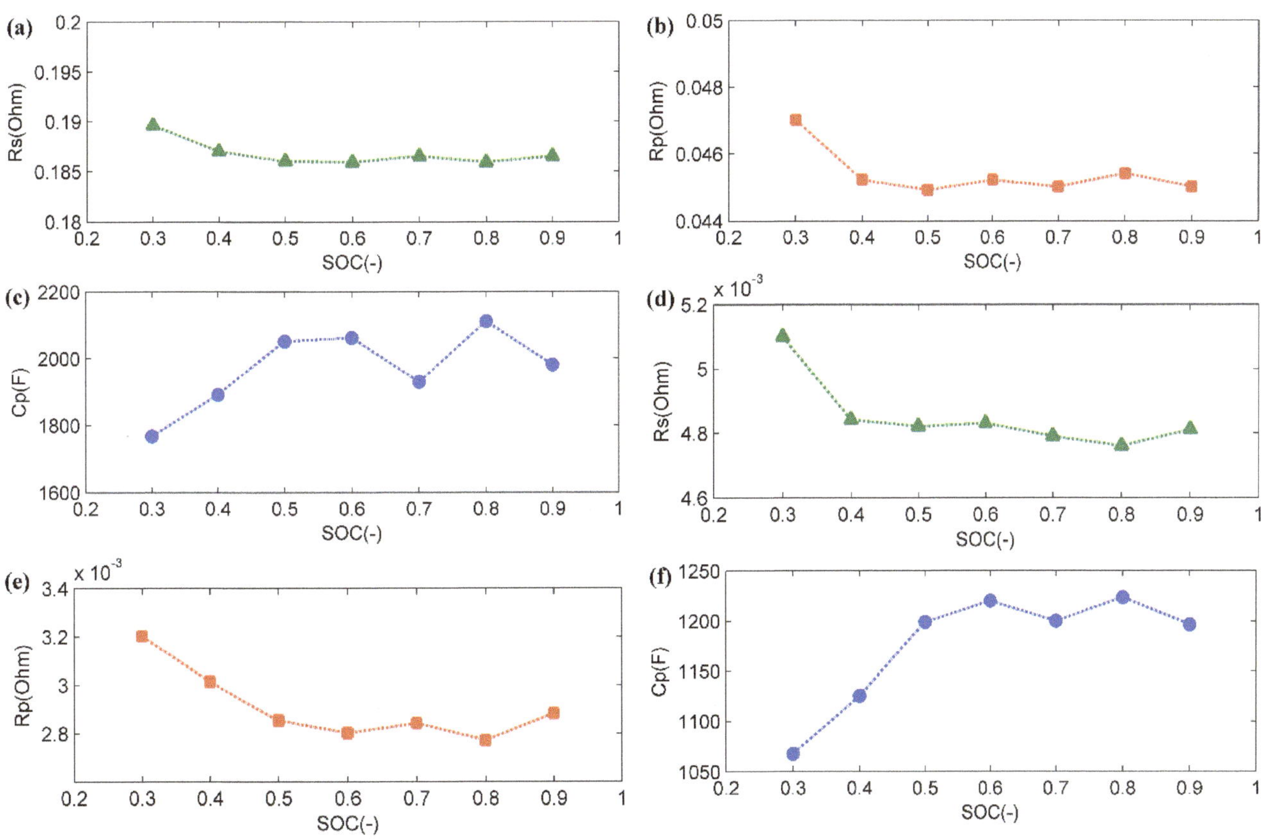

Figure 4. Parameters identification results: **(a)** R_s results of $LiCoO_2$ (LCO) cell; **(b)** R_p results of LCO cell; **(c)** C_p results of LCO cells; **(d)** R_s results of $LiFePO_4$ (LFP) cell; **(e)** R_p results of LFP cell; **(f)** C_p results of LFP cells.

(3) Open circuit voltage test: To calibrate the nonlinear SOC-OCV relationship, an open circuit voltage test is conducted as follows. The battery cell is discharged using 0.5 C constant current at 5% SOC interval from 100% SOC to 15% SOC. After each discharge period, the battery cell is rested for 3 hours to reach the close-to-equilibrium open-circuit potential for each SOC point. A similar procedure is conducted to get the SOC-OCV curve under the battery charge condition. Since the possible hysteresis voltage is neglected in this paper, the SOC-OCV relationship for the battery model is defined as the average of the equilibrium potentials of charging and discharging. The experimental results of the SOC-OCV curves for LCO and LFP battery cells are shown in Figure 5.

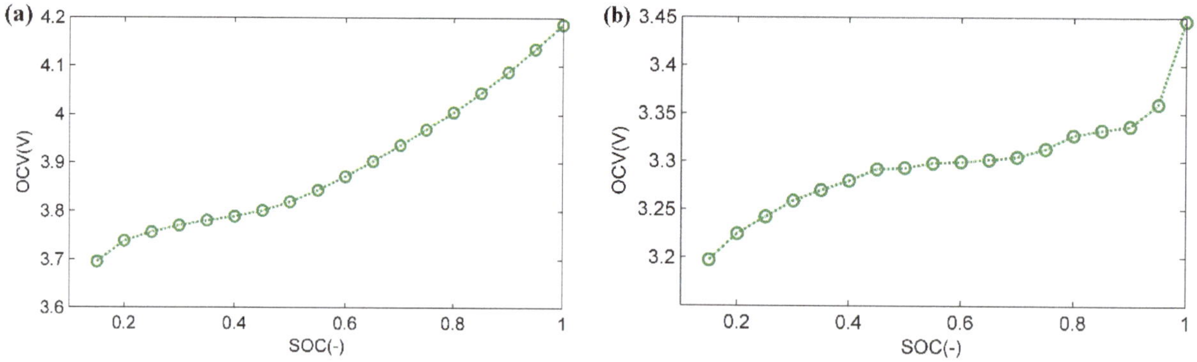

Figure 5. SOC-OCV curves: (**a**) LCO battery cell; (**b**) LFP battery cell.

Table 3. Model parameters.

LCO Battery Cell				
Model parameters	$R_s(\Omega)$	$R_p(\Omega)$	$C_p(F)$	$C_n(Ah)$
Values	0.187	0.046	1969	2.62
LFP battery cell				
Model parameters	$R_s(\Omega)$	$R_p(\Omega)$	$C_p(F)$	$C_n(Ah)$
Values	0.0048	0.0029	1186	4.29

2.3. Model Validation

In order to validate the battery models with identified parameters for LCO and LFP battery cells, the experimental test using Urban Dynamometer Driving Schedule (UDDS) driving cycle is conducted. UDDS is usually used for light duty vehicle testing under the city driving condition [32]. Therefore, the UDDS driving cycle is adopted in this study to simulate the battery dynamics under a realistic EV scenario. The model validation results for LCO and LFP battery cells are shown in Figures 6 and 7. Figure 6a,b shows the comparison profiles of the estimated terminal voltage and measured terminal voltage for LCO battery cell. Figure 6c shows the corresponding voltage error and it can be seen that the maximum model error is around 0.02 V. Figure 6d shows the root mean squared error (RMSE) of the terminal voltage, which is less than 0.007 V. Similar results for LFP battery cell are shown in Figure 7. The maximum model error is within 0.01 V for LFP battery cell, while the RMSE is less than 0.004 V. According to the results, it can be concluded that the first order battery model is able to accurately capture the dynamic behavior of both Li-ion battery cells.

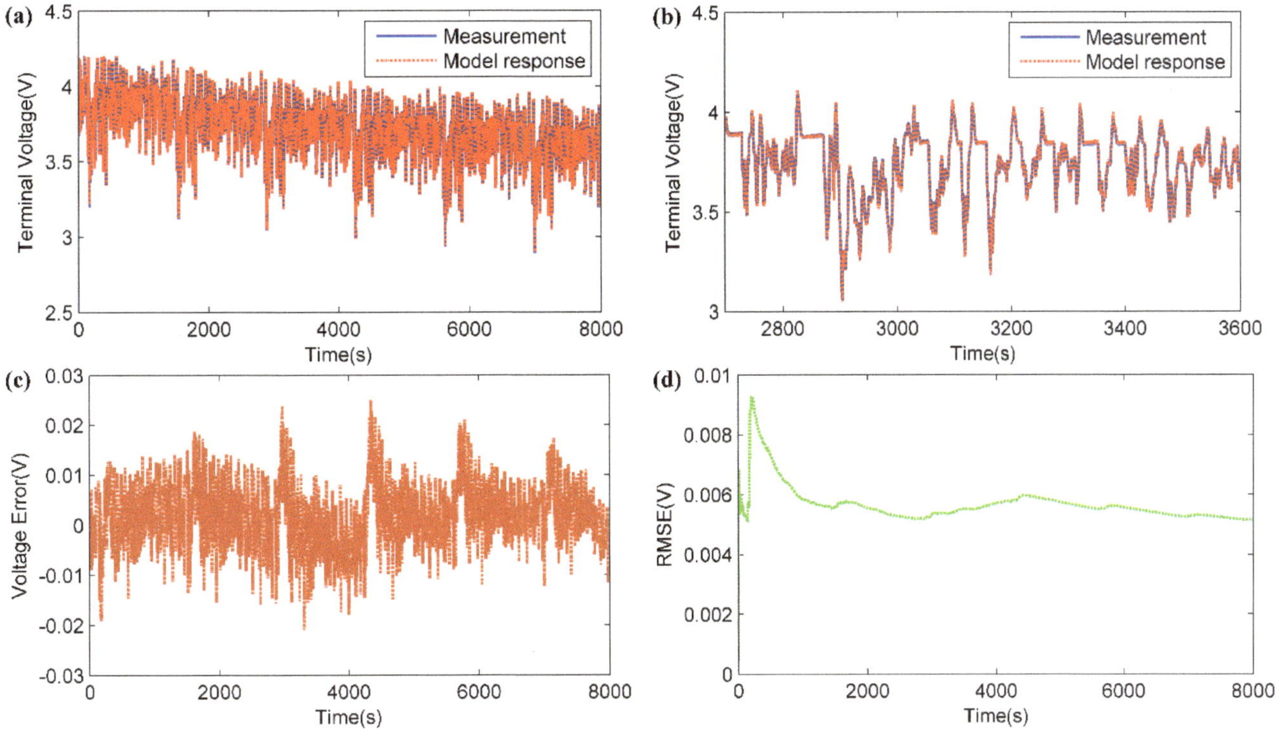

Figure 6. Model validation results of LCO cell: (**a**) Comparison of model output voltage and measured voltage; (**b**) Zoom plot; (**c**) Voltage error; (**d**) RMSE of voltage error.

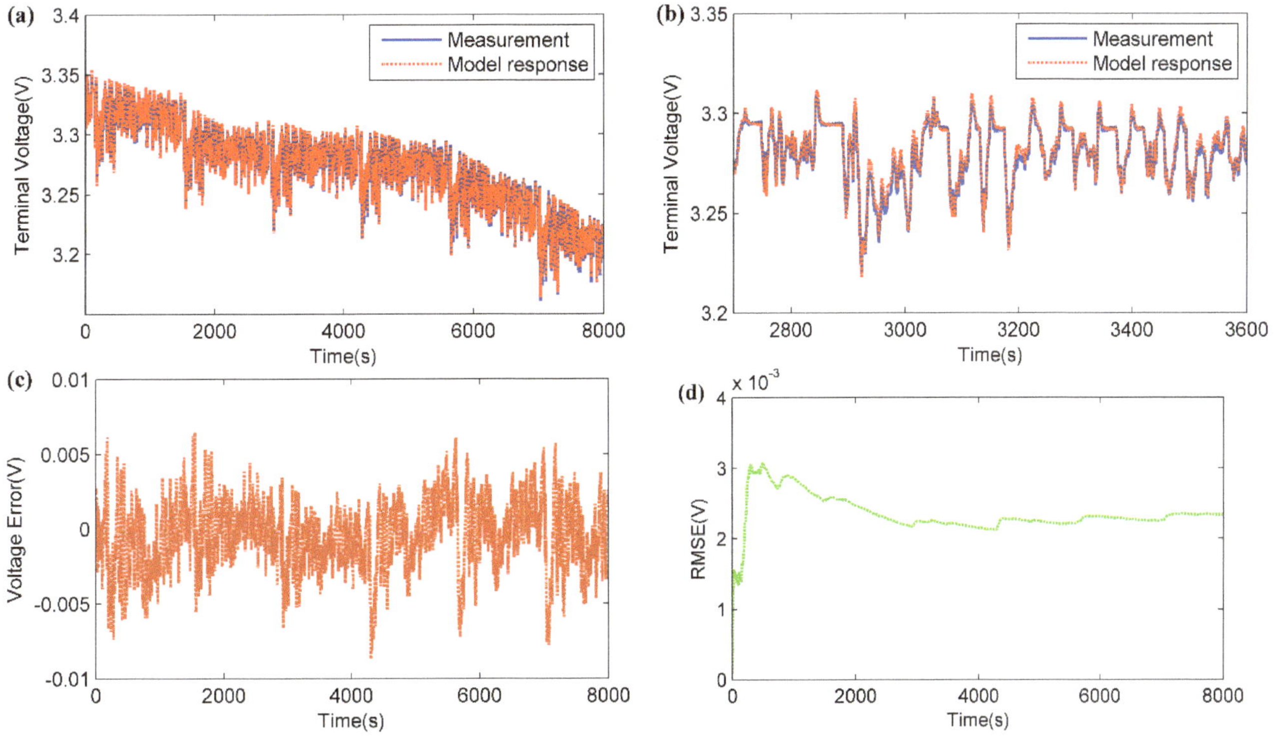

Figure 7. Model validation results of LFP cell: (**a**) Comparison of model output voltage and measured voltage; (**b**) Zoom plot; (**c**) Voltage error; (**d**) RMSE of voltage error.

3. Nonlinear Predictive Filter for SOC Estimation

In this section, the description of nonlinear predictive filter (NPF) is firstly introduced. Then, the NPF based SOC estimation using the first-order battery model is presented.

3.1. Nonlinear Predictive Filter

Nonlinear predictive filter (NPF) is a model based state estimation method implemented with nonlinear continuous-discrete time system. The state and measurement equations in NPF are given by Equations (7) and (8):

$$\dot{x}(t) = f[x(t),t] + g(t)d(t) \tag{7}$$

$$y(t_k) = h[x(t_k),t_k] + v(t_k) \tag{8}$$

where $x(t)$ is the state vector which needs to be estimated, f is system dynamic function, $d(t)$ is the model error which mainly represents the process noise and $g(t)$ is the model error distribution matrix. $y(t_k)$ is the system output sampled at time step t_k, h is the system measurement function, v is the measurement noise which is assumed to be an independent zero mean Gaussian white noise with:

$$E[v(t_k)] = 0, E[v(t_k)v(t_{k'})^T] = R\delta_{kk'} \tag{9}$$

where R is a positive-definite covariance matrix.

In order to estimate the system states, the model error $d(t)$ needs be obtained at every time step. A cost function consisting of the weighted sum square of the measurement-minus-estimate residuals plus the weighted sum square of the model correction term is established in order to get the mathematical expression of $d(t)$, as defined in Equaton (10):

$$J[d(t)] = 0.5[y(t+\Delta t) - \hat{y}(t+\Delta t)]^T R^{-1}[y(t+\Delta t) - \hat{y}(t+\Delta t)] + 0.5d(t)^T W d(t) \tag{10}$$

where $y(t) = y(t_k), y(t+\Delta t) = y(t_{k+1})$, Δt is the sampling interval, W is a positive semi-definite weighting matrix. A first-order Taylor expansion is used to approximate $\hat{y}(t+\Delta t)$, given in Equation (11):

$$\hat{y}(t+\Delta t) = \hat{y}(t) + \Delta t\dot{\hat{y}}(t) + 0.5\Delta t^2\ddot{\hat{y}}(t) + \cdots \approx \hat{y}(t) + Z(\hat{x}(t),\Delta t) + \Lambda(\Delta t)S(\hat{x}(t))d(t) \tag{11}$$

The solution of $d(t)$ is derived by minimizing the cost function $J[d(t)]$, given in Equation (12):

$$\begin{aligned} d(t) = &-\{[\Lambda(\Delta t)S(\hat{x}(t))]^T R^{-1}\Lambda(\Delta t)S(\hat{x}(t)) + W\}^{-1} \\ &\times[\Lambda(\Delta t)S(\hat{x}(t))]^T R^{-1}[Z(\hat{x}(t),\Delta t) + \hat{y}(t) - y(t+\Delta t)] \end{aligned} \tag{12}$$

where $S(\hat{x}(t_k)), Z(\hat{x}(t_k),\Delta t), \Lambda(\Delta t)$ are intermediate matrices, given as follows:

$$S(\hat{x}(t)) = \begin{bmatrix} L_{g_1}L_f^{r_1-1}h_1(\hat{x}(t)) & \cdots & L_{g_l}L_f^{r_1-1}h_1(\hat{x}(t)) \\ \vdots & \ddots & \vdots \\ L_{g_1}L_f^{r_m-1}h_m(\hat{x}(t)) & \cdots & L_{g_l}L_f^{r_m-1}h_m(\hat{x}(t)) \end{bmatrix} \tag{13}$$

$$\Lambda(\Delta t) = \begin{bmatrix} \lambda_{11} & \cdots & 0 \\ \vdots & \lambda_{ii} & \vdots \\ 0 & \cdots & \lambda_{mm} \end{bmatrix}, \lambda_{ii} = \frac{\Delta t^{r_i}}{r_i!}, i = 1,2,\cdots m \tag{14}$$

$$Z_i(\hat{x}(t),\Delta t) = \sum_{a=1}^{r_i}\frac{\Delta t^a}{a!}L_f^a h_i(\hat{x}(t)), i = 1,2,\cdots,m \tag{15}$$

Here, $L_{g_j} L_f^{r_i - 1} h_i(\hat{x}(t))$ is the Lie derivation, defined as:

$$L_f^{\,0} h_i(\hat{x}(t)) = h_i(\hat{x}(t)) \tag{16}$$

$$L_f^{\,n} h_i(\hat{x}(t)) = \frac{\partial L_f^{\,n-1} h_i(\hat{x}(t))}{\partial \hat{x}} f[\hat{x}(t), t] \tag{17}$$

$$L_{g_j} L_f^{\,n} h_i(\hat{x}(t)) = \frac{\partial L_f^{\,n} h_i(\hat{x}(t))}{\partial \hat{x}} g_j(t) \tag{18}$$

The index r_i is the relative degree, which satisfies the following two constrained equations:

$$L_{g_j} L_f^{\,n_i} h_i(\hat{x}(t)) = 0 (i = 1, 2, \cdots m; j = 1, 2, \cdots, l, n_i < r_i - 1) \tag{19}$$

$$L_{g_j} L_f^{\,r_i - 1} h_i(\hat{x}(t)) \neq 0 (i = 1, 2, \cdots m; j = 1, 2, \cdots, l) \tag{20}$$

Therefore, based on the measurement processed at time t_{k+1}, the new $d(t)$ in $[t_k : t_{k+1}]$ can be found. After that, the state estimates are propagated to time t_{k+1}.

The weighting matrix W in Equation (10) is derived as the inverse of the model error's covariance matrix D, given as follows:

$$W = D^{-1} = Cov[d(t)]^{-1} = E[(d(t) - E[d(t)])(d(t) - E[d(t)])^T]^{-1} \tag{21}$$

Assume that $d(t)$ is a stationary ergodic random process, then the covariance matrix D of the model error can be iteratively derived with a certain time interval; the steps of calculating W are listed as follows:

Step.1: Initialization: $W = E[(d(t_0) - E[d(t_0)])(d(t_0) - E[d(t_0)])^T]^{-1}$, for the total time length L, the time interval for updating W is defined as $l = L/r$, where r is the total iterations.

Step.2: For $t \in [t_{(k-1)l+1} : t_{kl}]$: estimate the model error $d(t)$ for $[t_{(k-1)l+1} : t_{kl}]$, and obtain the sequence of model error $[d(t_{(k-1)l+1}), \ldots, d(t_{kl})]$.

Step.3: For $t = t_{kl}$, update weighting matrix: $W = D_k^{-1}$, D_k is the covariance for $[d(t_{(k-1)l+1}), \ldots, d(t_{kl})]$.

Step.4: If $k < L/l$, return to step.2.

Based on the definitions and descriptions above, the workflow of NPF can be summarized as follows:

Step.1: Initialization: for $k = 0$:

(a) Set initial values: $\hat{x}(t_0) = E[x(t_0)]$, $W = E[(d(t_0) - E[d(t_0)])(d(t_0) - E[d(t_0)])^T]^{-1}$;

(b) Set weighting matrix update time interval: $l = L/r$.

Step.2: For every time step $k = 1, 2, \cdots$, doing the following:

(a) Estimate system output: $\hat{y}(t_k) = h[\hat{x}(t_k)]$;

(b) Calculate the intermediate parameter matrices $S(\hat{x}(t_k)), Z(\hat{x}(t_k), \Delta t), \Lambda(\Delta t)$;

(c) Estimate model error: $\begin{aligned} d(t_k) &= -\{[\Lambda(\Delta t) S(\hat{x}(t_k))]^T R^{-1} \Lambda(\Delta t) S(\hat{x}(t_k)) + W\}^{-1} \\ &\times [\Lambda(\Delta t) S(\hat{x}(t_k))]^T R^{-1} [Z(\hat{x}(t_k), \Delta t) + \hat{y}(t_k) - y(t_{k+1})] \end{aligned}$;

(d) Update state estimation from $\hat{x}(t_k)$ to $\hat{x}(t_{k+1})$ using discretized state equation: $\dot{\hat{x}}(t) = f[\hat{x}(t), t] + g(t) d(t)$.

Step.3: For $k = nl, 1 \le n \le r$

(a) Calculate covariance for $[d(t_{(k-1)l+1}),\ldots,d(t_{kl})]$: $D_k = Cov[d(t)]$, $t \in [t_{(k-1)l+1} : t_{kl}]$

(b) Update weighting matrix: $W = D_k^{-1}$

It can be seen that the process noise is determined online as a part of the solution without any prior assumption. As a result, the NPF method is robust to the process noise with any statistic distribution.

3.2. NPF Based SOC Estimation

In order to apply the NPF method to estimate the battery SOC, a continuous-discrete time model for the battery system is needed. According to the description of the equivalent circuit battery model in Section 2.1, the battery model in the continuous-discrete time form is formulated as:

$$\dot{x}(t) = \begin{pmatrix} S\dot{O}C(t) \\ \dot{U}_p(t) \end{pmatrix} = \begin{pmatrix} 0 \cdot SOC(t) + \dfrac{I_L(t)}{C_n} \\ -(R_p C_p)^{-1} U_p(t) + C_p^{-1} I_L(t) \end{pmatrix} + g \cdot \begin{pmatrix} d_1(t) \\ d_2(t) \end{pmatrix} \tag{22}$$

$$U_t(t_k) = U_{oc}[SOC(t_k)] + U_p(t_k) + I_L(t_k) R_s + v(t_k) \tag{23}$$

where the state vector $x(t)$ consists the components of $SOC(t)$ and $U_p(t)$, the model error vector $d(t)$ consists the components of $d_1(t)$ and $d_2(t)$ representing the errors of $SOC(t)$ and $U_p(t)$ respectively, and the battery's terminal voltage $U_t(t_k)$ is the system output sampled at t_k. An identity matrix $I_{2\times2}$ is assigned to the model error distribution matrix g. In addition, the input current $I_L(t)$ is treated as a known system parameter since we can sample the input current in every time step by the current sensor.

The expressions of the intermediate parameter matrices $S(\hat{x}(t_k)), Z(\hat{x}(t_k), \Delta t), \Lambda(\Delta t)$ for the nonlinear battery system are given in Equations (24)–(26), which are derived according to Equations (13)–(15). The relative order *r* is 1 which can be obtained following the definitions in Equations (19)–(20).

$$S(\hat{x}(t_k)) = [L_{g_1} h(\hat{x}(t_k)), L_{g_2} h(\hat{x}(t_k))] = \left[\frac{\partial U_{oc}[SOC(t_k)]}{\partial SOC(t_k)}, 1 \right] \tag{24}$$

$$Z(\hat{x}(t_k), \Delta t) = \Delta t \cdot L_f h(x(t_k)) = \Delta t \cdot \left(\frac{I_L(t_k)}{C_n} \cdot \frac{\partial U_{oc}[SOC(t_k)]}{\partial SOC(t_k)} - (R_p C_p)^{-1} U_p(t_k) + C_p^{-1} I_L(t_k) \right) \tag{25}$$

$$\Lambda(\Delta t) = \Delta t \tag{26}$$

The schematic diagram of the NPF based SOC estimation is shown in Figure 8. The $d_1(t)$ and $d_2(t)$ will be estimated at every iterative loop to predict the model errors of $SOC(t)$ and $U_p(t)$ in the next time step, then the battery state estimation of the next time step can be obtained based on the predicted model errors and battery dynamics.

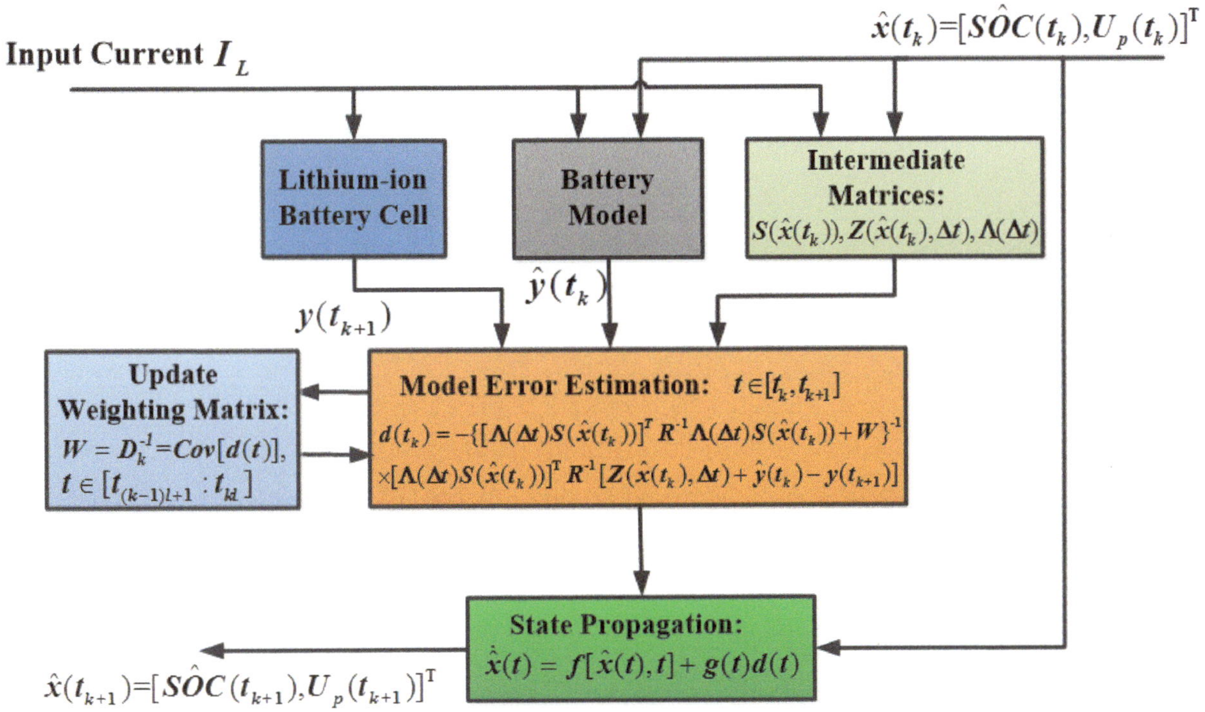

Figure 8. Schematic diagram of NPF based SOC estimation.

4. Results and Discussion

4.1. Part A: Evaluate the SOC Estimation of LCO Battery Cell

In this part, the experimental data is used to validate the performance of the NPF based estimation method for LCO battery cell. The model parameters in Table 3 are used in the estimation. The JC08 (Japanese Cycle 2008) driving cycle is applied as the loading profile to evaluate the proposed method. Figure 9 shows the SOC estimation results with the accurate initial SOC value, and Figure 10 shows the SOC estimation results with inaccurate initial SOC values.

Figure 9a shows the comparison profiles of the reference SOC and estimated SOC with the accurate initial SOC. Since the initial SOC is accurate and the integration time is not long, the SOC result obtained from coulomb counting is treated as an accurate estimation and thus taken as a reference. Figure 9b shows the estimation error between the reference SOC and estimated SOC and it can be seen that the maximum error is around 1%. Figure 9c,d shows the mean absolute error (MAE) and root mean squared error (RMSE) of SOC estimation, respectively. The results indicate that both of the MAE and RMSE are less than 0.4% at the end of estimation. Based on the discussion above, it can be concluded that the NPF method is able to accurately estimate the SOC of LCO battery cell.

Figure 10a shows the estimation results with inaccurate initial SOC values, while Figure 10b shows the corresponding errors. Two different initial SOC values (0.5 and 0.3) are used. It can be seen that the SOC estimates can quickly converge to the true solutions within several sampling steps for both initial SOC values. Therefore, the proposed NPF based SOC estimation method is robust to inaccurate initial values for LCO battery cell.

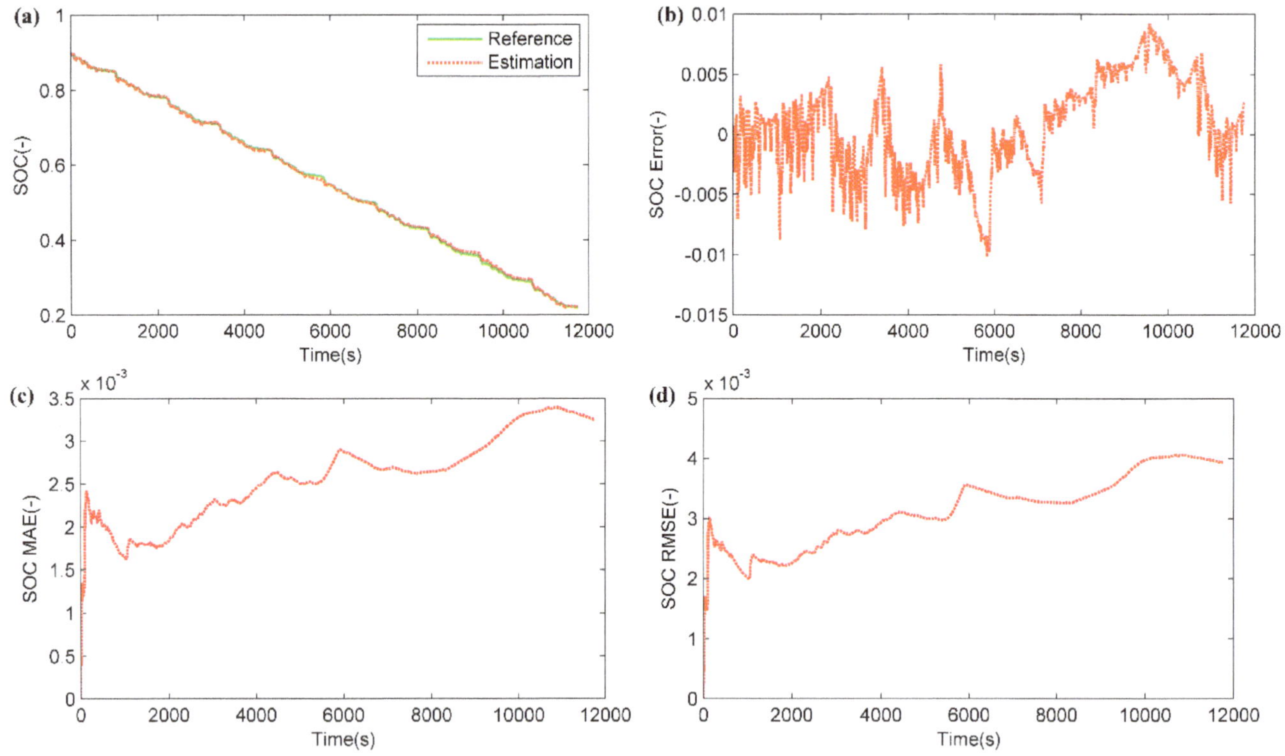

Figure 9. SOC estimation results with accurate initial SOC for LCO battery cell: (**a**) Comparative profiles of reference and estimated SOC; (**b**) Error of SOC estimation; (**c**) MAE of SOC estimation; (**d**) RMAE of SOC estimation.

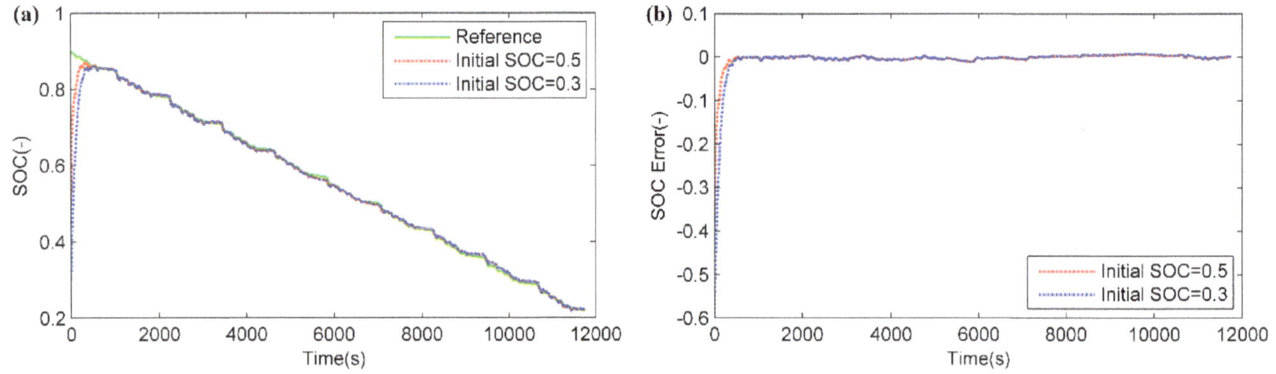

Figure 10. Estimation results with inaccurate initial SOCs for LCO cell: (**a**) Comparative profiles of reference and estimated SOC; (**b**) Error of SOC estimation.

4.2. Part B: Evaluate the SOC Estimation of LFP Battery Cell

In this section, the SOC estimation using NPF for LFP battery cell is conducted. Similarly, the JC08 driving cycle is applied as the loading profile. Figure 11 shows the SOC estimation results with the accurate initial SOC value. Figure 11a shows the comparison profiles of the estimated SOC and reference SOC, and Figure 11b shows the estimation error. It can be seen that the maximum estimation error is within 2%. Figure 11c,d shows the MAE and RMSE of SOC estimation, both of which are less than 0.9% at the end of estimation. Figure 12 shows the SOC estimation with inaccurate initial SOC values, where Figure 12a shows the comparison of the reference SOC and estimated SOC, and Figure 12b shows the estimation error. Similar to the LCO battery cell, two different initial SOC values are used. It can be seen from those results that the SOC estimation can also converge to the true solution within several sampling steps for both inaccurate initial values.

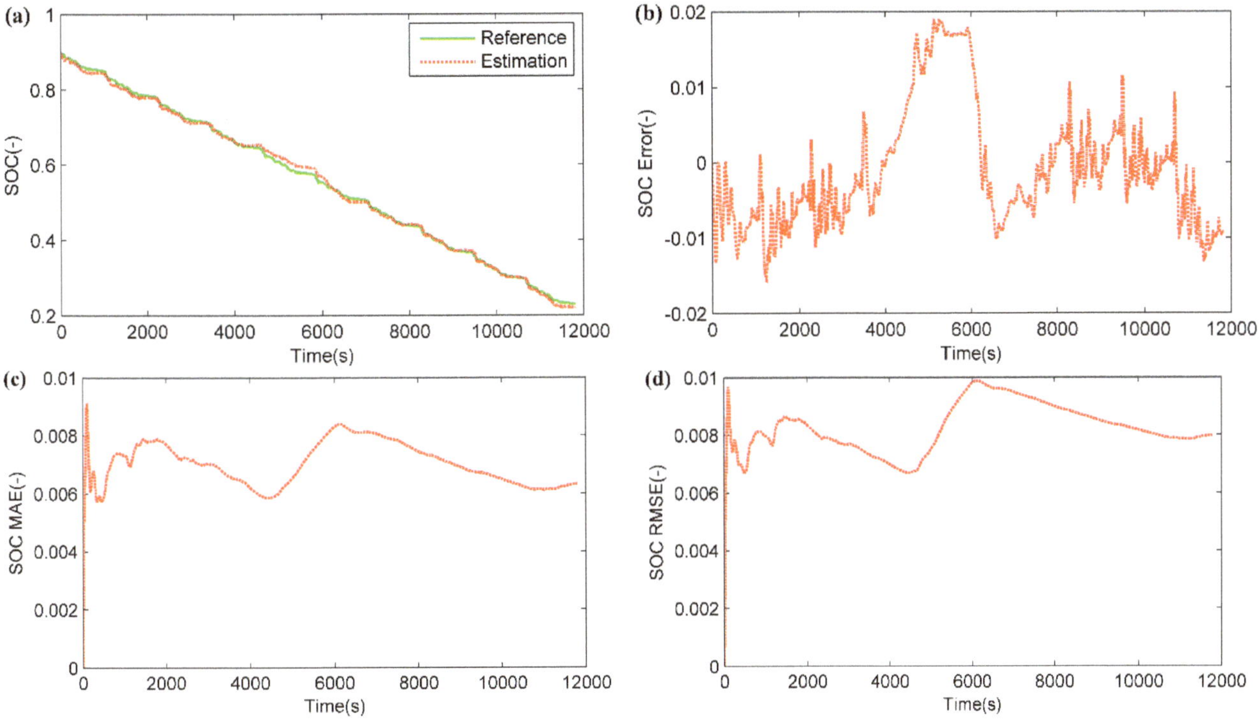

Figure 11. SOC estimation results with accurate initial SOC for LFP battery cell: (**a**) Comparative profiles of reference and estimated SOC; (**b**) Error of SOC estimation; (**c**) MAE of SOC estimation; (**d**) RMAE of SOC estimation.

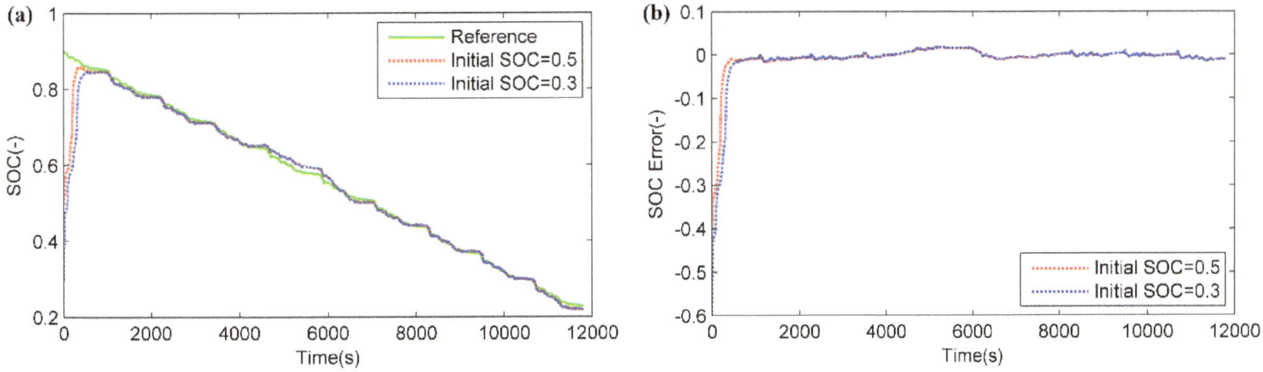

Figure 12. Estimation results with inaccurate initial SOC for LFP cell: (**a**) Comparative profiles of reference and estimated SOC; (**b**) Error of SOC estimation.

Based on the estimation results of LCO and LFP battery cells, a comparison study of SOC estimation for LCO and LFP cells is conducted. The comparison results are listed in Table 4. Compared to LFP battery cell, the SOC estimation of the LCO battery cell is benefited by a 0.91% improvement in terms of the maximum error, and by 0.31% and 0.42% improvements in terms of MAE and RMSE. With respect to the convergence rate, the SOC estimation of the LCO battery cell converges much faster than that of the LFP battery cell with inaccurate initial SOC values. It seems that the estimation performance of the LCO battery cell is better than that of the LFP battery cell. The difference in the estimation performance can be mainly explained by the difference of the SOC-OCV relationships between these two cells, as shown in Figure 13. The comparison results of SOC-OCV curves are shown in Figure 13a,b. It is obvious that, compared to the LCO battery cell, the LFP battery cell has a much flatter SOC-OCV curve with smaller derivation. In other words, for the same OCV difference ΔOCV, the corresponding SOC difference ΔSOC_{LFP} of LFP battery cell will be much larger than the corresponding SOC difference ΔSOC_{LCO} of LCO battery cell. Therefore, the observability of the LFP battery cell is weaker than that of the LCO battery cell. As a result, for similar model error, the SOC error of the LFP battery cell will be larger than the SOC error of the LCO battery cell. Thus, the performance of NPF based SOC estimation method for LFP battery cells is worse than that for LCO battery cells due to weaker observability.

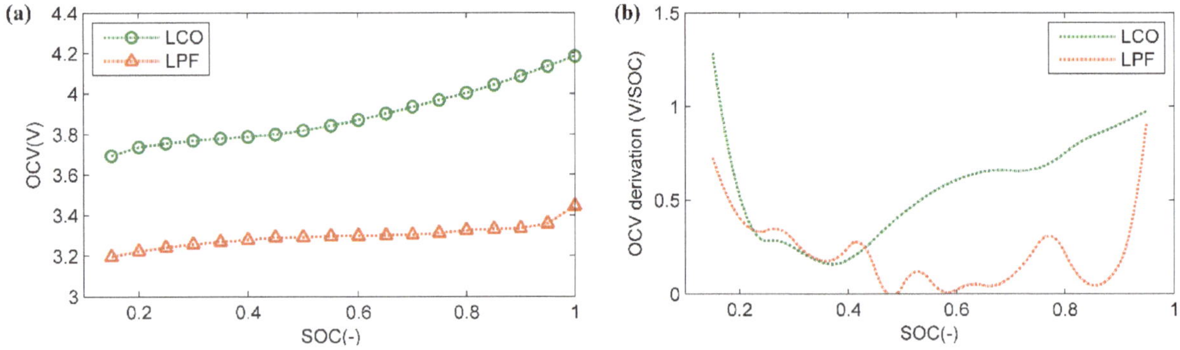

Figure 13. SOC-OCV curves for two cells: (**a**) Comparison of SOC-OCV; (**b**) Comparison of the derivation of SOC-OCV.

Table 4. Comparison of LCO and LFP battery cells.

Estimation accuracy	**Maximum error**	**MAE**	**RMSE**
LCO battery cell	1.01%	0.33%	0.39%
LFP battery cell	1.92%	0.64%	0.81%
Difference	0.91%	0.31%	0.42%
Convergence rate	**Convergence time (Initial SOC = 0.5)**	**Convergence time (Initial SOC = 0.3)**	
LCO battery cell	302 s	427 s	
LFP battery cell	392 s	556 s	
Difference	29.8%	30.2%	

4.3. Part C: Comparison with Extended Kalman Filter

In order to further evaluate the performance of proposed method, a comparison study is conducted between the NPF method and the well-established extended Kalman filter (EKF) method based on the same experimental conditions. The details of EKF for SOC estimation can be found in [15]. To facilitate the understanding of the SOC estimation using EKF, a summary of EKF is provided as follows:

$$\text{Nonlinear sate-space model:} \begin{cases} x_{k+1} = f(x_k, u_k) + w_k \\ y_k = h(x_k, u_k) + v_k \end{cases} \tag{27}$$

$$\text{Definitions}: \hat{A}_k = \left.\frac{\partial f(x_k, u_k)}{\partial x_k}\right|_{x_k=\hat{x}_k^+}, \hat{C}_k = \left.\frac{\partial h(x_k, u_k)}{\partial x_k}\right|_{x_k=\hat{x}_k^-} \tag{28}$$

Step.1: Initialization: For k = 0, set: $\hat{x}_0^+ = E[x_0]$, $p_0^+ = E[(x_0 - \hat{x}_0^+)(x_0 - \hat{x}_0^+)^T]$

Step.2: For k = 1,2, . . . n, do the following:

(a) State estimation time update: $\hat{x}_k^- = f(\hat{x}_{k-1}^+, u_{k-1})$

(b) Error covariance time update: $p_k^- = \hat{A}_{k-1} p_{k-1}^+ \hat{A}_{k-1}^T + R_w$

(c) Calculate the Kalman gain: $L_k = p_k^- \hat{C}_k^T \left[\hat{C}_k p_k^- \hat{C}_k^T + Q_v\right]^{-1}$

(d) State estimation measurement update: $\hat{x}_k^+ = \hat{x}_k^- + L_k[y_k - h(\hat{x}_k^-, u_k)]$

(e) Error covariance measurement update: $p_k^+ = (I - L_k \hat{C}_k) p_k^-$

where w_k and v_k are independent, zero-mean, Gaussian noise processes with covariance matrices R_w and Q_v.

In comparison with the NPF based method which uses a continuous-discrete time model, the battery system in EKF is modeled in a discrete time form, given as follows:

$$\text{State equations:} \begin{cases} SOC_{k+1} = SOC_k + \dfrac{\eta \Delta t}{C_n} I_{L,k} + w_{1,k} \\ U_{p,k+1} = (1 - (R_p C_p)^{-1} \Delta t) U_{p,k} + \dfrac{\Delta t}{C_p} I_{L,k} + w_{2,k} \end{cases} \tag{29}$$

$$\text{Measurement equations: } U_{t,k} = U_{oc}(SOC_k) + U_{p,k} + I_{L,k} R_s + v_k \tag{30}$$

In this paper, two main aspects, estimation accuracy and convergence rate, are studied to show the advantages of the NPF based method.

To compare the estimation accuracy of these two methods, the SOC estimation with the accurate initial SOC is evaluated. Figure 14 shows the comparison of SOC estimation for LCO battery cell. Figure 14a shows the comparison of the reference SOC and estimated SOCs by NPF and EKF, while Figure 14b shows the corresponding errors. It can be seen that the NPF based method has better estimation accuracy than the EKF based method. The MAE and RMSE of SOC estimation also indicate that the NPF based method has better accuracy (shown in Figure 14c,d). Similarly, an improvement of estimation accuracy by the NPF based method can also be found for LFP battery cells, as shown in Figure 15. A summary of the estimation accuracy for these two methods is listed in Table 5.

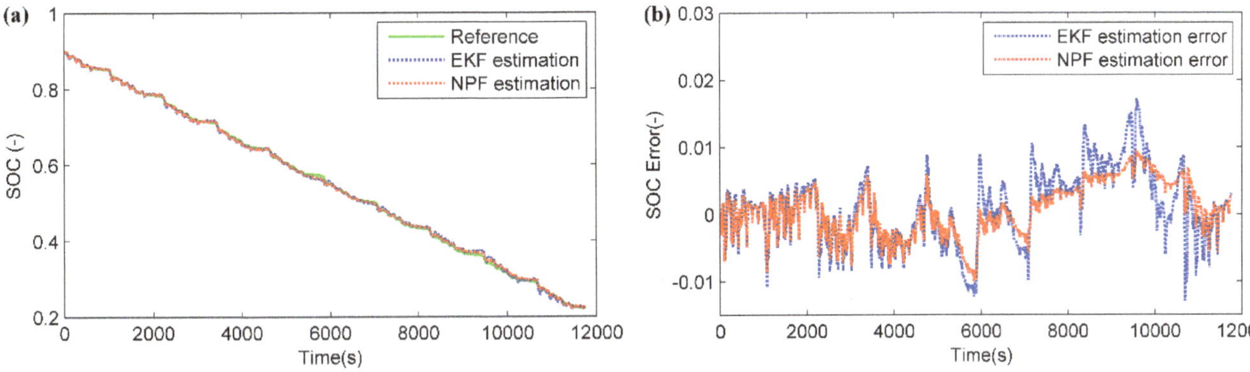

Figure 14. *Cont.*

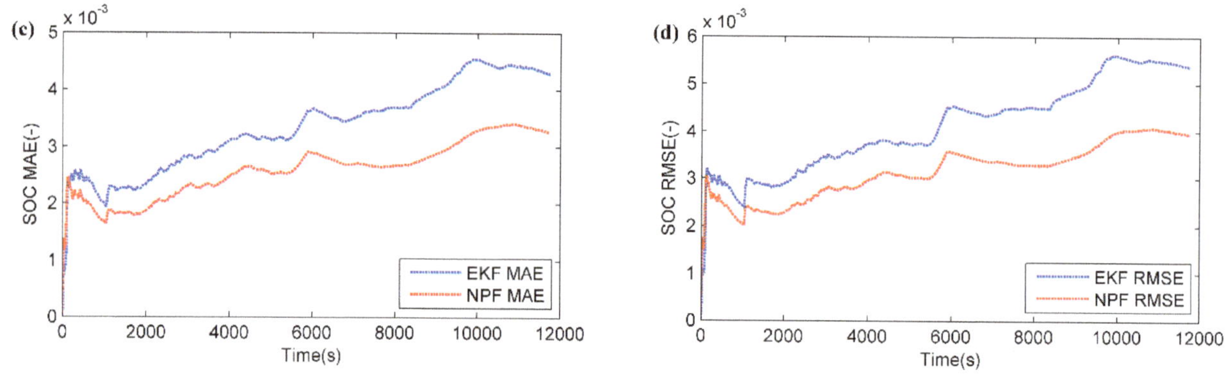

Figure 14. Comparison results of SOC estimation for LCO battery cell with accurate initial SOC: (**a**) SOC estimation; (**b**) Error of SOC estimation; (**c**) MAE of SOC estimation; (**d**) RMSE of SOC estimation.

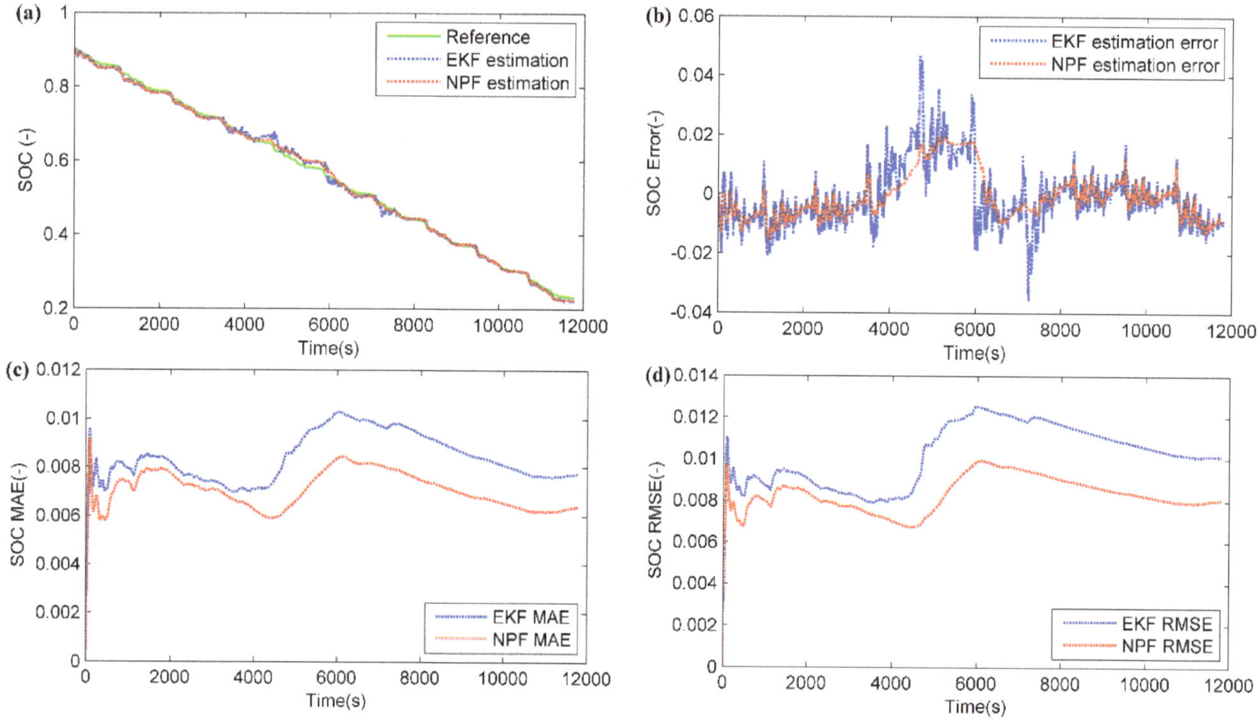

Figure 15. Comparison results of SOC estimation for LFP battery cell with accurate initial SOC: (**a**) SOC estimation; (**b**) Error of SOC estimation; (**c**) MAE of SOC estimation; (**d**) RMSE of SOC estimation.

Table 5. Comparison results of estimation accuracy.

Estimation Accuracy	Maximum Error	MAE	RMSE
	LCO battery cell		
EKF	1.64%	0.43%	0.54%
NPF	1.01%	0.33%	0.39%
Improvement	38.4%	17.9%	27.8%
	LFP battery cell		
EKF	4.66%	0.78%	1.01%
NPF	1.92%	0.64%	0.81%
Improvement	58.8%	17.9%	20.8%

To compare the convergence rate of these two methods, the SOC estimation with the inaccurate initial SOC is conducted. For convenience, the initial SOC value used in this study is set as 0.5 for both NPF and EKF. The results are shown in Figure 16 where Figure 16a,b shows the comparison for the LCO battery cell, and Figure 16c,d shows the comparison for the LFP battery cell. It can be seen that the NPF based method converges faster than the EKF based method for both cells. A summary of the convergence rate for these two methods is given in Table 6.

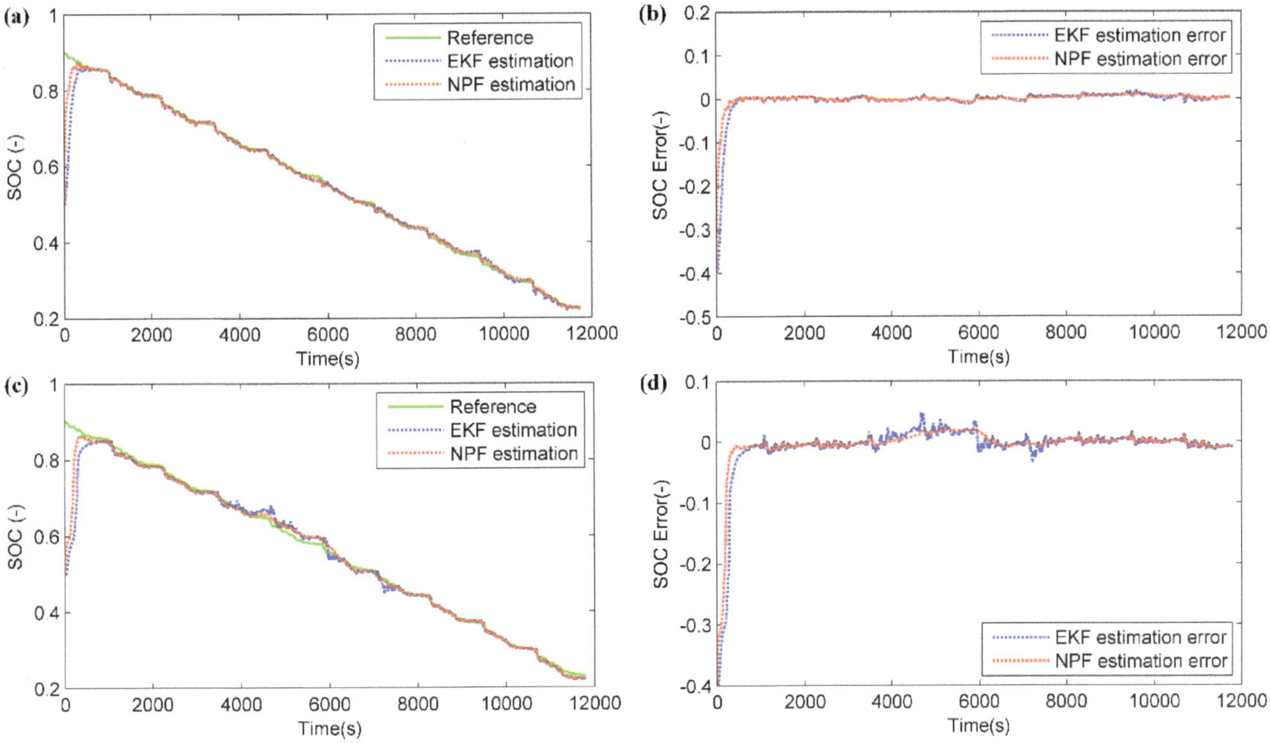

Figure 16. Comparison results of SOC estimation with inaccurate initial SOC: (**a**) SOC estimation of LCO battery cell; (**b**) SOC estimation error of LCO battery cell; (**c**) SOC estimation of LFP battery cell; (**d**) SOC estimation error of LFP battery cell.

Table 6. Comparison results of convergence rate.

Convergence Rate	Time
LCO battery cell	
EKF	439 s
NPF	302 s
Improvement	31.2%
LFP battery cell	
EKF	586 s
NPF	392 s
Improvement	33.1%

5. Conclusions

In this paper, the nonlinear predictive filter is developed to estimate the SOC of Li-ion batteries with two different chemistries. First, in consideration of the model accuracy and computational complexity, the first-order equivalent circuit battery model is adopted to characterize the dynamic performance of Li-ion battery cells. Second, a series of battery tests are conducted using two cells with different chemistries (LCO and LFP) to identify the model parameters. The model validation results based on UDDS driving cycle indicate that the battery models with identified parameters are able to accurately represent the behavior of LCO and LFP battery cells. Third, the nonlinear predictive filter is introduced and developed to estimate the battery SOC using the first-order battery model. The JC08 driving cycle is applied for these two cells to evaluate the proposed method. According to the results, the NPF based method is able to estimate the battery SOC accurately. Besides, the proposed SOC estimation method has good robust performance to inaccurate initial values (quickly converging to the true solution within several sampling steps). Finally, the comparison of EKF and NPF based methods indicates that the proposed method has better estimation accuracy and faster convergence rate than the EKF based method for both cells.

Acknowledgments

The authors gratefully acknowledge the support from Nippon Chemi-Con Corporation.

Author Contributions

Yin Hua and Min Xu planned and drafted the main part of the paper. Mian Li revised the paper. Chengbin Ma and Chen Zhao participated in the experiment.

Conflicts of Interest

The authors declare no conflict of interest.

References

1. Xi, J.; Li, M.; Xu, M. Optimal energy management strategy for battery powered electric vehicles. *Appl. Energy* **2014**, *134*, 332–341.
2. Wang, B.; Xu, M.; Yang, L. Study on the economic and environmental benefits of different EV powertrain topologies. *Energy Conver. Manag.* **2014**, *86*, 916–926.
3. Lukic, S.M.; Cao, J.; Bansal, R.C.; Rodriguez, F.; Emadi, A. Energy storage systems for automotive applications. *IEEE Trans. Ind. Electron.* **2008**, *55*, 2258–2267.
4. Hammond, G.P.; Hazeldine, T. Indicative energy technology assessment of advanced rechargeable batteries. *Appl. Energy* **2014**, *138*, 559–571.
5. Garche, J.; Jossen, A.; Döring, H. The influence of different operating conditions, especially over-discharge, on the lifetime and performance of lead/acid batteries for photovoltaic systems. *J. Power Sour.* **1997**, *67*, 201–212.
6. Alzieu, J.; Smimite, H.; Glaize, C. Improvement of intelligent battery controller: State-of-Charge indicator and associated functions. *J. Power Sour.* **1997**, *67*, 157–161.
7. Xing, Y.; He, W.; Pecht, M.; Tsui, K.L. State of charge estimation of lithium-ion batteries using the open-circuit voltage at various ambient temperatures. *Appl. Energy* **2014**, *113*, 106–115.

8. Eddahech, A.; Briat, O.; Vinassa, J.M. Adaptive voltage estimation for EV Li-ion cell based on artificial neural networks state-of-charge meter. In Proceedings of the 2012 IEEE International Symposium on Industrial Electronics (ISIE), Hangzhou, China, 28–31 May 2012; pp. 1318–1324.
9. Salkind, A.J.; Fennie, C.; Singh, P.; Atwater, T.; Reisner, D.E. Determination of state-of-charge and state-of-health of batteries by fuzzy logic methodology. *J. Power Sour.* **1999**, *80*, 293–300.
10. Anton, J.C.A.; Nieto, P.J.G.; Viejo, C.B.; Vilan, J.A. Support vector machines used to estimate the battery state of charge. *IEEE Trans. Power Electron.* **2013**, *28*, 5919–5926.
11. Tsang, K.M.; Sun, L.; Chan, W.L. Identification and modelling of Lithium ion battery. *Energy Convers. Manag.* **2010**, *51*, 2857–2862.
12. He, H.; Xiong, R.; Guo, H.; Li, S. Comparison study on the battery models used for the energy management of batteries in electric vehicles. *Energy Convers. Manag.* **2012**, *64*, 113–121.
13. Plett, G.L. Extended Kalman filtering for battery management systems of LiPB-based HEV battery packs: Part 1. Background. *J. Power Sour.* **2004**, *134*, 252–261.
14. Plett, G.L. Extended Kalman filtering for battery management systems of LiPB-based HEV battery packs: Part 2. Modeling and identification. *J. Power Sour.* **2004**, *134*, 262–276.
15. Plett, G.L. Extended Kalman filtering for battery management systems of LiPB-based HEV battery packs: Part 3. State and parameter estimation. *J. Power Sour.* **2004**, *134*, 277–292.
16. Plett, G.L. Sigma-point Kalman filtering for battery management systems of LiPB-based HEV battery packs: Part 1: Introduction and state estimation. *J. Power Sour.* **2006**, *161*, 1356–1368.
17. Plett, G.L. Sigma-point Kalman filtering for battery management systems of LiPB-based HEV Battery packs: Part 2: Simultaneous state and parameter estimation. *J. Power Sour.* **2006**, *161*, 1369–1384.
18. He, Z.; Gao, M.; Wang, C.; Wang, L.; Liu, Y. Adaptive state of charge estimation for Li-ion batteries based on an unscented kalman filter with an enhanced battery model. *Energies* **2013**, *6*, 4134–4151.
19. Han, J.; Kim, D.; Sunwoo, M. State-of-charge estimation of lead-acid batteries using an adaptive extended Kalman filter. *J. Power Sour.* **2009**, *188*, 606–612.
20. Sun, F.; Hu, X.; Zou, Y.; Li, S. Adaptive unscented Kalman filtering for state of charge estimation of a lithium-ion battery for electric vehicles. *Energy* **2011**, *36*, 3531–3540.
21. Schwunk, S.; Armbruster, N.; Straub, S.; Kehl, J.; Vetter, M. Particle filter for state of charge and state of health estimation for lithium-iron phosphate batteries. *J. Power Sour.* **2013**, *239*, 705–710.
22. Charkhgard, M.; Farrokhi, M. State-of-charge estimation for lithium-ion batteries using neural networks and EKF. *IEEE Trans. Ind. Electron.* **2010**, *57*, 4178–4187.
23. Tian, Y.; Xia, B.; Wang, M.; Sun, W.; Xu, Z. Comparison Study on two model-based adaptive algorithms for SOC estimation of lithium-ion batteries in electric vehicles. *Energies* **2014**, *7*, 8446–8464.
24. He, H.; Qin, H.; Sun, X.; Shui, Y. Comparison study on the battery SoC estimation with EKF and UKF algorithms. *Energies* **2013**, *6*, 5088–5100.
25. Zou, Z.; Xu, J.; Mi, C.; Cao, B.; Chen, Z. Evaluation of model based state of charge estimation methods for lithium-ion batteries. *Energies* **2014**, *7*, 5065–5082.
26. Yuan, S.; Wu, H.; Yin, C. State of charge estimation using the extended kalman filter for battery management systems based on the arx battery model. *Energies* **2013**, *6*, 444–470.

27. Xia, B.; Chen, C.; Tian, Y.; Sun, W.; Xu, Z.; Zheng, W. A novel method for state of charge estimation of lithium-ion batteries using a nonlinear observer. *J. Power Sour.* **2014**, *270*, 359–366.
28. Crassidis, J.L.; Markley, F.L. Predictive filtering for attitude estimation without rate sensors. *J. Guid. Control Dyn.* **1997**, *20*, 522–527.
29. Hua, Y.; Cordoba-Arenas, A.; Warner, N.; Rizzoni, G. A multi time-scale state-of-charge and state-of-health estimation framework using nonlinear predictive filter for lithium-ion battery pack with passive balance control. *J. Power Sour.* **2015**, *280*, 293–312.
30. Hu, X.; Li, S.; Peng, H. A comparative study of equivalent circuit models for Li-ion batteries. *J. Power Sour.* **2012**, *198*, 359–367.
31. Abu-Sharkh, S.; Doerffel, D. Rapid test and non-linear model characterisation of solid-state lithium-ion batteries. *J. Power Sour.* **2004**, *130*, 266–274.
32. US Environmental Protection Agency. EPA urban dynamometer driving schedule (UDDS). Available online: http://www.epa.gov/oms/standards/light-duty/udds.htm (accessed on 3 April 2015).

3

Application of a Central Composite Design for the Study of NO_x Emission Performance of a Low NO_x Burner

Marcin Dutka [1,*], Mario Ditaranto [2,†] and Terese Løvås [1,†]

[1] Department of Energy and Process Engineering, Norwegian University of Science and Technology, Kolbjørn Hejes vei 1b, Trondheim 7491, Norway; E-Mail: terese.lovas@ntnu.no

[2] SINTEF Energy Research, Sem Sælands vei 11, Trondheim 7034, Norway; E-Mail: mario.ditaranto@sintef.no

† These authors contributed equally to this work.

* Author to whom correspondence should be addressed; E-Mail: marcin.d.dutka@ntnu.no

Academic Editor: Jinliang Yuan

Abstract: In this study, the influence of various factors on nitrogen oxides (NO_x) emissions of a low NO_x burner is investigated using a central composite design (CCD) approach to an experimental matrix in order to show the applicability of design of experiments methodology to the combustion field. Four factors have been analyzed in terms of their impact on NO_x formation: hydrogen fraction in the fuel (0%–15% mass fraction in hydrogen-enriched methane), amount of excess air (5%–30%), burner head position (20–25 mm from the burner throat) and secondary fuel fraction provided to the burner (0%–6%). The measurements were performed at a constant thermal load equal to 25 kW (calculated based on lower heating value). Response surface methodology and CCD were used to develop a second-degree polynomial regression model of the burner NO_x emissions. The significance of the tested factors over their respective ranges has been evaluated using the analysis of variance and by the consideration of the coefficients of the model equation. Results show that hydrogen addition to methane leads to increased NO_x emissions in comparison to emissions from pure methane combustion. Hydrogen content in a fuel is the strongest factor affecting NO_x emissions among all the factors tested. Lower NO_x formation because of increased excess air was observed when the burner was fuelled by pure methane, but this effect diminished for hydrogen-rich fuel mixtures. NO_x emissions were slightly reduced when the burner head was shifted closer to the burner outer tube, whereas a secondary fuel stream provided to the burner was found to have no impact on NO_x emissions over the investigated range of factors.

Keywords: burner; central composite design; hydrogen enrichment; nitrogen oxides

---

1. Introduction

The latest trends in environmental legislation associated with carbon dioxide (CO_2) emission reductions are a great challenge for the industrial sector. It can be assumed that a competitive position for businesses utilizing fossil fuels will be strongly conditioned on the ability to meet strict environmental regulations. Therefore, the necessity of reducing CO_2 emissions has led to the development of various technological solutions, called carbon capture and storage technologies. One of these technologies is associated with hydrogen or hydrogen-enriched fuel combustion. The pre-combustion CO_2 capture technology can be applied to power plants, but also to industrial processes as in refineries [1–3]. In this case, the fossil fuel is processed by gasification or reforming and water-gas shift reaction to generate a fuel composed mainly of hydrogen and CO_2. CO_2 is then captured leaving a hydrogen-rich fuel, what allows for CO_2 emission reduction. However, the use of such fuels in engineering applications is associated with corresponding changes in nitrogen oxides (NO_x) emissions, which in turn are affected by many factors.

In both science and engineering, determining which factors of a complex system are significant and how they affect the response of the system is often difficult. In such cases, usually full factorial design of the experiment is used to test all possible combinations of various factors. Full factorial design is often the only choice when one is interested in accurate measurement results under various operating conditions or when the response is expected to change in unforeseen ways. Note that this approach often requires a large number of experimental trials, because the number of trials increases geometrically with the number of factors to be tested. Furthermore, the interpretation of a large set of measurement data is difficult and may be unnecessary, because especially in engineering and practical applications, focus on only trends in how factors affect system response can be sufficient. Therefore, response surface methodology combined with central composite design (CCD) [4] is an efficient technique for experimentally exploring relationships between investigated factors and system response.

Most practical industrial burner designs are much more sophisticated combustion systems than those used in the scientific investigations of combustion processes. Many factors may affect pollutant emissions from these burners, while in cases of high-burner thermal power, conducting measurements in such industrial facilities is often difficult and expensive [5]. Statistically cognizant design of experiment facilitates an understanding of the influence of the factors tested and the interactions between these factors on system response by using a minimum number of experimental trials. Such an approach applied to testing of burners might be also helpful in reducing the costs of large industrial scale experiments.

The purpose of this paper is to present the use of the high performance and predictive strength of CCD to study NO_x emissions from a Partially Premixed Bluff Body (PPBB) burner. The PPBB burner was chosen because it allows testing several factors including fuel composition, excess air, fuel distribution method and the factor that determines burner geometry. Understanding the influence of these factors on NO_x emissions is considerably important for burner development.

2. PPBB Burner

Bluff body stabilized flames have been an important research topic for a long time, mainly because of the fact that some form of flame stabilization is often required in industrial applications. When bluff body is used as a flame holder, the incoming fuel and oxidizer mixture is ignited by the hot combustion products recirculating in a wake structure formed behind the bluff body. This stabilization mechanism has been investigated extensively by many researchers [6–12]. Many studies associated with bluff body stabilization mechanism focus on premixed combustion systems, where premixed fuel and oxidizer are provided to the combustion zone and thus complex physics that controls mixing process can be neglected. However, it is worth noting that premixed combustion systems may suffer from flashback phenomena, what in turn can lead to burner damage and it may have potentially dangerous consequences, especially in industrial applications. Risk of flashback can be eliminated by using partially premixed flames.

The PPBB burner described in Figure 1 functions by partially mixing the fuel with air rapidly and stabilizing the flame by a bluff body. The air flow is suddenly accelerated when approaching the burner exit section, also called burner throat, because of the progressively reduced cross section formed between the diverging burner head and the outer tube. Gaseous fuel is distributed from the burner head, *i.e.*, fuel lance, in the accelerating air flow and is partially mixed with air before stabilization. A flame stabilization zone is created at the front of the lance by recirculation initiated by the shape of the lance, as shown in Figure 2.

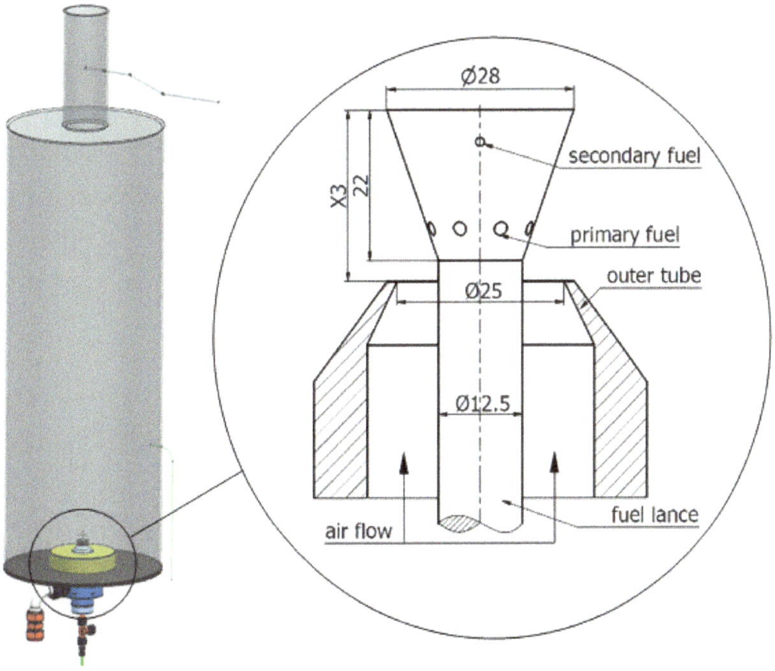

Figure 1. Schematic representation of the test rig and the PPBB burner used in the experiments (dimensions in mm).

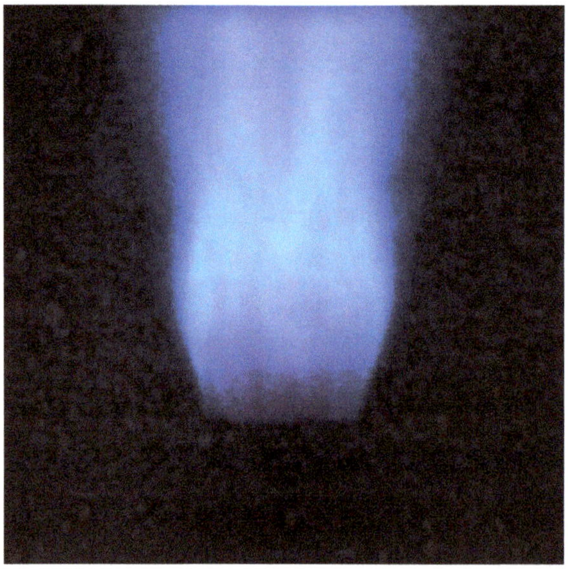

Figure 2. Photography of hydrogen-enriched methane flame stabilized behind the lance of the PPBB burner.

The PPBB burner design allows for controlling the combustion process using two design parameters for a given fuel composition. The first parameter is the distance between the lance and the burner exit section (labelled X3 in Figure 1). The distance can be changed by moving the fuel lance relative to the burner throat section. The lance position determines the point of fuel injection into the accelerated air, relative to the outer tube, and affects the fuel and air mixing process. The fuel distribution method is the second parameter affecting burner operation. Fuel provided to the burner is divided into a primary fuel stream selectively distributed from ports at the bottom of the lance and a secondary fuel stream from ports arranged around at the top of the lance. The fuel distribution method affects gas composition in the reaction zone. Consequently, it can be stated that the design parameters control mixing process in the burner, which involves changes of flow pattern behind the lance, fuel jets-air stream interaction and primary fuel jet-secondary fuel jet interactions. These phenomena may have significant impact on NO_x emissions and flame stability [13,14].

3. Central Composite Design of the Experiment

CCD enables estimation of the regression parameters to fit a second-degree polynomial regression model to a given response. A polynomial, as given by Equation (1), quantifies relationships among the measured response y and a number of experimental variables $X_1 \ldots X_k$, where k is the number of factors considered, β are regressors and ε is an error associated with the model:

$$\begin{aligned} y = {} & \beta_0 + \beta_1 X_1 + \beta_2 X_2 + \ldots + \beta_k X_k \\ & + \beta_{11} X_1^2 + \beta_{22} X_2^2 + \ldots + \beta_{kk} X_k^2 \\ & + \beta_{12} X_1 X_2 + \ldots + \beta_{k-1,k} X_{k-1} X_k + \varepsilon \end{aligned} \tag{1}$$

The regressors (β_1, β_2, β_3…) in the various terms of Equation (1) provide a quantitative measure of the significance of linear effects, curvilinear effects of factors and interactions between factors. It is worth noting that the model presented by Equation (1) is not a model in purely physical sense, but rather it should be understood as a statistical model, *i.e.*, a correlation developed based on regression analysis.

However, this nomenclature is widely used in the field of design of experiments and statistics, and therefore it is used hereafter.

CCD requires three types of trials, *i.e.*, 2^k factorial trials, $2k$ axial trials and n_c center point trials, where k is number of factors studied in the experiment [15]. As an example, this is illustrated in Figure 3, where each point defines a unique set of values of experimental trials for the three factors ($k = 3$) tested in an experiment.

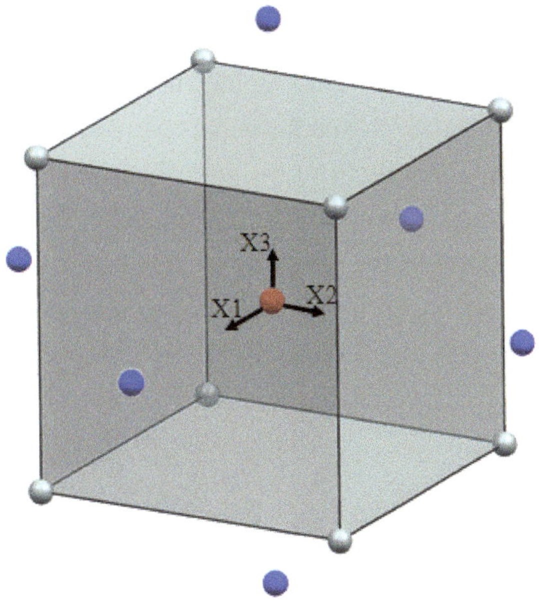

Figure 3. Visualization of original type rotatable CCD for three factors: X1, X2 and X3.

Values at the center point (red point with coordinates 0, 0, 0) that is located in the center of the cube are used to detect curvature in the response, *i.e.*, they contribute to the estimation of the coefficients of quadratic terms. Axial points (six blue points located at a distance α from the center point) are also used to estimate the coefficients of quadratic terms, while factorial points (eight grey points located at corners of the cube with a side length equal to 2) are used mainly to estimate the coefficients of linear terms and two-way interactions. For testing four or more factors in an experiment, Figure 3 should be extended to the fourth or more dimensions.

In CCD, factors are tested at a minimum of three levels: minimum, middle and maximum, equivalent to levels −1, 0 and 1, which are called *coded units*. If X_{min} and X_{max} are respectively minimum and maximum absolute, *i.e.*, *uncoded*, values of a factor, the absolute values X corresponding the respective coded values can be obtained by a simple linear transformation of the original measurement scale, namely:

$$X = b \cdot coded\,value + a \tag{2}$$

where:

$$a = \frac{X_{max} + X_{min}}{2}, b = \frac{X_{max} - X_{min}}{2} \tag{3}$$

To obtain the *rotatability* of a design, each experimental factor must be represented at the five levels of coded units: −α, −1, 0, 1, α. This property ensures constant variance at points that are equidistant from the center point, and therefore provides equal precision of response estimation in any direction of the design. For a full factorial CCD, it can be shown [16] that a design is rotatable if:

$$\alpha = \left(2^k\right)^{0.25} \tag{4}$$

A geometric region restricted by levels −1 and 1 on factor setting is defined as a region of interest to the experimenter. Therefore, based on the positions of factorial and axial points in the design one can distinguish the following main varieties of CCD: circumscribed CCD, face-centered CCD and inscribed CCD. Circumscribed CCD, shown in Figure 3, is the original type of CCD, where axial points are located at distance α, defined by Equation (4), from the center point. In face-centered CCD that is presented in Figure 4, axial points are located at a distance 1 from the center point, *i.e.*, at the face of the design cube, if the design involves three experimental factors. In turn, inscribed CCD is characterized by that axial points are located at factor levels −1 and 1, while factorial points are brought into the interior of the design space and located at distance 1/α from the center point, as illustrated in Figure 4. Geometrically, inscribed CCD reminds of circumscribed CCD. This is because factorial points are set at a distance from the center point so that these distances between the factorial points, the axial points and the center point are in the same proportion as in circumscribed CCD.

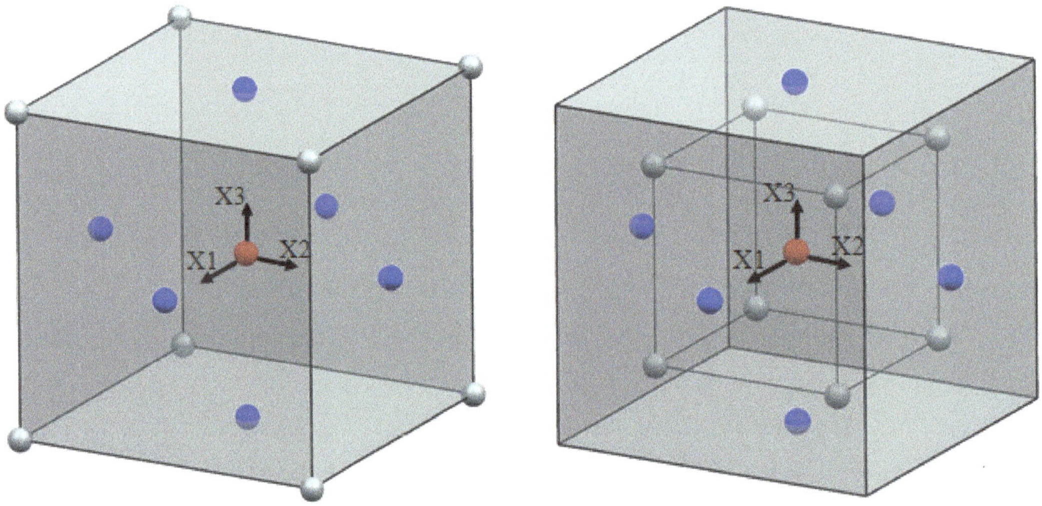

(a) (b)

Figure 4. Visualization of face-centered CCD (**a**) and inscribed CCD (**b**) for three factors: X1, X2 and X3.

All the varieties of CCD enable the experimenter to study the same region of interest, but they differ in respect of combinations of factor level settings to be tested. However, this affects properties of the prediction variance of a design. Design performance can be evaluated using variance dispersion graph (VDG) [17].

VDG displays the maximum, the minimum and the averaged scaled prediction variance (SPV) against the radius of the spherical space with the center at the center point of the design. SPV can be estimated with the Equation (5):

$$v(x) = \frac{N \cdot Var[y(x)]}{\sigma^2} = \frac{N\sigma^2 \cdot x^T (X^T X)^{-1} x}{\sigma^2} = N \cdot x^T (X^T X)^{-1} x \tag{5}$$

where N is the number of points in the experimental design; $Var[y(x)]$ is the variance of a predicted value at a point x; σ^2 is the variance of the experimental error; X is the design matrix; and x is a vector valued function of the coordinates whose elements correspond to the columns of the design matrix X. According to Equation (5), SPV is independent of the response data and quality of a design can be assessed *a priori*.

The main varieties of CCD were compared using VDGs, assuming that three factors are tested in the experiment and second-order polynomial model will be developed based on the design. VDGs for the CCDs are shown in Figure 5. SPV for the circumscribed CCD is relatively constant over the distance 1 from the center point of the design. The minimum, the maximum and the averaged SPV are equal at the same distance from the origin and it is evidence that the design is rotatable.

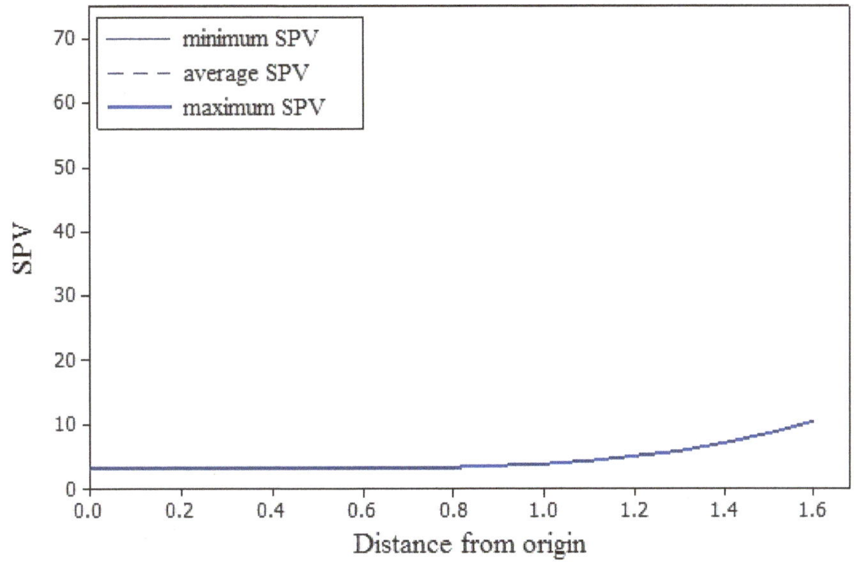

(a)

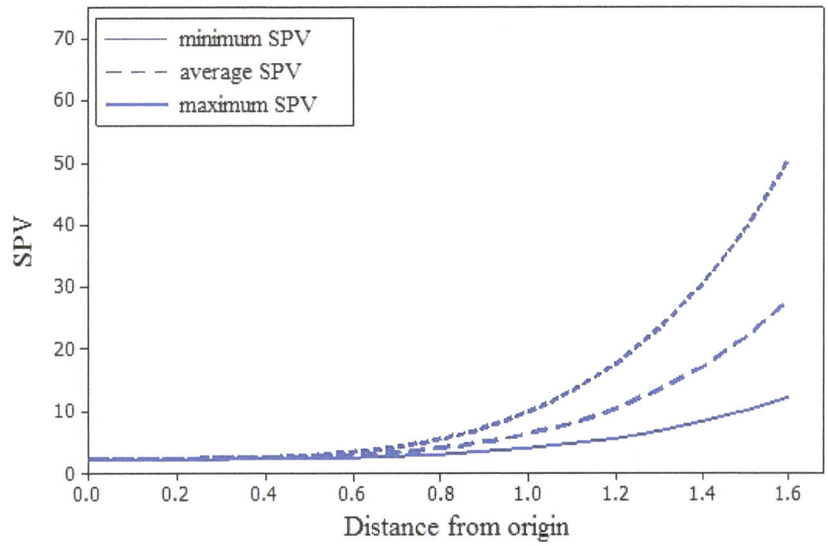

(b)

Figure 5. *Cont.*

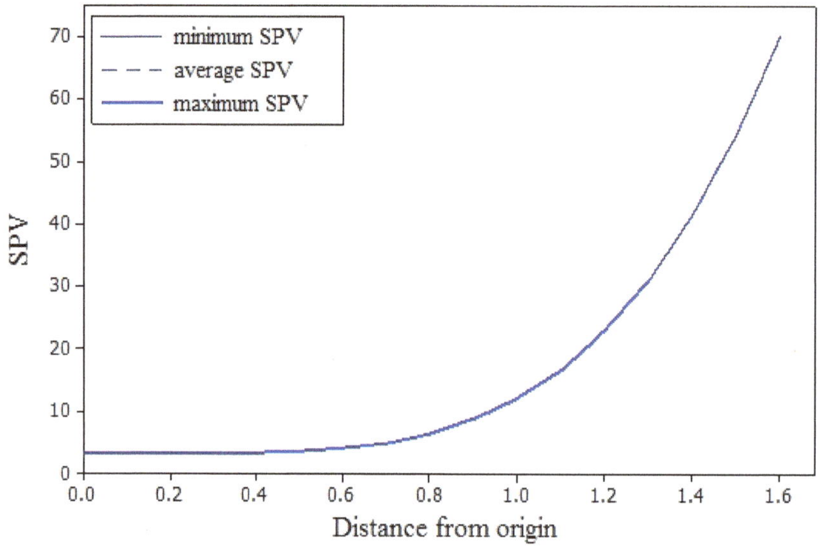

(c)

Figure 5. VDGs for: (**a**) circumscribed CCD; (**b**) face-centered CCD; (**c**) inscribed CCD. SPVs plotted in the graphs have been calculated assuming that full quadratic model for 3 factors is tested when running the experiment and each of the designs consists of 20 experimental trials, including 6 center point trials.

The face-centered CCD ensures lower SPV near the center point compared to the circumscribed CCD, however, at a certain point of the design SPV increases to a level higher than for circumscribed CCD. Furthermore, the maximum and the minimum SPV at the same distance from the origin are not equal to each other, hence the design is not rotatable. VDG for the inscribed CCD shows the same trend as for the circumscribed CCD, but SPV value is higher. This design is also rotatable, similarly to the circumscribed CCD. Despite the fact inscribed CCD does not require testing factors at as extreme levels as the remaining types of CCD, the penalty for this is seen in the variance, because the further is the distance from the center point, the higher is the variance. It is worth noting that each type of CCD strictly defines design points. Therefore, selection of type of CCD should be made taking into account region of experiment operability. If a design point cannot be tested due to various experiment constraints, the experimenter should reduce factor ranges, what affects the region of interest, or generate face-centered CCD or inscribed CCD.

In practice, no model can be fit perfectly to measured values because of measurement errors or relationships between factors and response that cannot be described by a second-order polynomial. This results in residual values at the design points, *i.e.*, deviations from the measured values. The quality of the model is assessed by the coefficients of determination: R^2, *adjusted* R^2 and *predicted* R^2.

R^2 represents a pure correlation between measured and predicted values and is indicative of the response variation explained by a model [18]. However, every term added to the model equation will improve the model fit to the measured data. Therefore, adjusted R^2 is used to compare the explanatory power of models, and its value increases only when an added term improves the model more than by chance [18]. Both these coefficients are calculated using data that were themselves used for model development. A model's predictive capability for new observations is assessed using predicted R^2. This coefficient is calculated by systematically removing each observation from the data set, estimating the regression equation and determining the model's capability in predicting the removed observations [19]. Predicted residual sums of squares statistic [20] is used to calculate the value of predicted R^2.

The statistical significance of the terms of the model defined by Equation (1) can be evaluated using the analysis of variance (ANOVA) [21]. ANOVA is based on partitioning the variation in the data into components. For all the terms of the model equation, values characteristic of a so-called ANOVA table are calculated individually. These values will be important in subsequent discussions and are thus defined here.

The *adjusted sum of squares* (Adj SS) for a specific term calculates reduction in the residual sum of squares resulting from the inclusion of this term to the model. *Adjusted mean squares* (Adj MS) are calculated by dividing Adj SS by the number of degrees of freedom (DF) for the respective term.

Variation in the data unexplained by the model is represented by the *Residual Error* (RE), for which Adj SS is calculated as the residual sum of squares and Adj MS value of the RE is calculated as explained above.

Ratios of the Adj MS for all terms of the model equation and Adj MS of the RE are calculated. Because the ratios of variances follow an F-distribution [16], a statistical F-test [18] is employed to identify statistically significant terms of the model. One can obtain *p*-values from this test for each term of the model, which are a measure of the probability of obtaining data at least as extreme as the data from the model, assuming that the *null hypothesis* is true, *i.e.*, in this case, a particular term does not provide an effect on the results from the model [22]. Therefore, the lower the *p*-values for the analyzed terms, the greater effect these terms have on the response predicted by the model.

Pure error lack-of-fit test [18] is used to assess whether the model is adequate to describe the functional relationships between the experimental factors and the response. The test is based on partitioning the RE sum of squares into two components: lack-of-fit sum of squares, which is associated with variation due to factors other than measurement error, and pure error sum of squares, resulting from random variation caused by measurement error. The ratio of mean squares for lack-of-fit and pure error follows F-statistic, and similar to the aforementioned description, low *p*-value for lack-of-fit in ANOVA table means that the analyzed model does not fit to the experimental data.

4. Experimental Apparatus and Approach

The experimental campaign involved NO_x emission measurements from the PPBB burner, shown in Figure 1, fuelled by pure methane and hydrogen-enriched methane mixtures. All the measurements were conducted at a constant thermal load equal to 25 kW (calculated based on lower heating value, LHV). The fuel mixture was supplied to the burner using two separate fuel supply lines for primary and secondary fuel ports. Mass flow controllers were used to obtain accurate mass flow rates of methane, hydrogen and air to the burner. Maximum measurement error of the mass flow controllers for the investigated burner settings was ±1 nL/min for each fuel supply line and ±10 nL/min for air.

The combustion process took place in a stainless steel combustion chamber at atmospheric pressure. The height of the combustion chamber was 1000 mm with an outer diameter of 360 mm; while the wall thickness was 5 mm. The chamber was cooled only by natural convection and radiation to ambient conditions. The wall temperature was measured on the side of the chamber at a height 300 mm from the bottom to be 390 °C–400 °C. The exhaust gas composition was determined by probing a sample of exhaust gases at the combustion chamber outlet section. The sample was transported through a heated hose to the cooler to avoid uncontrolled water condensation in the gas sampling line and dissolving of nitrogen dioxide (NO_2) in water droplets formed in the hose. NO_x; CO; CO_2 and O_2 were measured with a pre-calibrated Horiba PG-250 gas analyzer. The NO_x concentration in the dry exhaust gas (cooled to 5 °C) was measured by chemiluminescence at a precision of 1 ppmv (part per million by volume); and concentrations of CO_2 and CO were measured using non-dispersive infrared technique. In addition; a Fourier transform infrared gas analyzer was used to measure methane and water concentration in the wet exhaust gas to ensure that all fuel provided to the chamber was burned. The influence of the following four factors on NO_x emissions from the PPBB burner was experimentally investigated:

- hydrogen addition to methane
- amount of excess air
- lance position
- secondary fuel fraction

Levels for the investigated factors were specified as follows: hydrogen content in the fuel was measured by hydrogen mass fraction (X1), excess air was measured using the global air/fuel equivalence ratio (X2), lance position was defined as the distance between the top of the lance and the burner throat (X3) and the amount of secondary fuel was determined as the ratio of the secondary fuel flow rate to the total fuel flow rate (X4).

All the tested factors affect the flow field in the burner and may change the mixing process, flow pattern or flame stability because of complex fluid dynamics. Furthermore, hydrogen addition to methane influences chemical processes occurring in the reaction zone. Interactions between the tested parameters obscure the impact of individual parameters on NO_x emissions from the burner, and the influence of these factors on the emissions cannot be easily described or predicted based on theoretical analyses alone.

Thus, response surface methodology and CCD were employed to identify the significance of each of these factors and develop a second-degree polynomial correlation for NO_x emission prediction. A rotatable CCD was used as the experimental design. Because four factors were studied in the experiment, the number of factorial runs was equal to $2^4 = 16$, and to maintain rotatability in accordance with Equation (4), α was set to 2.

The operating ranges for all the factors were chosen by a series of preliminary measurements to ensure complete fuel combustion and to avoid flame instability, extinction phenomena and combustion chamber acoustic issues at the measurement points. The experimental variables and the levels at which they were tested are shown in Table 1.

Table 1. Tested levels of experimental factors X1, X2, X3 and X4.

Factor	Short Name	Unit	Coded Levels and Corresponding Absolute Levels				
			−1	$-1/\alpha$	0	$1/\alpha$	1
X1	Hydrogen fraction	%	0	3.75	7.5	11.25	15
X2	Air/fuel equivalence ratio	-	1.05	1.1125	1.175	1.2375	1.3
X3	Lance position	mm	20	21.25	22.5	23.75	25
X4	Secondary fuel fraction	%	0	1.5	3	4.5	6

Emission measurement at the center point, *i.e.*, at level zero in coded units for each factor, was replicated four times at various stages of the experiment to obtain an estimate of experimental error. Therefore, the total number of experimental trials, based on the number of design factors $k = 4$, was equal to $N = 2^k + 2k + n_c = 28$. Full factorial design represents a possible alternative approach, but it would require a minimum of $3^4 = 81$ experimental trials, while measurement error could not be evaluated and the model would be unnecessarily made far more complicated.

The entire experiment was conducted in random order without replication. The approach enabled the identification of important parameters for NO_x emission reduction and trends in NO_x emissions from the PPBB burner.

5. Results and Discussion

Statistical analysis of the experimental results was performed using the Minitab® 16 software (Minitab, Inc., State College, PA, USA) [19]. A least squares method was used to derive a mathematical correlation by fitting a response surface to the measured values of NO_x emissions at specific points of the experimental design matrix presented in Table 2. NO_x emissions from the burner have been analyzed for all fuel mixtures investigated, various air/fuel equivalence ratios and burner design parameters.

Table 2. Experimental matrix.

Factor	X1	X2	X3	X4	NO_x	CO	NO_x	Predicted NO_x	Residual
	%	–	mm	%	ppmvd	ppmvd	mg/kWh	mg/kWh	mg/kWh
1	3.75	1.1125	21.25	1.5	17.9	0	35.2	35.2	−0.04
2	3.75	1.2375	21.25	1.5	15.5	5.5	34.3	34.1	0.18
3	11.25	1.1125	21.25	1.5	20.1	0	37.9	37.6	0.35
4	11.25	1.2375	21.25	1.5	17.7	0	37.7	37.7	0.00
5	3.75	1.1125	21.25	4.5	17.9	0	35.3	35.2	0.03
6	3.75	1.2375	21.25	4.5	15.5	7.9	34.2	34.1	0.15
7	11.25	1.1125	21.25	4.5	20.0	0	37.7	37.6	0.13
8	11.25	1.2375	21.25	4.5	18.0	0	38.2	37.7	0.54
9	3.75	1.1125	23.75	1.5	18.5	0	36.3	35.7	0.59
10	3.75	1.2375	23.75	1.5	15.6	4.5	34.6	34.6	0.00
11	11.25	1.1125	23.75	1.5	20.2	0	38.2	38.1	0.13
12	11.25	1.2375	23.75	1.5	17.8	0.1	37.8	38.2	−0.37
13	3.75	1.1125	23.75	4.5	18.0	0	35.4	35.7	−0.29
14	3.75	1.2375	23.75	4.5	15.5	4.7	34.2	34.6	−0.32
15	11.25	1.1125	23.75	4.5	19.9	0	37.5	38.1	−0.53
16	11.25	1.2375	23.75	4.5	17.8	0.1	37.8	38.2	−0.37
17	7.5	1.05	22.5	3	20.2	0	36.4	36.6	−0.17
18	7.5	1.3	22.5	3	15.6	10.0	35.7	35.6	0.12
19	0	1.175	22.5	3	15.4	3.3	32.8	32.9	−0.13
20	15	1.175	22.5	3	19.7	0	38.9	38.8	0.09
21	7.5	1.175	22.5	0	17.8	0	36.4	36.8	−0.44
22	7.5	1.175	22.5	6	17.8	0	36.3	36.8	−0.48

Table 2. *Cont.*

Factor	X1	X2	X3	X4	NO_x	CO	NO_x	Predicted NO_x	Residual
	%	–	mm	%	ppmvd	ppmvd	mg/kWh	mg/kWh	mg/kWh
23	7.5	1.175	20	3	17.6	0	36.0	36.3	−0.29
24	7.5	1.175	25	3	18.7	0	38.3	37.3	0.96
25	7.5	1.175	22.5	3	18.0	0	36.8	36.8	−0.02
26	7.5	1.175	22.5	3	18.1	0	37.1	36.8	0.28
27	7.5	1.175	22.5	3	17.7	2.3	36.2	36.8	−0.61
28	7.5	1.175	22.5	3	18.2	0	37.3	36.8	0.51

5.1. Statistical Analysis of the Model

Due to the fact that mixtures of methane and hydrogen were tested during the experiment, it was found that it is useful to express NO_x emissions measured in parts per million by volume on dry basis (ppmvd) as mass of emitted NO_x per unit of heat released during the combustion process. Such definition is independent of O_2 concentration in flue gas, amount of water present in flue gas that is different for methane and hydrogen combustion, and also takes into account various lower heating values of fuels used in the experimental campaign. Therefore, the formula defined by Equation (6) was used to recalculate NO_x emissions to mg/kWh:

$$NO_x[\mathrm{mg/\,kWh}] = \frac{r_{NO_x} V_{dry\,gas} \rho_{NO_x}}{LHV_{fuel}} 3.6 \cdot 10^6 \quad (6)$$

In Equation (6) r_{NOx} is measured NO_x emissions expressed in ppmvd; $V_{dry\ gas}$ is volume of dry flue gas at standard conditions produced from combustion of 1 kg of fuel in Nm^3/kg, assuming that the flue gas consists of CO_2, N_2 and O_2; ρ_{NO_x} is density of NO_2 at standard conditions equal to 2.05 kg/m^3 and LHV_{fuel} is the lower heating value of fuel mixture used during the burner test in J/kg.

Based on the experimental results of NO_x emissions expressed in mg/kWh and regression analysis for CCD, the full quadratic model is given by Equation (7):

$$\begin{aligned} NOx = {} & -30.011 \\ & -0.418X_1 + 122.783X_2 - 0.494X_3 + 0.400X_4 \\ & -0.017X_1^2 - 49.402X_2^2 + 0.053X_3^2 - 0.052X_4^2 \\ & +1.357X_1X_2 - 0.024X_1X_3 + 0.009X_1X_4 \\ & -1.097X_2X_3 + 1.247X_2X_4 - 0.074X_3X_4 \end{aligned} \quad (7)$$

This model takes into account linear effects, quadratic effects and two-way interactions between the studied factors. The empirical correlation represented in Equation (7) must use factors expressed in uncoded units, *i.e.*, actual values of these factors. NO_x emissions obtained using this correlation are given in mg/kWh.

The coefficient of determination R^2, presented in Table 3, indicates that the model approximates the data at the design points. The predictive power of the developed model for new observations may be 85%, based on the predicted R^2 value. The calculated p-value for lack-of-fit is greater than 0.05; therefore, there is no statistically significant evidence that the model does not represent the data at a 95% confidence level.

Table 3. Model fitting test results.

Model Parameter	Full Quadratic Model	Enhanced Model
R^2	96.63%	94.06%
Adjusted R^2	93.01%	92.36%
Predicted R^2	84.99%	90.19%
p-value for lack-of-fit	0.75	0.195

Based on the results of ANOVA, shown in Table 4, the model was improved by removing terms (one-by-one) with *p*-values greater than 0.05, considered as statistically insignificant at a 95% confidence level. The removed terms were not taken into account in the regression analysis.

Table 4. ANOVA table for the full quadratic model.

Source of Variation	DF	Adj SS	Adj MS	F-Ratio	*p*-Value
X1	1	53.2362	53.2362	322.24	<0.001
X2	1	1.5645	1.5645	9.47	0.009
X3	1	1.4741	1.4741	8.92	0.010
X4	1	0.1050	0.1050	0.64	0.440
X1·X1	1	1.4259	1.4259	8.63	0.012
X2·X2	1	0.8937	0.8937	5.41	0.037
X3·X3	1	0.1648	0.1648	1.00	0.336
X4·X4	1	0.3233	0.3233	1.96	0.185
X1·X2	1	1.6182	1.6182	9.80	0.008
X1·X3	1	0.2071	0.2071	1.25	0.283
X1·X4	1	0.0410	0.0410	0.25	0.627
X2·X3	1	0.1176	0.1176	0.71	0.414
X2·X4	1	0.2187	0.2187	1.32	0.271
X3·X4	1	0.3088	0.3088	1.87	0.195
Residual Error	13	2.1477	0.1652		
Lack-of-fit	10	1.4444	1.4444	0.62	0.754
Pure error	3	0.7033	0.2344		
Total	27				

The enhanced model is defined by Equation (8):

$$\begin{aligned} NOx = &-20.678 \\ &-0.944X_1 + 97.837X_2 + 0.198X_3 \\ &-0.017X_1^2 - 47.701X_2^2 \\ &+1.357X_1X_2 \end{aligned} \quad (8)$$

This model predicts NO_X emissions in mg/kWh based on three linear effects, two quadratic effects and one two-way interaction. Secondary fuel fraction (X4) has been found to be negligible. Again, the empirical correlation given in Equation (8) must be used with the actual, uncoded factor values.

Although R^2 and adjusted R^2, shown in Table 3, decreased compared with those of the full quadratic model given by Equation (7), the enhanced model offers a higher prediction capability of 90.19% for new observations. Simultaneously, with a *p*-value for lack-of-fit greater than 0.05, it can be assumed that the model adequately represents the experimental results.

A summary of ANOVA applied to the enhanced model is presented in Table 5. The estimated standard deviation of the error in the model is 0.42 mg/kWh. Since all the terms are statistically significant and the model gives reasonable response estimation, it was used to analyze the NO_x emissions from the PPBB burner.

Table 5. ANOVA table for the enhanced model.

Source of Variation	DF	Adj SS	Adj MS	F-Ratio	*p*-Value
X1	1	53.2362	53.2362	295.01	<0.001
X2	1	1.5645	1.5645	8.67	0.008
X3	1	1.4741	1.4741	8.17	0.009
X1·X1	1	1.4992	1.4992	8.31	0.009
X2·X2	1	0.9259	0.9259	5.13	0.034
X1·X2	1	1.6182	1.6182	8.97	0.007
Residual Error	21	3.7895	0.1805		
Lack-of-fit	8	1.9270	0.2409	1.68	0.195
Pure error	13	1.8625	0.1433		
Total	27				

5.2. *Effects of the Factors*

Figure 6 shows the sorted absolute values of the coefficients for each term of the second-degree polynomial based on factors expressed in coded units. Coding of factors removes any pseudo effects due to the use of different measurement scales. Therefore, coefficients in the model equation developed on the basis of coded units are a measure of the magnitude of the response resulting from one unit change in a factor in one specific term, with all other terms held constant. It should be stressed that this interpretation applies over the entire investigated range of factors to linear effects only. Because the model includes an interaction involving two factors (X1 and X2), the effect of a change in one factor associated with the interaction term varies depending on the value chosen for the other factor. In the case of quadratic terms, the response to a change in the value of a factor depends on the value of the factor itself.

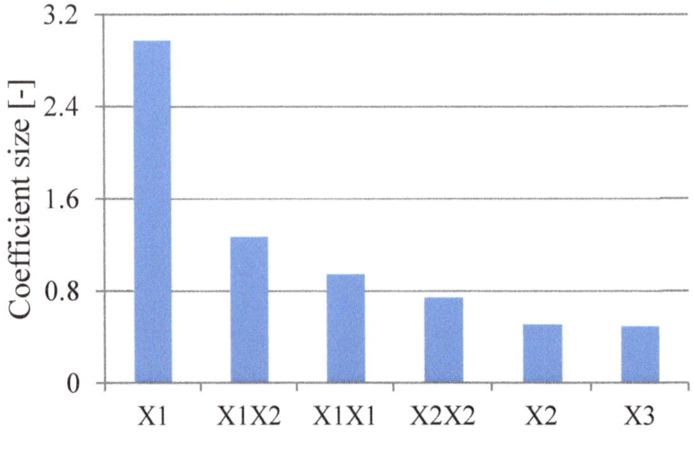

Figure 6. Comparison of the absolute values of the coefficients for each term of the enhanced model, which is expressed in coded units.

The most significant factor for NO_x emissions is clearly hydrogen fraction in the fuel (X1). The coefficient of the linear term determining fuel composition causes this term to dominate the other terms. Adiabatic flame temperatures, assuming chemical equilibrium of products, for various stoichiometric mixtures of methane and hydrogen and air as oxidizer are presented in Figure 7. Significance of the factor X1for the NO_x emission prediction can as expected be attributed to the increased flame temperature that results from the added hydrogen and thus enhanced thermal NO_x formation.

However, it is worth noting that the above brief explanation does not show the overall picture of the problem and the complexity of NO_x chemistry. When firing hydrocarbon fuels, NO_x are largely formed via thermal or prompt mechanism and, for simplicity, it can be assumed that total NO_x emissions are a combination of these two mechanisms.

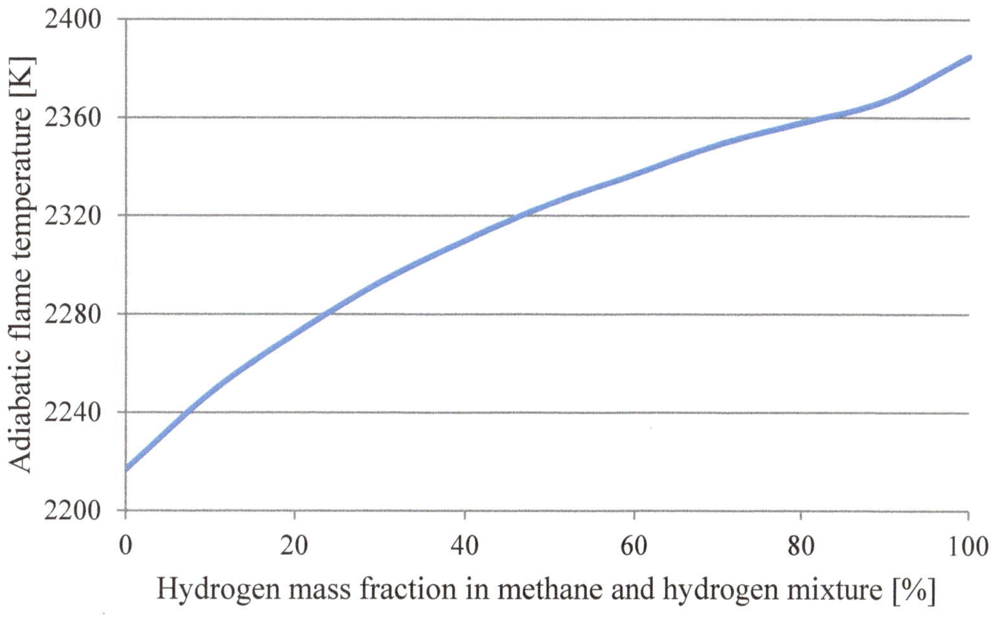

Figure 7. Adiabatic flame temperature for various mixtures of methane and hydrogen and air as oxidizer at stoichiometric conditions at pressure equal to 1 atm and temperature of reactants equal to 293 K, calculated assuming chemical equilibrium of products.

The NNH mechanism [23] plays a role in NO_x formation in hydrogen-enriched flames, but it is not discussed here. As it is well-known that hydrogen addition to methane increases flame temperature, it is also known that prompt NO_x formation mechanism requires carbon-containing radicals in the reaction zone. Concentration of these radicals decreases when hydrogen is added to methane and then prompt NO_x formation is inhibited. In the light of the above and despite the fact that NO_x increase is explained here by the higher adiabatic flame temperature of hydrogen-enriched flames, if a low NO_x burner that effectively reduces peak flame temperature is used, the situation when total NO_x emissions from hydrogen combustion are lower than NO_x emissions from pure methane combustion cannot be ruled out [24].

Linear terms related to excess air (X2) and the lance position (X3) are considerably less significant than the hydrogen fraction in the fuel. The effect of the interaction between fuel composition and excess air (X1X2) varies depending on the values of these factors, but for all values of excess air, the interaction affects the response less than hydrogen enrichment of methane. Comparable values for the coefficients of quadratic terms of hydrogen fraction and excess air are evidence that there is a curvature in the response surface describing NO_x emissions. These terms may significantly change the response of the model over the investigated range of factors, depending on the values of the respective factors, but never more than the linear term determining fuel composition. Because the fuel composition is the most important factor, the value of X1 was always taken into account while evaluating the influence of the other experimental variables on NO_x emissions.

5.3. Excess Air Effect

Significant change in NO_x emissions was observed with increasing air flow to the burner in comparison with stoichiometric air consumption in the combustion process. However, this relationship is visible only for pure methane and fuel mixtures containing hydrogen of a few percent mass fraction. As illustrated in Figure 8, with increasing hydrogen content, NO_x emissions approach a state independent of excess air, or even slightly increase with increasing excess air for hydrogen content approaching 15% mass fraction. The lowest NO_x emissions are approximately 30 and 38 mg/kWh for pure methane and hydrogen-rich fuel, respectively. Calculated 95% confidence intervals for these values are between 29.0 and 31.7 mg/kWh, and 37.5 and 40.2 mg/kWh.

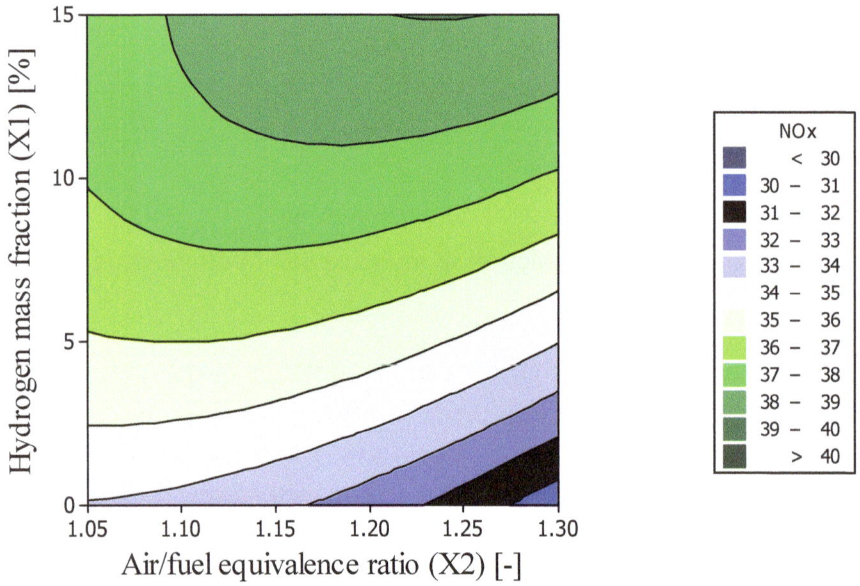

Figure 8. NO_x emissions in mg/kWh *versus* X1 and X2 for the fuel lance positioned at 22.5 mm.

The relationship shown in Figure 8 can be explained by that additional air plays the role of an inert gas and suppresses NO_x formation via the thermal mechanism. However, for methane enriched with 15% mass fraction hydrogen, this suppressing effect is diminished. Despite the fact that the fuel mixture is partially mixed with air before combustion and the flame is stabilized behind the lance, NO_x emissions are not reduced. It can be assumed that the higher adiabatic flame temperature of hydrogen-enriched methane entails increased thermal NO_x formation, and this mechanism cannot be inhibited by partially premixed air, even at global excess air values approaching 30%. Simultaneously, this effect can be caused by the fact that hydrogen-rich methane-hydrogen-air flame is short and it is stabilized just behind the trailing edge of the lance, while in case of hydrogen-lean methane-hydrogen-air flame or pure methane-air flame, the reaction zone extends over a certain distance further downstream the lance to a better mixed region, for example, due to turbulence. Therefore, in the latter case, higher amount of excess air enables effective reduction of peak flame temperature, what in turn results in lower NO_x emissions.

Note that because the density of hydrogen is approximately eightfold lower than that of methane, hydrogen addition results in increased fuel mixture velocity at the fuel ports. Consequently, air and fuel mixing occurs differently for different fuel mixtures due to different both physical properties such as diffusivity and flow pattern. The amount of excess air provided to the combustion zone behind the lance is probably not linearly dependent on the global excess air value. Therefore, it should be kept in mind when comparing the results that the measured emissions are affected by the different flow fields created by both different fuel composition and excess air. Thus, the extent of NO_x formation may result from either a modified chemical kinetic mechanism, when hydrogen is added, or altered fluid dynamics.

5.4. PPBB Burner Design Parameters Effect

Modifications of the distance between the lance and the burner exit section create variations in air velocity in the air/fuel mixing region. The location of the fuel distribution ports is different for various lance positions. According to Figure 9, for all the investigated fuel compositions, NO_x emissions are reduced with the lance shifted upstream relative to air flow. Minimum NO_x emissions estimated based on 95% confidence interval are between 28.5 mg/kWh and 31.2 mg/kWh, and are observed at a lance position of 20 mm when the burner is fuelled by pure methane. This trend is confirmed for hydrogen-enriched methane as well. However, measurements were conducted with lance positions from 20 to 25 mm because the further shifting of the lance upstream results in incomplete combustion of pure methane. For the investigated settings, primary fuel ports are always located above the top edge of the air outlet, and the lance shape does not substantially affect air velocity at the burner throat, *i.e.*, air outlet section. Therefore, the effect of lance position on NO_x emissions is not strong.

All the terms associated with secondary fuel provided to the burner (X4) were determined to be statistically insignificant over the range of burner settings investigated and have therefore not been included in the enhanced model. The influence of this factor was investigated over rather low mass flow rates ranging from 0% to 6% of total fuel stream. The reason for this was flame extinction, which occurs when a higher amount of secondary fuel is provided to the burner and the burner is fuelled by pure methane. The low settings resulted in the unstable operation of mass flow controllers, and as a result, the influence of this factor may have been underestimated. ANOVA results of the full quadratic model indicated that these terms contributed to the response of the model, with a magnitude lower than the inaccuracy of the model. Quantitatively, the full quadratic model would give approximately the same value for NO_x emissions as the enhanced model. However, the quality of the enhanced model has been improved in that the model facilitates an understanding of the impact of the tested factors on NO_x emissions and has a better predictive capability.

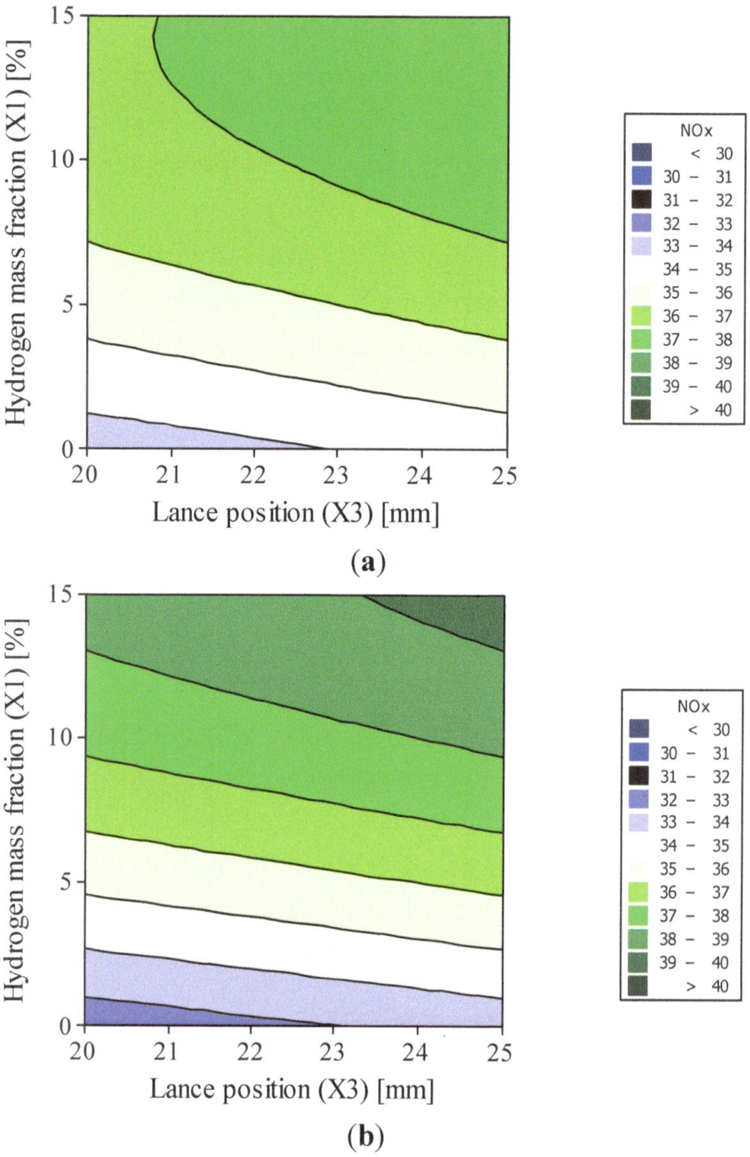

Figure 9. NO_x emissions in mg/kWh *versus* X1 and X3 for (**a**) X2 = 1.05; (**b**) X2 = 1.3.

5.5. Limitations of the CCD Approach

CCD is perhaps the most popular class of experimental designs, which allow for efficient estimation of second-order response surfaces. Nevertheless, some important limitations have been identified when it comes to application of CCD in the field of combustion.

Second-order polynomial model is unable to describe real behavior of the explanatory variable as a function of the investigated factors, if significant non-linearities or multimodal distribution are present within the tested range of factors. Expert knowledge or additional measurements are necessary in order to ensure that the explanatory variable can be estimated with second-order response surface and higher-order effects do not occur. In contrast to NO_x emissions that can be described by the model developed based on CCD, it was found during preliminary measurements that CCD approach could not be used for CO emission prediction. CO emission trends often exhibit exponential dependence on various factors, e.g., due to so-called CO breakthrough at low amounts of excess air, and thus the predicted values of CO emission do not fit to the experimental data. One possible solution to this problem is to conduct burner tests at a firebox temperature high enough to completely oxidize CO formed in the reaction zone, and focus only on NO_x emissions, as it was done in the experiment described above. Reasonableness of this approach is, however, conditioned by the target application of the tested burner, *i.e.*, whether it will be used in high-temperature furnaces or low-temperature furnaces [25]. It is worth noting that NO_x emissions are dependent on a firebox temperature [26], but as soon as combustion process is complete, means all CO is reburned to CO_2, optimal burner design parameters determined at a certain firebox temperature are still optimal in terms of emissions at a higher firebox temperature.

Another important limitation of the CCD is the fact that this approach requires symmetrical design space. In other words minimum and maximum values of tested factors must be equidistant from the respective values of these factors at a center point. It may be a significant disadvantage when CCD is used for burner development, because conducting measurements at certain points of the design may not be possible due to operational reasons. At certain combinations of factors tested within the design, burner operation may not be stable or even possible due to difficulties such as: unacceptably high CO emission, flame instability or extinction, chamber acoustic issues, and flame stabilization at surfaces, which are not designed for this purpose. These issues naturally affect region of experiment operability and they may cause that it can be highly asymmetrical. An easy way to deal with these difficulties is to resign from testing the most limiting factor or narrow ranges of the factors tested in an experiment. However, one may be also interested in NO_x emissions beyond the symmetrical design space of the reduced CCD, which is fitted into a wider region of operability. If it is the case, an alternative approach to experimental design is D-optimal or I-optimal design. The former aims to minimize the variance of the factor-effect estimates, while the latter minimizes the average variance of prediction over the region of experimentation and it makes I-optimal design more appropriate for prediction than D-optimal design [27]. These methods offer a great deal of flexibility to reach a viable compromise between number of experimental trials, design choices, constraints, and usable results. On the other hand, the disadvantage of the optimal designs is that they are model-dependent. It means that they require knowledge about form of a model, *i.e.*, knowledge of terms of the second-order polynomial which should be used for a proper description of the relationship between the response and the factors. This information is necessary to generate optimal design and the form of the model must be specified by the experimenter *a priori*. Testing of all the unique combinations of factors overcomes all the aforementioned difficulties, but reasonableness of this approach is, of course, dependent on its intended purpose and available resources.

6. Conclusions

NO_x emissions from a laboratory-scale PPBB burner were investigated using a CCD of the experiment. Experimental measurements were conducted at a constant thermal load equal to 25 kW (calculated based on LHV). The influence of hydrogen enrichment of methane, excess air and burner design parameters on NO_x emissions was analyzed. The developed model allowed for an evaluation of linear and quadratic effects of analyzed factors, as well as significant interactions between these factors. The significance of the model terms was assessed by ANOVA. The values predicted by the second-order polynomial model are well-aligned with the measured NO_x emissions. Limitations of the used approach to experimental design have been identified and discussed.

Based on the developed model, it was found that addition of hydrogen to methane is accompanied by an increase of NO_x emissions. While this effect is very strong and cannot be avoided, it can be reduced by properly selecting burner operation settings. Increase in excess air results in significant NO_x emissions reduction, but only when the burner is fuelled by pure methane or fuel mixtures containing small amounts of hydrogen. However, NO_x formation is largely insensitive to changes in global excess air when the burner is fuelled by methane enriched with about 15% mass fraction hydrogen. This can be attributed to the location of hydrogen-enriched reaction zone close to the lance in a poorly mixed region and a higher adiabatic flame temperature, resulting in an enhanced thermal NO_x formation mechanism. A proper fuel lance position enables slight reduction of NO_x emissions. Secondary fuel fraction was found to have no significant effect on NO_x emissions over the investigated range of burner operation settings.

The developed correlation demonstrates that response surface methodology combined with a CCD of the experiment can be used as an effective method for investigating the influence of various factors on NO_x emissions and finding optimal burner operation settings, if such exist. The technique is also useful in a context where numerous factors affecting the results are involved, and factors warranting deeper insight can be highlighted.

Future Work

Future work involves flow field measurements using particle imaging velocimetry (PIV) technique in order to correlate NO_x emission characteristics of the PPBB burner with the respective changes of the flow field behind the lance. The measurements will be performed for both non-reactive and reactive flows, and influence of the burner design parameters and fuel composition on the flow field will be investigated. In addition, NO_x emission measurements for the PPBB operating with pre-heated combustion air are planned.

Acknowledgments

This publication has been produced with support from the BIGCCS Centre, performed under the Norwegian research program Centers for Environment-friendly Energy Research (FME). The authors acknowledge the following partners for their contributions: ConocoPhillips, Gassco, Shell, Statoil, TOTAL, GDF SUEZ and the Research Council of Norway (193816/S60).

Author Contributions

Marcin Dutka analyzed available design of experiments techniques and developed the original direction and methodology behind the work. Marcin Dutka designed the experiment, conducted experimental measurements, developed the regression model for NO_x emission prediction from the PPBB burner and analyzed the experimental results. The work was performed under supervision of Terese Løvås and Mario Ditaranto. All the authors discussed the results, revised the paper and approved its publication.

Conflicts of Interest

The authors declare no conflict of interest.

References

1. *CO_2 Abatement in Oil Refineries*: *Fired Heaters*; Technical Report PH3/31, 2000. Available online: http://ieaghg.org/docs/General_Docs/Reports/Ph3_31%20Oil%20refinery%20fired%20heaters.PDF (accessed on 27 April 2015).
2. Miracca, I.; Asen, K.I.; Assink, J.; Coulter, C.; Curran, L.; Lowe, C.; Moure, G.T.; Schlasner, S. The CO_2 capture project (CCP): Results from phase II (2004–2009). *Energy Procedia* **2009**, *1*, 55–62.
3. Ditaranto, M.; Anantharaman, R.; Weydahl, T. Performance and NO_x emissions of refinery fired heaters retrofitted to hydrogen combustion. *Energy Procedia* **2013**, *37*, 7214–7220.
4. Box, G.; Wilson, K. On the experimental attainment of optimum conditions. *J. R. Stat. Soc. Ser. B* **1951**, *13*, 1–45.
5. Bělohradský, P.; Kermes, V. Experimental study on NO_x formation in gas-staged burner based on the design of experiments. *Chem. Eng. Trans.* **2012**, *29*, 79–84.
6. Shanbhogue, S.J.; Husain, S.; Lieuwen, T. Lean blowoff of bluff body stabilized flames: Scaling and dynamics. *Prog. Energy Combust. Sci.* **2009**, *35*, 98–120.
7. Park, N.S.; Ko, S.C. Large eddy simulation of turbulent premixed combustion flow around bluff body. *J. Mech. Sci. Technol.* **2011**, *25*, 2227–2235.
8. Kopp-Vaughan, K.M.; Jensen, T.R.; Cetegen, B.M.; Renfro, M.W. Analysis of blowoff dynamics from flames with stratified fueling. *Proc. Combust. Inst.* **2013**, *34*, 1491–1498.
9. Düsing, M.; Kempf, A.; Flemming, F.; Sadiki, A.; Janicka, J. Combustion LES for premixed and diffusion flames. *Prog. Comput. Fluid Dyn.* **2005**, *5*, 363–374.
10. Dawson, J.R.; Gordon, R.L.; Kariuki, J.; Mastorakos, E.; Masri, A.R.; Juddoo, M. Visualization of blow-off events in bluff-body stabilized turbulent premixed flames. *Proc. Combust. Inst.* **2011**, *33*, 1559–1566.
11. Chaudhuri, S.; Kostka, S.; Renfro, M.W.; Cetegen, B.M. Blowoff dynamics of bluff body stabilized turbulent premixed flames. *Combust. Flame* **2010**, *157*, 790–802.
12. Giacomazzi, E.; Battaglia, V.; Bruno, C. The coupling of turbulence and chemistry in a premixed bluff-body flame as studied by LES. *Combust. Flame* **2004**, *138*, 320–335.
13. Sobiesiak, A.; Rahbar, S.; Becker, H.A. Performance characteristics of the novel low-NO_x CGRI burner for use with high air preheat. *Combust. Flame* **1998**, *115*, 93–125.
14. Grandmaison, E.W.; Yimer, I.; Becker, H.A.; Sobiesiak, A. The strong-jet/weak-jet problem and aerodynamic modeling of the CGRI burner. *Combust. Flame* **1998**, *114*, 381–396.
15. Montgomery, D.C. *Design and Analysis of Experiments*; John Wiley & Sons, Inc.: Hoboken, NJ, USA, 2001; p. 456.
16. Box, G.E.P.; Hunter, J.S.; Hunter, W.G. *Statistics for Experimenters: Design, Innovation, and Discovery*; Wiley-Interscience: Hoboken, NJ, USA, 2005; pp. 47, 455.
17. Giovannitti-Jensen, A.; Myers, R.H. Graphical assessment of the prediction capability of response surface designs. *Technometrics* **1989**, *31*, 159–171.
18. Mason, R.L.; Gunst, R.F.; Hess, J.L. *Statistical Design and Analysis of Experiments*, 2nd ed.; John Wiley & Sons: New York, NY, USA, 2003; pp. 194–195, 504–507.
19. *Minitab*, Version 16, Statistical Software; Minitab, Inc.: State College, PA, USA, 2010.

20. Allen, D.M. The Relationship between variable selection and data agumentation and a method for prediction. *Technometrics* **1974**, *16*, 125–127.
21. Doncaster, C.P.; Andrew, J.H.D. *Analysis of Variance and Covariance: How to Choose and Construct Models for the Life Sciences*; Cambridge University Press: Cambridge, UK, 2007; pp. 1–14, 238.
22. Andersson, O. *Experiment!: Planning, Implementing and Interpreting*; John Wiley & Sons: Chichester, UK, 2012; p. 164.
23. Konnov, A.A.; Colson, G.; de Ruyck, J. NO formation rates for hydrogen combustion in stirred reactors. *Fuel* **2001**, *80*, 49–65.
24. Waibel, R.T.; Athens, L.; Claxton, M. Effect of fuel composition on emissions from ultra low NO_x burners. In Proceedings of the International Symposium Combustion Research and Industrial Practice: From Equations to Equipment, Monterey, CA, USA, 15–18 October 1995.
25. Bussman, W.R.; Baukal, C.E., Jr.; French, K.W. Variable test furnace cooling. In Proceedings of the ASME Summer Heat Transfer Conference, San Francisco, CA, USA, 17–22 July 2005; Volume 4, pp. 291–304.
26. Bussman, W.; Baukal, C.; Colannino, J. The effect of firebox temperature on NO_x emissions. In Proceedings of the Air and Waste Management Association's Annual Conference and Exhibition, AWMA, Indianapolis, IN, USA, 22–25 June 2004.
27. Jones, B.; Goos, P. I-optimal *versus* D-optimal split-plot response surface designs. *J. Qual. Technol.* **2012**, *44*, 85–101.

4

Pyroelectric Harvesters for Generating Cyclic Energy

Chun-Ching Hsiao * and Jia-Wai Jhang

Department of Mechanical Design Engineering, National Formosa University, No. 64, Wunhua Rd., Huwei Township, Yunlin County 632, Taiwan; E-Mail: a8334573@gmail.com

* Author to whom correspondence should be addressed; E-Mail: cchsiao@nfu.edu.tw

Academic Editor: Shi Xue Dou

Abstract: Pyroelectric energy conversion is a novel energy process which directly transforms waste heat energy from cyclic heating into electricity via the pyroelectric effect. Application of a periodic temperature profile to pyroelectric cells is necessary to achieve temperature variation rates for generating an electrical output. The critical consideration in the periodic temperature profile is the frequency or work cycle which is related to the properties and dimensions of the air layer; radiation power and material properties, as well as the dimensions and structure of the pyroelectric cells. This article aims to optimize pyroelectric harvesters by matching all these requirements. The optimal induced charge per period increases about 157% and the efficient period band decreases about 77%, when the thickness of the PZT cell decreases from 200 μm to 50 μm, about a 75% reduction. Moreover, when using the thinner PZT cell for harvesting the pyroelectric energy it is not easy to focus on a narrow band with the efficient period. However, the optimal output voltage and stored energy per period decrease about 50% and 74%, respectively, because the electrical capacitance of the 50 μm thick pyroelectric cell is about four times greater than that of the 200 μm thick pyroelectric cell. In addition, an experiment is used to verify that the work cycle to be able to critically affect the efficiency of PZT pyroelectric harvesters. Periods in the range between 3.6 s and 12.2 s are useful for harvesting thermal cyclic energy by pyroelectricity. The optimal frequency or work cycle can be applied in the design of a rotating shutter in order to control the heated and unheated periods of the pyroelectric cells to further enhance the amount of stored energy.

Keywords: pyroelectric harvester; cyclic heating; waste heat energy; PZT

---

1. Introduction

Energy harvesting is the process of converting, processing and storing environmental energy. The ambient energy comes in the form of wind, water, sun, heat, *etc*. Energy harvesting is also called energy scavenging, and it is employed for recovering energy that otherwise would be wasted. Sources for energy harvesting are mainly categorized as mechanical, thermal and solar energy. A significant amount of waste heat is lost as a by-product of power, refrigeration, or heat pump cycles. Direct energy conversion using thermoelectric generators mainly relies on the Seebeck effect to convert a steady-state temperature difference between two dissimilar metals or semiconductors into electrical energy. Thermoelectric generators need a temperature difference, as they cannot work in an environment with spatially uniform and time-dependent temperature fluctuations [1]. Alternatively, pyroelectric devices directly convert temperature fluctuations into electrical energy [1–7]. Pyroelectric materials have the potential to operate with high thermodynamic efficiency and, compared to thermoelectric generators, do not require bulky heat sinks to maintain the temperature gradient. Pyroelectric materials respond to changes in temperature which cause an internal strain, and this, in turn, results in electrical charges on the material's surface. Advances in materials and in thermal-electrical cycling methods are expected to provide low cost and high-power-density electrical generators [8]. When a pyroelectric element is subjected to a power density radiation (W) causing a temperature variation ($\mathrm{d}T$), the induced charge ($\mathrm{d}Q$, units: μC) is released by the electrode area (A) of the element due to a decrease in polarization, as presented by [1,9]:

$$\mathrm{d}Q = \eta \times P \times A \times \mathrm{d}T \quad (1)$$

where η is the absorption coefficient of radiation; A is the electrode area; $\mathrm{d}T/\mathrm{d}t$ is the temperature variation rate of the pyroelectric materials; Q is the induced charge; and P (units: $\mu\mathrm{Cm}^{-2}\mathrm{K}^{-1}$) is the pyroelectric coefficient of the pyroelectric materials, as presented by [1,9]:

$$P = \mathrm{d}P_s/\mathrm{d}T \quad (2)$$

where P_s is the magnitude of the electrical polarization vector. Pyroelectric elements, such as flat-plate capacitors, are sandwiched in between the top and bottom electrodes, and poled along the axis perpendicular to the plates. P_s is perpendicular to the electrode surface and its magnitude equals the electrode charge density. Therefore, pyroelectric elements with high P values and applications with large temperature variations over time should be considered. Moreover, increasing the electrode area will result in enhanced current under parity of incident thermal power density per unit area. When the dimensions and the materials of pyroelectric elements are determined, creating a larger temperature variation rate in the pyroelectric element is useful for harvesting pyroelectric energy, however, the temperature variation rate is difficult to extract from pyroelectric layers by experimental measurements. Some finite element models were built using the commercial software package COMSOL to explore the temperature variation rate in pyroelectric cells, with some designs of cavities created by wet etching, trenches made with a precision dicing saw, and grooves generated by a sandblast etching technique, in order to improve the energy conversion efficiency of PZT cells by pyroelectricity [5–7].

Cyclic heat energy harvesting mainly uses the pyroelectric elements to convert temperature fluctuations into electricity. Periodic heating leads to continuous power generation over time, and provides a means of continuously harvesting energy. Therefore, it was necessary to design a periodic temperature profile and apply it to the pyroelectric elements to generate temperature variation rates. The factors affecting the periodic temperature profiles for improving the temperature variation rate in pyroelectric elements include the frequency or work cycle, radiation power, properties and dimensions of the air layer, and the material properties, dimensions and structure of the pyroelectric elements, *etc.* This research proposes a complete analysis of all these factors for a thorough comprehension of the relationships among the abovementioned considerations.

2. Materials and Methods

2.1. Cyclic Heating System for Pyroelectricity

A commercial PZT pyroelectric cell with the dimensions of 9 mm × 9 mm × 0.214 mm was used. The cell comprised a 0.2 mm thick PZT sheet sandwiched between a top and a bottom electrode. The electrodes were 7 μm thick silver films. PZT samples were provided by Eleceram Technology Co. (Taoyuan, Taiwan). Table 1 shows the material properties of the commercial PZT pyroelectric cell. A thick PZT sheet with a low electrical capacitance is more likely to generate higher voltages than a thin PZT film with a high electrical capacitance under a given temperature variation [1], but a thick pyroelectric material has high thermal capacity, which hinders quick temperature variations. Obviously, thin pyroelectric materials have a high temperature variation rate. Although thin film deposition technology can be used to grow films with lesser thickness, the pyroelectric properties usually decrease quickly. Commercial PZT bulk materials possessing excellent pyroelectric properties, together with the designs of trench and groove structures fabricated using a dicing saw apparatus, as well as a sandblast etching technique, were useful for improving the temperature variation rate and thus enhancing the efficiency of pyroelectric harvesters [6,7].

Table 1. Properties of the commercial PZT pyroelectric cell.

Sample ID	Thickness (μm)	Area (mm^2)	Size (mm × mm)	Relative dielectric constant ($\varepsilon_{33}{}^T/\varepsilon_0$)	Density (g/cm^3)	Poling field (V/μm)	Pyroelectric coefficient ($10^{-4}cm^{-2}K^{-1}$)
KA	200	81	9 × 9	2100	7.9	3.5	6.97

The cyclic heating system was mainly meant to achieve a periodic temperature profile for generating temperature fluctuations in the pyroelectric cells. Figure 1 shows a schematic diagram of a pyroelectric harvester for generating cyclic heat energy and the measurement setup. This system consisted of a rotating disk with apertures, a gear motor, a radiating heat lamp (Heat Plus 250W 110V, Philips, Amsterdam, The Netherlands), thermocouple sensors and the PZT pyroelectric cells placed on a printed circuit board (PCB). A cavity was dug on the PCB to hold the PZT cell along the edges in order to expose it completely to the air. The distance between the heat lamp and the PZT cell was about 15 cm. The rotating speed of the disk could be changed by controlling the voltage applied to the gear motor. Type K (Chromel/Alumel) thermocouple sensors were used to probe the temperature variations in the pyroelectric cells; they were attached to the bottom electrode at the center of the PZT cell, thereby ensuring good thermal contact.

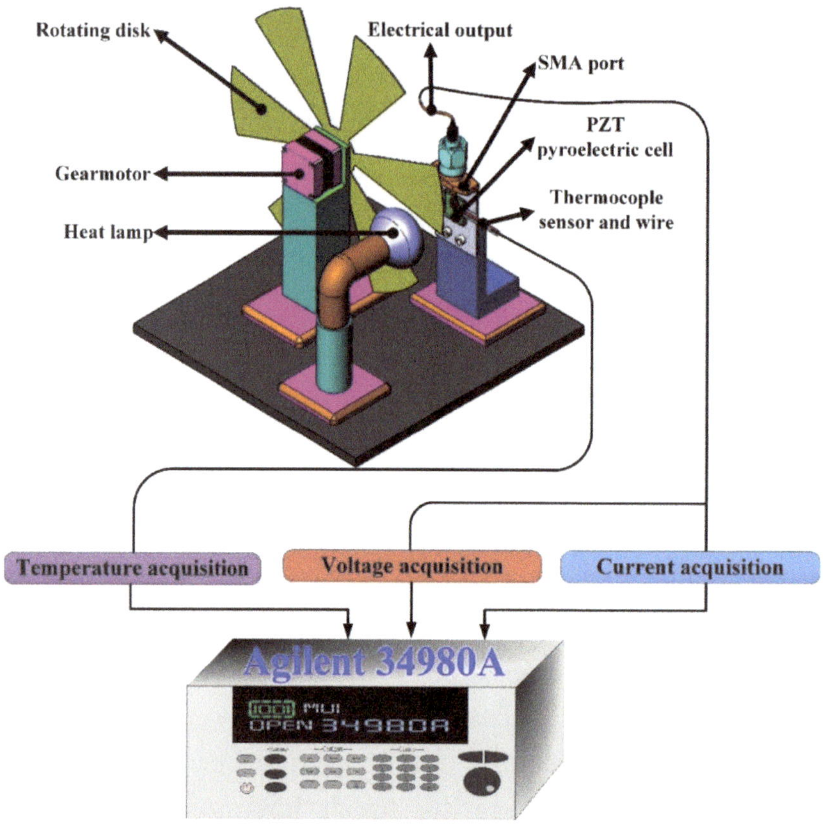

Figure 1. Schematic diagram of the pyroelectric harvester with cyclic heating and measurement setup.

The PZT pyroelectric cells with thermocouple sensors were attached to the PCB for measuring the electrical outputs via the SMA connectors. The radiating heat lamp was the heat source; it was projected onto the PZT pyroelectric cells through the apertures of the disk. Therefore, the disk could prevent the PZT cells from absorbing the heat from the radiating heat lamp, except for the apertures in the disk. When the disk rotated, the PZT cells were periodically heated. Finally, the output data of temperature, current and voltage were measured simultaneously with a computer-controlled data acquisition apparatus (Agilent 34980A, Santa Clara, CA, USA). The variables in the periodic temperature profiles included the frequency or work cycle, properties and dimensions of the air layer, material properties, and the dimension and structure of the pyroelectric PZT cells. The work cycle was defined as the heated and unheated periods of the pyroelectric cells, as depicted in Figure 2.

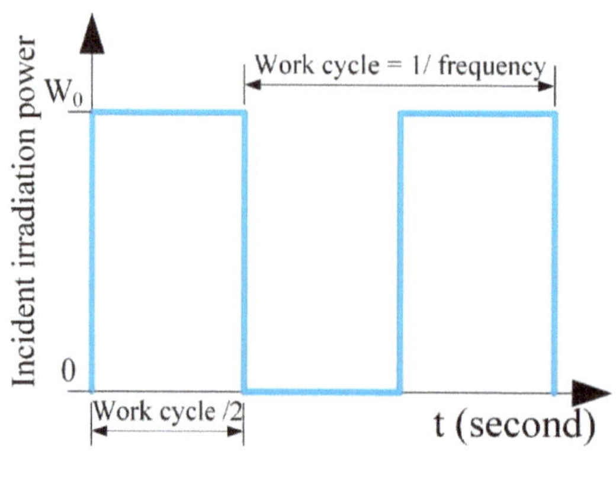

Figure 2. The incident radiation power with a work cycle applied to the PZT cells.

Moreover, the charge generated from thermal oscillations was stored in a capacitor C_L as a storage element using the full-wave rectifier circuit, as depicted in Figure 3. The measured forward voltage drop of the diodes (Model 1N4148, IHS, Englewood, CO, USA) was 0.62 V. C_L was a 4.7 μF electrolytic capacitor with a 50 V maximum voltage. During the heating cycle, the two forward-biased diodes (D_1 and D_3) allowed the generated current flow i_P through to charge C_L. The other diodes (D_2 and D_4) were reverse-biased, and blocked the current flow. During the cooling cycle, the direction of i_P was reversed, and C_L was charged through D_2 and D_4. The temperature in the PZT sheets and the output voltage of the circuit (V_0) were also measured with the Agilent 34980A data acquisition unit.

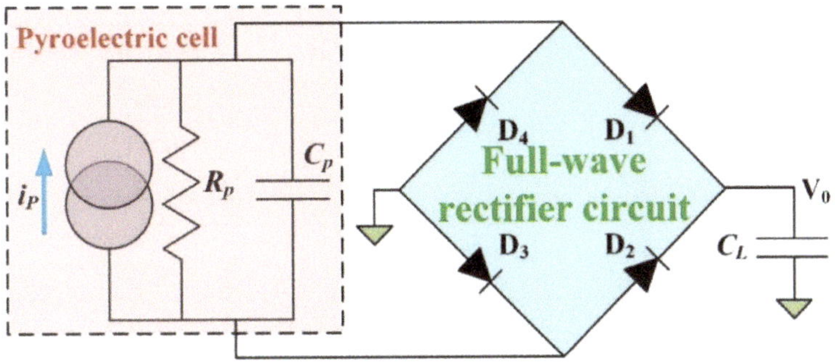

Figure 3. Full-wave rectifier circuit used to store the induced charge from the pyroelectric cell.

2.2. Simulation of Pyroelectric Harvesters from Cyclic Temperature Fluctuations

When the dimensions and materials of the pyroelectric cells are settled, the performance of the pyroelectric harvesting system can be estimated by means of the temperature variation rate of the pyroelectric cells. However, the difficulty is in extracting the temperature variation rate along the thickness direction of the pyroelectric cells. The transient temperature field in the cells is also not easy to extract by means of experimental measurements. In the present study, a two-dimensional finite element model was generated by commercial multiphysics software COMSOL Multiphysics®4.2 to explore the temperature variation rates in the PZT pyroelectric cells with various considerations. These considerations were derived from the frequency or work cycle, radiation power, properties and dimensions of the air layer, and the material properties, dimensions and structure of the pyroelectric PZT cells. The periods of rising and dropping temperature in the PZT pyroelectric cell determined the electrical output of the pyroelectric harvester, which was affected by the thermal capacity and electrical capacitance of the PZT pyroelectric cells. The various angular velocities of the disk and dimensions of the aperture determine the work cycles, resulting in the periods of rising and dropping temperature in the PZT pyroelectric cells. The distance between the PZT pyroelectric cells and the radiating heat lamp could determine the heat flux applied on the PZT cells. The dimensions and structure of the pyroelectric PZT cells could set the thermal capacity and electrical capacitance of the PZT pyroelectric cells. Hence, the performance of pyroelectric harvesters could be optimized by matching the conditions related to the various angular velocities of the disk, dimensions of the aperture, and distances between the PZT pyroelectric cells and the radiating heat lamp, when a PZT pyroelectric cell with constant dimensions and material is adopted. The thermal properties of the materials used in the numerical analysis model are listed in Table 2. There was an isotropic and time-independent assumption regarding the PZT cell, the electrodes and the PCB in the model. The thermal conductivity, heat capacity and air density were all temperature-dependent material properties, which were according to the settings in the COMSOL software. The model was meshed by regular mesh, as shown in Figure 4. Furthermore, thermal isolation was applied to the rear side of the PZT pyroelectric cells, and the two lateral sides of the model were symmetric, in order to serve as boundary conditions. The top border of the model was applied to the incident irradiation power of about 1.228×10^{-12} W/μm^2 [5–7], with various unit step functions for the resulting various frequencies or work cycles.

Table 2. Material properties used for finite element analysis.

Material	Thermal conductivity ($Wm^{-1}K^{-1}$)	Specific heat ($Jg^{-1}K^{-1}$)	Density (gcm^{-3})	Thickness (mm)
PCB	0.12	1.38	0.51	0.4
Silver electrode	429	0.235	10.53	0.007
PZT cell	2.1	0.36	7.9	0.2 and 0.05
Air	TC(T)	SH(T)	DS(T)	0.2

Note: (T: Absolute temperature); $TC(T) = -0.00227583562 + 1.15480022 \times 10^{-4} \times T - 7.90252856 \times 10^{-8} \times T^2 + 4.11702505 \times 10^{-11} \times T^3 - 7.43864331 \times 10^{-15} \times T^4$; $SH(T) = 1.04763657 - 3.72589265 \times 10^{-4} \times T + 9.45304214 \times 10^{-7} \times T^2 - 6.02409443 \times 10^{-10} \times T^3 + 1.2858961 \times 10^{-1}3 \times T^4$; $DS(T) = 101.325 \times 103 \times 0.02897/8.314/T$.

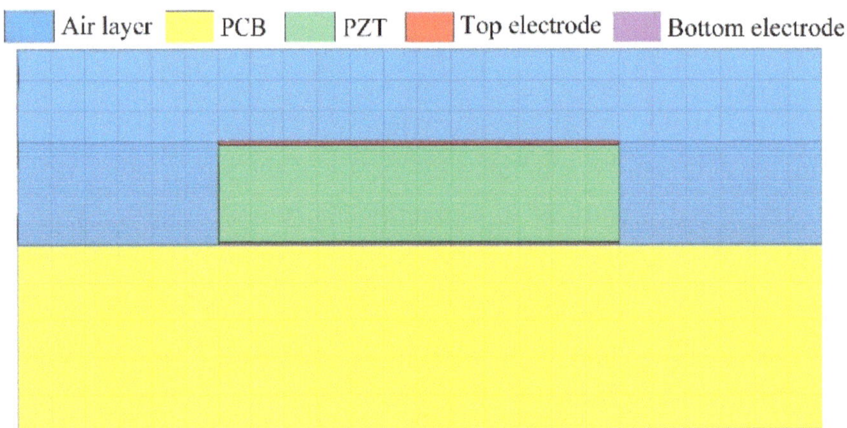

Figure 4. Regular meshed finite element model for the PZT pyroelectric harvester.

In fact, the PZT pyroelectric cells used the current responsivity and induced charge in the performance of the harvesters. Hence, the current responsivity and induced charge were calculated from the transient temperature fields, so as to estimate the electrical outputs of the harvesters, and for further evaluating the output voltage and stored energy in a load capacitor. Moreover, the dynamic pyroelectric response current (i_p, unit: μA) produced by the induced charge can be expressed as [1,9]:

$$i_p = \eta \times P \times A \times dT/dt \tag{3}$$

Moreover, the induced charge (Q, unit: μC) can be expressed as:

$$Q = \int i_p \, dt, \tag{4}$$

Finally, the current and induced charge resulting from the cyclic temperature fluctuations were rectified and stored in a 1 μF load capacitor (C_L), using the full-wave rectifier circuit depicted in

Figure 3. Then, the output voltage of the stored capacitor (V_0) across C_L and the stored energy would increase. Cuadras *et al.* [1] derived V_0 as:

$$V_0(N) = \frac{Q - Q_{leak} - 4V_D C_P}{2C_P}\left[1 - \left(\frac{C_L - C_P}{C_L + C_P}\right)^N\right] \tag{5}$$

where Q_{leak} is the charge leakage; V_D is the voltage drop across the diodes when forward biased; C_P is the electrical capacitance of the pyroelectric cell and N is the heating/cooling cycle. When N approaches infinity; V_0 converges to:

$$V_{0,\max} = \frac{Q - Q_{leak} - 4V_D C_P}{2C_P} \tag{6}$$

Assuming $C_L > C_P$, and considering the forward voltage drop of the diode of 0.62 V and no leakage current, $V_{0,\max}$ (unit: V) can be expressed as:

$$V_{0,\max} = \frac{Q - 2.48C_P}{2C_P} \tag{7}$$

It is obvious that $V_{0,\max}$ does not depend on C_L, but only the internal capacitance of the pyroelectric cell (C_P). The stored energy (E_C, unit: nJ) in C_L can be expressed as:

$$E_C = \frac{1}{2}C_L V_{0,\max}^2 \tag{8}$$

The relative conditions for calculating the current (i_p), induced charge (Q), output voltage of the stored capacitor (V_0) and stored energy (E_C) are shown in Table 3.

Table 3. Relative conditions for calculating i_p, Q, V_0 and E_c.

P **(10^{-4} C/m²·K)**	***A*** **(mm²)**	**C_L (μF)**	**η (Unit Free)**
6.97	81	1	1

3. Results and Discussion

The cyclic heating system was designed to periodically heat and cool the pyroelectric elements to produce periodic temperature changes and thus harvest the resulting pyroelectric energy. With cyclic heating, the temperature changes were smaller; however, the total energy harvested could accumulate and become larger over time [10]. The present study is focused on optimizing the efficiency of PZT pyroelectric harvesters with consideration of the frequency or work cycle, radiation power, properties and dimensions of the air layer, and the dimensions and structure of the pyroelectric PZT cells.

The temperature variation rate increased when the point approached the top side of the ZnO layer because incident radiation power approached the top electrode. Using the temperature variation rate at the point near the top electrode could not discriminate the properties of the pyroelectric sensors with various thicknesses of the ZnO films. Therefore, a conservative consideration was to adopt the properties at the point near the bottom electrode for inspecting the sensors because the temperature variation rate at this point could display differences of the ZnO films with various thicknesses [5,6]. As the ZnO films at the point near the bottom electrode possess the lowest temperature variation rates, this position could certainly enhance the responsivity of the pyroelectric devices. Hence, the temperature variation rates of the ZnO films at the point near the bottom electrode were adopted to calculate the current responsivity and estimate the induced charge.

Figure 5 shows the relationships among the temperature variation rate, the current and time at five periods of 2×10^{-2} s, 4×10^{-2} s, 6×10^{-2} s, 8×10^{-2} s and 10×10^{-2} s when a 200 μm thick PZT cell used. The current curve was calculated from the temperature variation rate data. Therefore, the shapes of the curves related to the current and temperature variation rate were almost identical. It was also obvious that these curves had the best performance for the period of about 6×10^{-2} s.

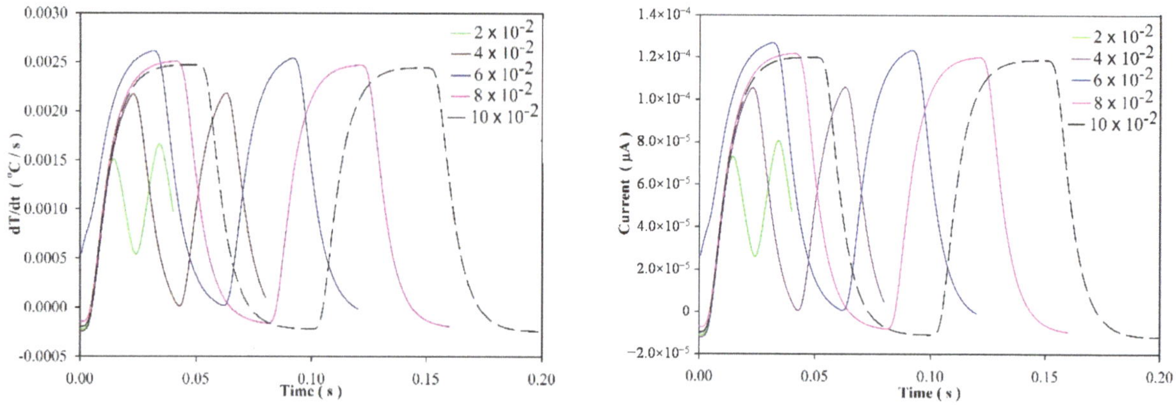

(a) (b)

Figure 5. Relationships among the temperature variation rate (**a**); the current (**b**) and time at five periods of 2×10^{-2} s, 4×10^{-2} s, 6×10^{-2} s, 8×10^{-2} s and 10×10^{-2} s for the 200 μm thick PZT cell used.

The charge was inferred from the integration of the area under the current curves by a numerical integration of Simpson's rule. Figure 6 shows the relationships between the induced charge per period, the output voltage per period, the stored energy per period and various periods when the 200 μm thick PZT cell was used to construct the model for estimating the efficiency of pyroelectric harvesters. The induced charge per period, the output voltage per period and the stored energy per period were the largest when the work cycle or the period approached 6×10^{-2} s. Moreover, a larger period over 6×10^{-2} s proved more beneficial for harvesting the pyroelectric energy than a smaller period under 6×10^{-2} s because the 200 μm thick PZT cell needed sufficient time to absorb and conduct the incident radiation power. A smaller period under 6×10^{-2} s would reduce the thermal energy absorbed in the PZT cell. However, a much larger period over 6×10^{-2} s rapidly reduced the pyroelectric energy harvesting because the temperature variation rate rapidly decreased and the period quickly increased.

Figure 7 shows the relationships among the temperature variation rate, the current and time at five periods of 9×10^{-3} s, 1×10^{-2} s, 2×10^{-2} s, 3×10^{-2} s and 4×10^{-2} s for the 50 μm thick PZT cell used. Figure 8 shows the relationships between the induced charge per period, the output voltage per period, the stored energy per period and various periods when the 50 μm thick PZT cell was used. The induced charge per period, the output voltage per period and the stored energy per period were the largest when the work cycle or the period approached 2×10^{-2} s. The optimal period for harvesting the pyroelectric energy was reduced when the thickness of the PZT cell decreased. The induced charge per period, the output voltage per period and the stored energy per period were greatly reduced over the period of 0.8 s.

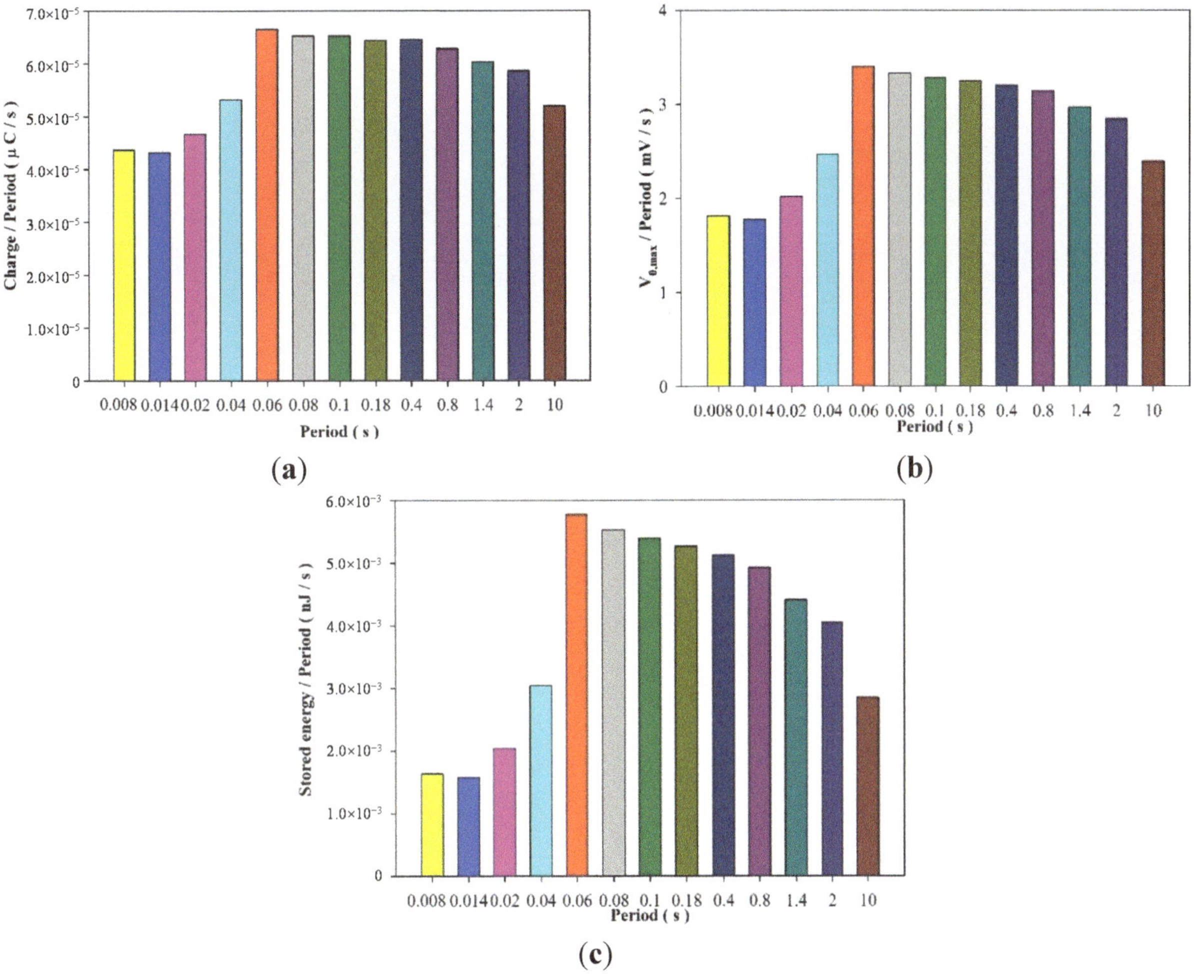

Figure 6. Relationships among (**a**) the induced charge per period; (**b**) the output voltage per period; (**c**) the stored energy per period and various periods for the 200 μm thick PZT cell used for constructing the analysis model.

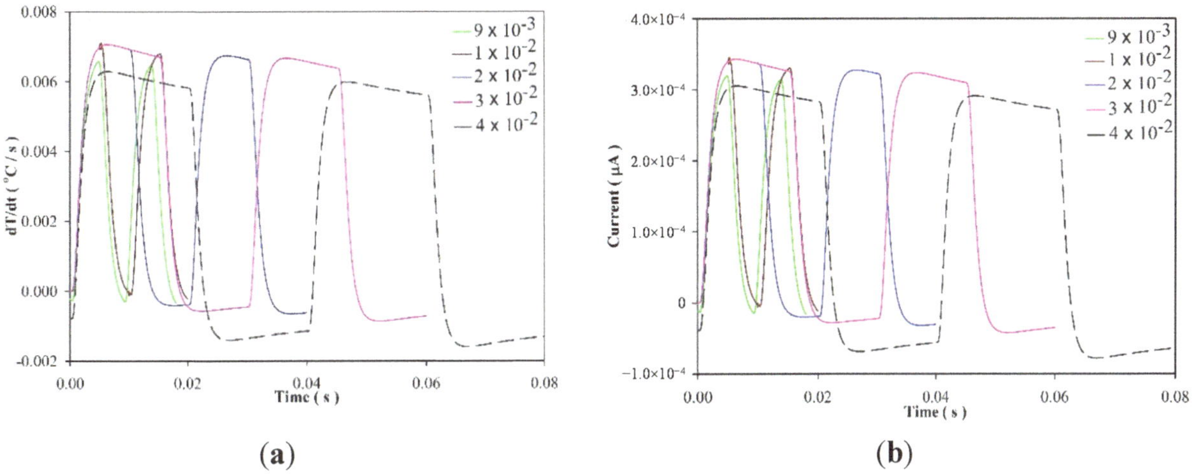

Figure 7. Relationships among (**a**) the temperature variation rate; (**b**) the current and time at five periods of 9×10^{-3} s, 1×10^{-2} s, 2×10^{-2} s, 3×10^{-2} s and 4×10^{-2} s for the 50 μm thick PZT cell used.

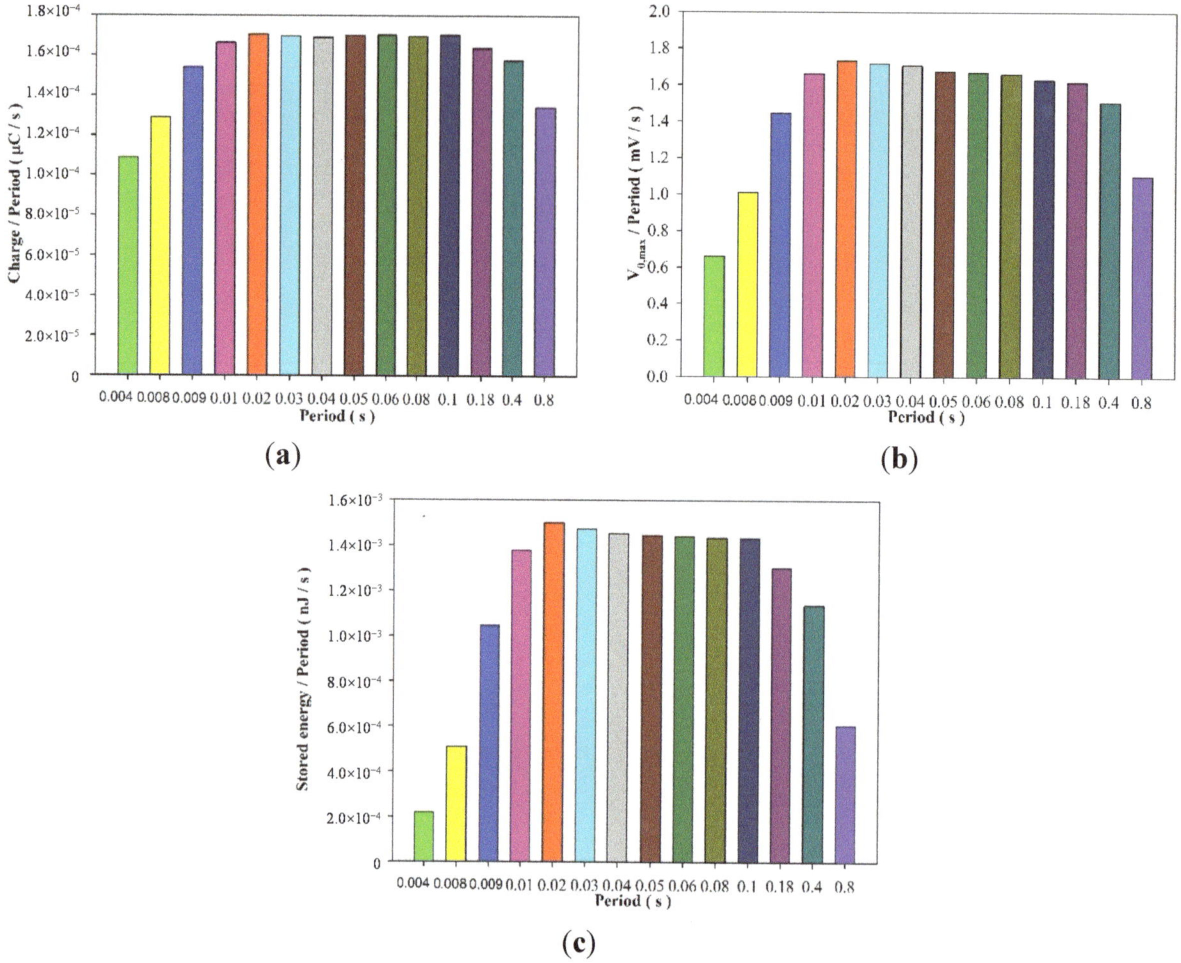

Figure 8. Relationships among (**a**) the induced charge per period; (**b**) the output voltage per period; (**c**) the stored energy per period and various periods for the 50 μm thick PZT cell used for constructing the analysis model.

Moreover, the 50 μm thick PZT cell showed superior harvesting of the pyroelectric energy compared to the 200 μm thick PZT cell during a smaller period under 4×10^{-3} s. This could be attributed to the fact the thinner PZT cell possesses a lower thermal capacity for enhancing the temperature variation rate. The period band of the 50 μm thick PZT cell for efficiently harvesting the cyclic heating energy was narrower than that of the 200 μm thick PZT cell. The efficient period band of the 50 μm thick PZT cell was about 0.01 to 0.18 s, while that of the 200 μm thick PZT cell was about 0.06 to 0.8 s. The thicker PZT cell had a wider period band than the thinner PZT cell did. The efficient period band decreased about 77% when the thickness of the PZT cell decreased from 200 μm to 50 μm, about 75%. Therefore, in using the thinner PZT cell for harvesting the pyroelectric energy, it was not easy to focus on a narrow band of the efficient period. The pyroelectric energy harvesting with the thinner PZT cell needed to accurately use the period to achieve the best performance. Moreover, harvesting the pyroelectric energy transformed into the induced charge in the thinner PZT cell was greater than that in the thicker PZT cell during the optimal period. The optimal induced charge per period increased about 157% when the thickness of the PZT cell decreased about 75%. However, the optimal output voltage and stored energy per period decreased about 50% and 74%, respectively, when the thickness of the PZT cell decreased about 75%. This phenomenon could be attributed to the thinner pyroelectric cell possessing greater electrical capacitance (C_P). Moreover, the electrical capacitance of the 50 μm thick pyroelectric cell was about four times greater than that of the 200 μm thick pyroelectric cell. Because the stored energy was proportional to the square of the output voltage, the stored energy experienced a larger decline than the output voltage. An experimental setup was used to verify and observe the work cycle to be able to affect the efficiency of the pyroelectric harvesters. The PZT cells were heated with the heat lamp, and naturally cooled with the rotating disk to obstruct the heat source. The temperature fluctuation was controlled on a range between 406 K and 403 K. Figure 9 shows the relationships between the current and time at five periods of 1 s, 1.9 s, 4 s, 12.2 s and 25.2 s for the commercial PZT pyroelectric cell with dimensions of 9 mm × 9 mm × 0.214 mm used. The response current was bumpier when the periods increased to 25.2 s due to the environmental temperature noise which was greater in the larger period.

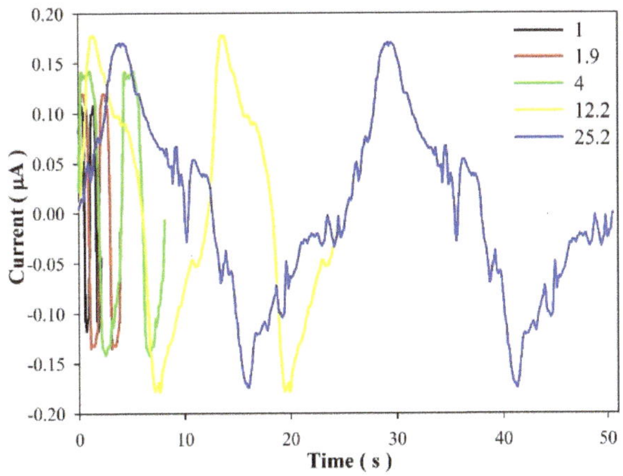

Figure 9. Relationships between the current and time at five periods of 1 s, 1.9 s, 4 s, 12.2 s and 25.2 s under the commercial PZT pyroelectric cell used.

Figure 10 shows the relationships between the stored energy per period and various periods when the commercial PZT pyroelectric cell was used. The efficient period band of the commercial PZT pyroelectric cell was about 3.6 s to 12.2 s. Figure 11 shows the relationships between the stored energy in the 4.7 μF electrolytic capacitor and time at various periods.

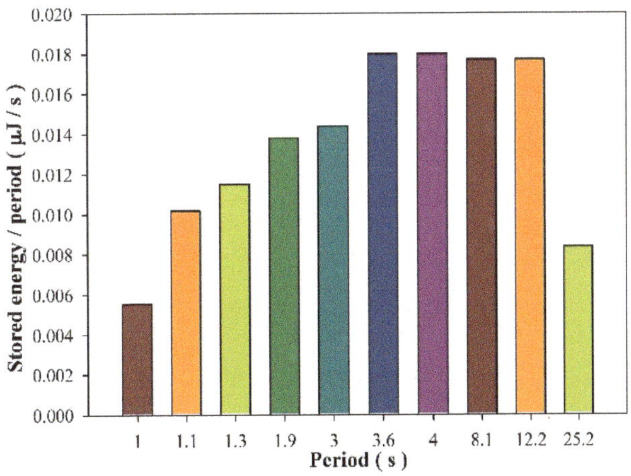

Figure 10. Relationships between the stored energy per period and various periods under the commercial PZT pyroelectric cell used.

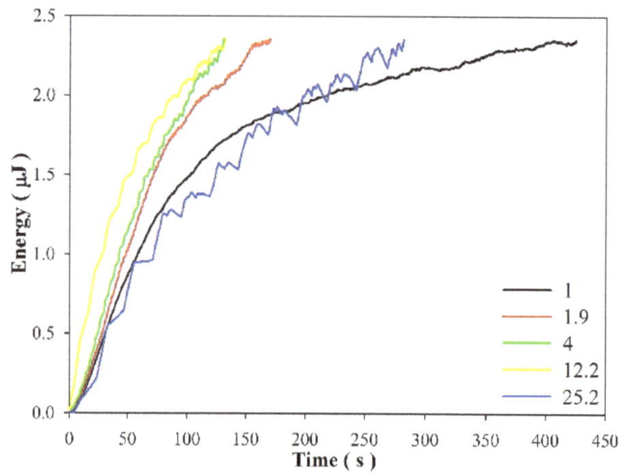

Figure 11. The stored energy over time at various periods during consecutive heating and cooling cycles.

Obviously, the period in the efficient period band revealed the best performance for harvesting thermal cyclic energy by pyroelectricity. Moreover, the stored energy in the capacitor increased according to expression (5). The smaller and the larger period could not efficiently harvest thermal cyclic energy. The pyroelectric element conversion of thermal cyclic energy into electricity was feeble in the smaller period due to the thermal capacity of pyroelectric elements. In other words, the pyroelectric element needed a thermal time constant ($\tau_T = H/G_T$, H: thermal capacity, G_T: thermal conductance) to sufficiently harvest thermal energy with the pyroelectric material. Moreover, the pyroelectric element conversion of thermal cyclic energy into electricity was also weak in the larger period due to the lower temperature variation rate over the period. Although the temperature variation rate was larger for harvesting thermal cyclic energy in the initial step, the large temperature variation rate could not be maintained over the period, therefore the temperature variation rate decreased, the induced charge decreased, and then the output voltage decreased. These resulted in a decrease of the stored energy, although the stored energy would increase until the next period. Therefore, the curve of the stored energy over time revealed some ripples when a larger period was used to harvest thermal cyclic energy. The ripples indicated that the larger period was unsuitable for harvesting thermal cyclic energy. The periods of 4 s and 12.2 s presented the best performance when the commercial PZT pyroelectric cell with dimensions of 9 mm × 9 mm × 0.214 mm was used. Hence, a period in the suitable range was beneficial for using the pyroelectric harvesters. The estimation of the frequency or work cycle could be applied to the design of the aperture sizes and number of the rotating disk, and the angular velocities for controlling the heated and unheated periods of the pyroelectric cells.

4. Conclusions

Energy harvesting using a cyclic heating system can effectively transform waste heat energy into electricity by pyroelectricity. The cyclic heating system is developed for varying the temperature in the pyroelectric cells between the hot and cold regions. A disk with apertures for radiation heating and natural cooling is used as the cyclic heating system. The critical consideration in a periodic temperature profile for optimizing the pyroelectric harvesters is the frequency or work cycle. This factor is related to the properties and dimensions of the air layer, radiation power, and the material properties, dimensions and structure of the pyroelectric cells. This study proposes a convenient method for estimating the frequency or work cycle in pyroelectric cells via the cyclic heating system. It is obvious that the optimal period decreased when the thickness of the PZT cell was reduced. Moreover, the optimal induced charge per period increased about 157% when the thickness of the PZT cell decreased from 200 μm to 50 μm, about 75%; however, the optimal output voltage and stored energy per period decreased about 50% and 74%, respectively. This phenomenon could be attributed to the thinner pyroelectric cell possessing greater electrical capacitance. In addition, an experiment was used to verify the work cycle to be able to critically affect the efficiency of the PZT pyroelectric harvesters. The periods in the range between 3.6 s and 12.2 s were optimal for harvesting thermal cyclic energy by pyroelectricity when a commercial PZT pyroelectric cell with dimensions of 9 mm × 9 mm × 0.214 mm was used. The optimal frequency or work cycle can be applied to the design of the rotating disk for controlling the heated and unheated periods of the pyroelectric cells, further improving the energy storage.

Acknowledgments

The authors are thankful for the financial support from the National Science Council of Taiwan through Grant No. MOST 103-2221-E-150-038, and the experimental support from the Common Laboratory for Micro and Nano Science and Technology at the National Formosa University.

Author Contributions

Jia-Wai Jhang was involved in the data collection and experimental work under supervision of Chun-Ching Hsiao who carefully edited the paper, provided technical support and guidance, and directly contributed the results of this article.

Conflicts of Interest

The authors declare no conflict of interest.

References

1. Cuadras, A.; Gasulla, M.; Ferrari, V. Thermal energy harvesting through pyroelectricity. *Sens. Actuators A* **2010**, *158*, 132–139.
2. Lee, F.Y.; Navid, A.; Pilon, L. Pyroelectric waste heat energy harvesting using heat conduction. *Appl. Therm. Eng.* **2012**, *37*, 30–37.
3. Nguyen, H.; Navid, A.; Pilon, L. Pyroelectric energy converter using co-polymer P(VDF-TrFE) and Olsen cycle for waste heat energy harvesting. *Appl. Therm. Eng.* **2010**, *30*, 2127–2137.
4. Sebald, G.; Lefeuvre, E.; Guyornar, D. Pyroelectric energy conversion: Optimization principles. *IEEE Trans. Ultrason. Ferroelect. Freq. Control* **2008**, *55*, 538–551.
5. Hsiao, C.C.; Ciou, J.C.; Siao, A.S.; Lee, C.Y. Temperature field analysis for PZT pyroelectric cells for thermal energy harvesting. *Sensors* **2011**, *11*, 10458–10473.
6. Hsiao, C.C.; Siao, A.S.; Ciou, J.C. Improvement of pyroelectric cells for thermal energy harvesting. *Sensors* **2012**, *12*, 534–548.

7. Hsiao, C.C.; Siao, A.S. Improving pyroelectric energy harvesting using a sandblast etching technique. *Sensors* **2013**, *13*, 12113–12131.
8. Olsen, R.B.; Bruno, D.A.; Briscoe, J.M. Pyroelectric conversion cycles. *J. Appl. Phys.* **1985**, *58*, 4709–4716.
9. Whatmore, R.W. Pyroelectric devices and materials. *Rep. Progr. Phys.* **1986**, *49*, 1335–1386.
10. Mane, P.; Xie, J.; Leang, K.K.; Mossi, K. Cyclic energy harvesting from pyroelectric materials. *IEEE Trans. Ultrason. Ferroelectr. Freq. Control* **2011**, *58*, 10–17.

5

Thermal Performance of Ventilated Double Skin Façades with Venetian Blinds

Jordi Parra [1], Alfredo Guardo [1,*], Eduard Egusquiza [1] and Pere Alavedra [2]

[1] Center for Industrial Diagnostics and Fluid Dynamics, Polytechnic University of Catalonia *BARCELONATECH* (UPC-CDIF), Av. Diagonal 647, ETSEIB, 08028 Barcelona, Spain; E-Mails: jordi.parra@estudiant.upc.edu (J.P.); egusquiza@mf.upc.edu (E.E.)

[2] International University of Catalonia, C\Immaculada 22, 08017 Barcelona, Spain; E-Mail: pere.alavedra@uic.cat

* Author to whom correspondence should be addressed; E-Mail: alfredo.guardo-zabaleta@upc.edu;

Academic Editor: Hossam A. Gabbar (Gaber)

Abstract: Venetian blinds (VB) are shading devices of widespread use in residential and corporate buildings. They can reflect or transmit light into buildings and at the same time allow daylighting and exterior views. They can also efficiently block radiative heat from entering the building, and if combined with a heat dissipation system such as forced ventilation, they can improve the thermal performance of double skin façades (DSF). Computational Fluid Dynamics (CFD) has proven to be a useful tool for modeling flow and heat transfer in DSF, including conduction, convection and radiation heat transfer phenomena. The aim of this work is to evaluate, by means of CFD, the influence of several optical, construction and operation parameters of a DSF (such as optical properties of the materials, geometrical relations of the VB or flow stream conditions) in terms of energy savings, measured as a reduction of the solar load entering the building. Results obtained show that parameters such as the proximity of the VB to the exterior skin of the façade or a differentiated surface treatment for the exterior and interior faces of the VB louvers can notably affect the thermal performance of the DSF and hence the heat gains experienced by the building.

Keywords: active transparent façades; double skin façades; venetian blinds; building thermal performance; computational fluid dynamics (CFD)

---

1. Introduction

A double-skin facade (DSF) is a building typology consisting of two skins (a glazed outer layer and either a glazed or mixed inner layer) placed in such a way that air flows in the intermediate cavity. The cavity air ventilation (either natural or mechanical) is used for evacuating the radiative heat absorbed by the façade elements. The outer glazed skin can be single or double glazing units with a distance from 20 cm up to 2 m from the inner skin. Sometimes, for radiation protection, solar shading devices are placed inside the cavity. DSF are becoming of widespread use in commercial buildings due not only to the unique aesthetics they can offer, but also because they can help increase the comfort and efficiency of the building by reducing external noise and wind loads [1].

The use of a DSF has a direct impact on thermal comfort indoors. Gratia and de Herde [2] modeled the thermal behavior and the heating or cooling loads (depending on the season) necessary to obtain certain pre-set indoor comfort parameters for a building located in Uccle (Belgium) and using the external climatic conditions of representative sunny spring, summer, fall and winter days. Their analysis of the modeled cases showed that the use of a DSF on the selected location decreases the heating loads during winter and increases the cooling loads during summer. In Mediterranean climates the use of DSF can be problematic as large solar gains and moderate to warm temperatures are a permanent condition throughout year, even during winter [3]. Strategies used to improve the thermal performance of a DSF include shading devices, double/triple glazing and/or enhanced natural or mechanical ventilation.

Shading devices such as venetian blinds (VB) mounted indoors of glazed surfaces are of general use for daylighting control. These devices also modify the thermal performance of the façade. Early studies in the subject dealt with the effect of VB on the free convection heat transfer at indoor surfaces with and without solar irradiance to simulate daytime and nighttime conditions [4–6]. Energy performance studies attempted to develop models to evaluate the influence of VB in energy cooling loads and building energy consumption [7–10]. Other authors have focused on the effects of VB in daylight transmission and interior illuminance [11,12]. Nowadays several authors work in developing automated control strategies for the blinds, maximizing indoor daylighting and minimizing solar heat gains, thus reducing the building energy consumption and improving its sustainability (see e.g., [13,14]).

Predicting the thermal performance of a DSF is not an easy task. The solid/fluid temperatures and cavity airflow patterns are the result of several flow, thermal and optical processes happening simultaneously. These processes are highly dynamic and in constant interaction with each other, and they depend on the geometric, thermo-physical, optical and aerodynamic characteristics of the different DSF elements. The indoor and outdoor temperatures, outdoors wind speed and direction, radiation levels and weather forecast—some of which are highly transient—control the heat transport driving forces. In order to achieve a suitable computational model of such complex phenomena, it is necessary to use a numerical code that allows transient analyses of mass, momentum and energy conservation equations together with optical/radiation models.

Computational fluid dynamics (CFD) has been validated as a numerical simulation tool in the study of DSF by several authors for natural and mechanical ventilation scenarios without shading devices [15–18]. The good agreement between numerical and experimental results reported by all authors cited demonstrates the viability of CFD to study the thermal performance of a DSF. Previous work from this research group [3,18–21] presented numerical simulations of a story-high DSF located in Barcelona (Spain) and evaluated the influence of construction and operation parameters as well as the ventilation method in the thermal performance of the façade. Several ventilation and injection methods were analyzed, and upper-crossed horizontal ventilation using Coandă nozzle injectors was found to be an interesting ventilation option because of its good compromise between solar heat gain reductions, flow rate injection and overall costs.

Few numerical studies have been performed for DSF with VB. The small thickness of the louvers and the large number of louvers present in a VB make it necessary to use a highly refined mesh in order to properly capture the flow and heat transfer around the VB. Manz [15] studied a DSF with a metallic shading screen. Baldinelli [22] modeled a DSF with an integrated movable shading device. Safer *et al.* [23] and Zeng *et al.* [24] used a porous media approach to simplify the flow around the VB louvers and reduce the computational cost of simulations with acceptable results for flow patterns but high deviations in VB temperature predictions, which excludes its application to thermal performance studies.

This works aims to contribute on the numerical modeling of DSF with VB by presenting a numerical investigation of the influence of several construction and operation parameters (such as optical properties of the materials, geometric characteristics or ventilation conditions) on the thermal performance of a DSF with VB in Mediterranean climates measured as a reduction on solar heat gains for a façade module located in Barcelona (Spain). Results were obtained by means of a commercial CFD solver (Ansys Fluent® v15). Numerical models, meshing and modeling strategy used were validated against an open literature experimental database in previous work [18]. The parametric study presented show the effects of VB louvers surface treatments, VB position and ventilation flow rate on cavity temperatures and heat gains.

2. Theoretical Considerations

A dimensionless analysis under work conditions of the set of equations used was carried out to determine the relative importance of the transport mechanisms present in the study case. The dimensionless equations corresponding to the mass, momentum and energy conservation equations for the fluid phase are shown in Equations (1)–(3) respectively. Equation (4) shows a dimensionless energy conservation equation for the solid phases:

$$\mathrm{Sr}\left(\frac{\partial\hat{\rho}}{\delta\hat{t}}\right) + (\hat{\boldsymbol{u}}\cdot\hat{\nabla})\hat{\rho} = \mathrm{Ma}^2\cdot\left(-\hat{\rho}\hat{\nabla}\cdot\hat{\boldsymbol{u}}\right) \tag{1}$$

$$\mathrm{Sr}\left(\frac{\partial\hat{\boldsymbol{u}}}{\delta\hat{t}}\right) + (\hat{\boldsymbol{u}}\cdot\hat{\nabla})\hat{\boldsymbol{u}} = \mathrm{Eu}\left(-\frac{1}{\hat{\rho}}\hat{\nabla}\hat{\rho}\right) + \frac{1}{\mathrm{Re}}\hat{\nabla}\cdot\left\{\frac{\hat{\mu}}{\hat{\rho}}\left[\hat{\nabla}\hat{\boldsymbol{u}} + (\hat{\nabla}\hat{\boldsymbol{u}})^T\right]\right\} + \frac{1}{\mathrm{Re}_T}\hat{\nabla}\cdot\left\{\frac{\hat{\mu}_T}{\hat{\rho}}\left[\hat{\nabla}\hat{\boldsymbol{u}} + (\hat{\nabla}\hat{\boldsymbol{u}})^T\right]\right\} + \frac{1}{\mathrm{Fr}}(\hat{F}_b) \tag{2}$$

$$\mathrm{Sr}\left(\frac{\partial\hat{T}}{\delta\hat{t}}\right) + (\hat{\boldsymbol{u}}\cdot\hat{\nabla})\hat{T} = \frac{\mathrm{Ec}}{\mathrm{Re}}\cdot\left[\frac{\hat{\mu}}{\hat{\rho}}(\hat{\tau}:\hat{\nabla}\hat{\boldsymbol{u}})\right] + \frac{1}{\mathrm{Re}\cdot\mathrm{Pr}}\left(\frac{\hat{k}_f}{\hat{\rho}}\hat{\nabla}\hat{T}\right) + \frac{1}{\mathrm{Re}\cdot\mathrm{Pr}_T}\left(\frac{\hat{k}_{f,T}}{\hat{\rho}}\hat{\nabla}\hat{T}\right) + \frac{aL}{\mathrm{Th}}(\hat{\xi}\hat{a}\hat{\sigma}\hat{T}^4) \tag{3}$$

$$\hat{\rho}_w\hat{c}_{p,w}\left(\frac{\partial\hat{T}}{\delta\hat{t}}\right) + (\hat{k}_w\cdot\hat{\nabla})\hat{T} = \frac{L_{0,w}}{k_{o,w}T_{o,w}}q \tag{4}$$

A dimensional analysis is not valid unless it takes full account of the boundary and initial conditions, since these conditions affect the solution of the problem directly. For non-slip solid surfaces, in Equation (3) the boundary conditions become: $\hat{\boldsymbol{u}} = 0; \hat{T} = 1$ or $\mathrm{Nu} = \hat{k}\left(d\hat{T}/d\hat{n}\right)_w$ where $\hat{n}$ is the normal direction from the surface. Thus, if the wall heat flux is specified, a new parameter appears: the Nusselt number (Nu). In this case it is one of the driving parameters that affect the solution. In Equation (4) the only parameter is dimensionless heat generation. The prescribed boundary conditions in this case are a radiating boundary and a convective heat flux for both the external and internal walls. For the radiating boundary a sort of Biot number (Bi_{rad}) could be defined as $\mathrm{Bi}_{rad} = (h_{rad} \cdot L_o/k_w)$, where the heat transfer coefficient is $h_{rad} = \left(\sigma \cdot F_{w,\infty} \cdot T^3\right)$. For the convective boundary condition a standard Biot number (Bi) is obtained, $\mathrm{Bi} = (h \cdot L_o/k_w)$. In this case, the initial conditions do not introduce new non-dimensional parameters.

The orders of magnitude of the dimensionless groups found were estimated by using the thermo-physical property values for all the fluid and solid phases under different DSF operating conditions. Reynolds analogy was used to estimate Pr_T from Re_T. The results obtained are shown in Table 1.

Table 1. Dimensionless groups' orders of magnitude.

Dimensionless group	Magnitude order			
Re	10^3	10^4	10^5	10^6
Bi	10^{-4}	10^{-3}	10^{-2}	10^{-1}
Bi_{rad}	10^{-2}	10^{-3}	10^{-3}	10^{-3}
Ec	10^{-9}	10^{-7}	10^{-5}	10^{-3}
Eu	10^3	10^1	10^{-1}	10^{-3}
Fr	10^{-3}	10^{-2}	10^{-1}	10^0
Ma	10^{-5}	10^{-4}	10^{-3}	10^{-2}
Nu	10^0	10^1	10^2	10^3
Pr	10^{-1}	10^{-1}	10^{-1}	10^{-1}
Pr_t	10^{-1}	10^{-2}	10^{-3}	10^{-4}
Re_t	10^{-1}	10^{-2}	10^{-3}	10^{-4}
Sr	10^0	10^{-1}	10^{-2}	10^{-3}
Th	10^0	10^1	10^2	10^3

Dimensionless analysis allows the problem to be identified as heat transfer by conduction/radiation for solid surfaces and convection in laminar (natural ventilation) or turbulent (mechanical ventilation) flow for the fluid phase, depending on the flow inlet velocity. Analyzing the obtained results for the energy conservation equation, it can be seen that convection is an important transport mechanism, and turbulent mixing becomes an important transport mechanism as the flow rate increases. The contribution of viscous dissipation to the energy balance is negligible. The radiation term vanishes independently of the value of Th because the absorption coefficient *a* for transparent gases is zero. Sr has a similar order of magnitude than the group $(\mathrm{Re} \cdot \mathrm{Pr}_T)^{-1}$, suggesting that a transient thermal analysis should be performed. Nu is the most important parameter for the fluid phase energy balance and therefore the surface convective heat transfer coefficients must be estimated carefully. For the solid phases (VB louvers, glazed surfaces, wall) Bi_{rad} is greater than Bi at low Re, which indicates that the contribution of radiative transport in these surfaces is greater than the contribution of the conduction and convection to the overall heat transport phenomena. At high Re the opposite behavior is expected.

3. Modeled Case

The numerical setup used for this study was set to replicate the numerical setup used in previous work published by this research group [18–20], in order to establish a common reference frame for thermal performance comparisons between a horizontally ventilated DSF with VB and other DSF configurations studied in the past.

3.1. Geometry

An isometric view and detail of the base geometric configuration used for this numerical study is shown in Figure 1. The control volume comprises (from outdoors to indoors) an external glazing, a flow cavity with a venetian blind, an internal double glazing (52.5% of the internal façade area) and an internal wall (47.5% of the internal façade area). The flow cavity dimensions were set to 0.8 m wide, 4 m height and 24 m depth following fire hazard regulations of mandatory compliance in Spain [25].

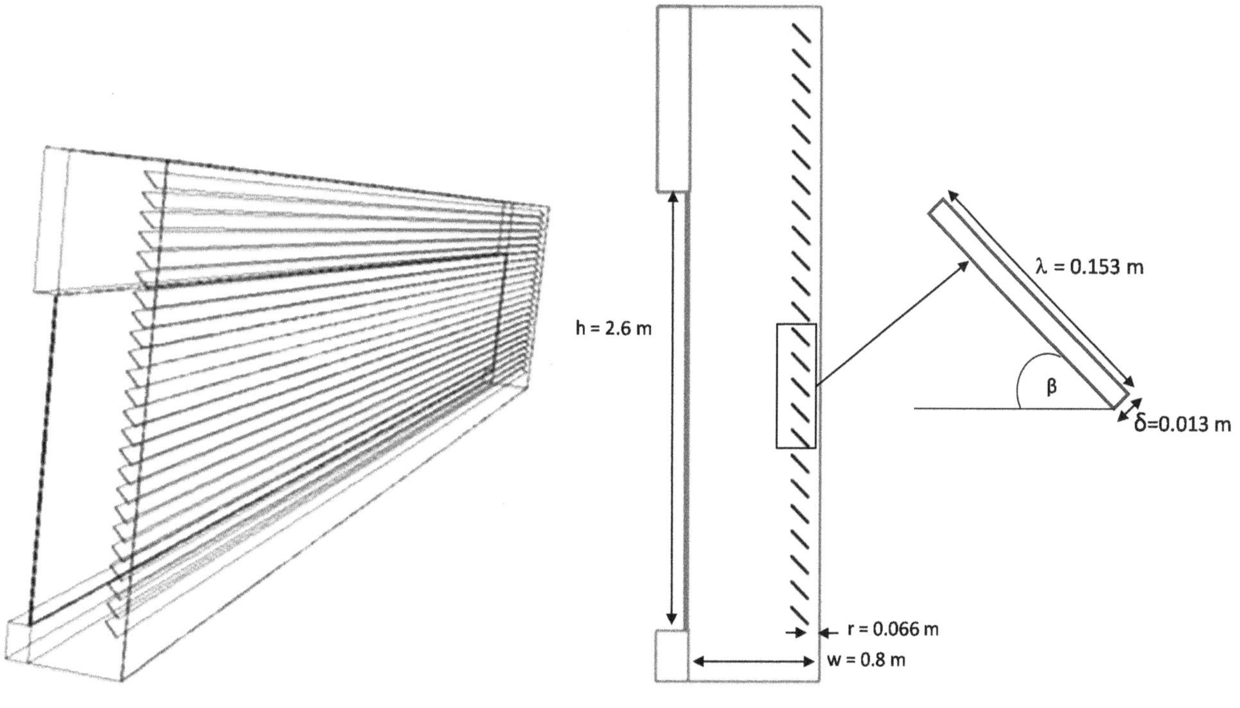

Figure 1. Base geometry configuration for the studied DSF.

VB louvers tilt angle (β) was set to 45° in order to avoid glare risk due to direct daylighting into the building at high sun elevation angles. Previous studies by Gratia and de Herde [26] suggest that the most efficient positions of shading devices in a DSF is in the middle of the cavity but that would be considered a fire hazard according to the local regulations [25] as it blocks accessibility to the cavity in case of emergencies. For this study the VB was placed next to the external glazing, suggested by the aforementioned study as the second best location in terms of the thermal efficiency of the DSF [26]. The geometrical model also distinguishes between the upper (facing outdoors) and the lower (facing indoors) surfaces of the VB louvers, which were created into the computational model as two separated wall boundary conditions for purposes of the parametric study presented.

3.2. Materials

Table 2 shows the physical and optical properties for all the DSF elements used in the parametric study base case. Thicknesses for the different layers were assigned following common practices in construction.

Table 2. Thermo-physical and optical properties of the construction materials.

Façade Element	ρ (kg/m³)	c_p (J/kg·K)	k (W/m·K)	Absorptivity	Transmissivity	Thickness
Glass	2.500	795	1.16	0.15	0.78	6 mm
Interior walls	720	1.069	0.1	0.8	-	200 mm
VB louvers (Aluminum)	2719	878	169	0.18	-	δ_{bc} = 13 mm

3.3. Location and Climatic Conditions of the DSF

It was assumed that the studied DSF was located in Barcelona, Spain (41.23° N–2.11° E) and facing south. Theoretical maximum values for solar irradiance (global and diffuse) and climatic data (for outdoors temperature) for a warm, sunny summer day (17th July) were taken from open-access databases [27]. Time-dependency of the outdoors temperature for the selected operation date was fitted into a 6th-degree polynomial and introduced as a dynamic boundary condition to the CFD solver via user defined functions.

3.4. Construction and Operation Parameters Studied

The influence of several design and operation parameters in the thermal performance of a horizontally ventilated DSF with VB (measured in terms of solar gain reductions) was evaluated using CFD simulations. Details on the definition of the studied parameters, its base values and the parametric variations studied can be found in Table 3. In order to properly assess the parameter effects on solar load gains all studied variables are presented as dimensionless quantities.

Table 3. Parameters selected for numerical study.

Parameter	Definition	Base case value	Parametric variations
Absorptivity Louvers inner face	α_i	$\alpha_{i,bc} = 0.18$	$\alpha_i = 0.3/0.6/0.9$
Emissivity Louvers outer face	ζ_o	$\zeta_{o,bc} = 0.1$	$\zeta_o = 0.4/0.6/0.8$
Louvers distance to outer glass	$R = \frac{r}{w}$ (Figure 1)	$R_{bc} = \frac{0.066\ m}{0.8\ m} = 0.0825$	$R = 0.0450/0.0575/0.0700$
Cavity flow rate	$\dot{m}_r = \frac{\text{Mass flow rate}}{\text{Maximum mass flow rate}}$	$\dot{m}_{r,bc} = \frac{1.21\ kg/s}{3.32\ kg/s} = 0.3645$	$\dot{m}_r = 0.5663/0.7922/1.0000$

3.5. Meshing and Boundary Conditions

A structured hexahedral grid was selected as the computational mesh for all modeled cases (e.g., Figure 2). Mesh-sensitivity tests were performed in previous work [3], selecting an optimal hexahedral mesh size of 2 × 12.5 × 1 mm for all solid phases. Boundary layers were defined around all solid surfaces in contact with a fluid to properly model the heat transfer phenomena. For the fluid phase (cavity), a new mesh-sensitivity test was performed considering the VB. Three different mesh densities were tested (2500, 3500 and 4500 kel) obtaining similar results in terms of heat fluxes for the two finer meshes tested. A cavity mesh of 3500 kel was selected for the purposes of this work.

Boundary conditions were set as close as possible to DSF regular operating conditions. Horizontal surfaces were set to adiabatic, reflective boundaries and the mass flow rate at the inlet was set to a constant value for each case. Thermal boundary conditions for the exterior and interior layers were imposed as convective + radiative heat fluxes and convective heat flux respectively. Convection coefficients were set to 12 W/m^2K and 8 W/m^2K for the outdoors and indoors glazing surfaces respectively. Radiative heat flux was computed using a radiation model.

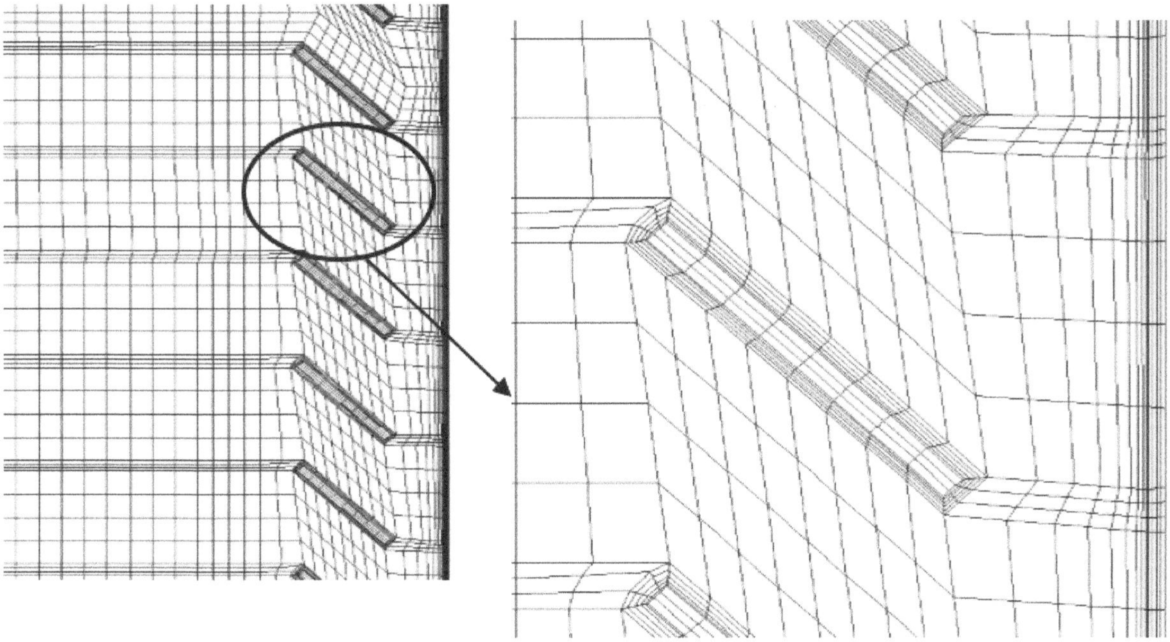

Figure 2. Detail of the mesh generated for the cavity and VB.

3.6. Solver Set-up

RNG k-epsilon turbulence model and P1 radiation model were selected as numerical sub-models for all cases tested. Navier–Stokes equations together with the energy conservation equation were solved using the CFD code Ansys Fluent® v15. A pressure-based double-precision transient solver was selected in order to solve the set of equations used. Second order upwind discretization schemes were imposed on all the transport equations. PISO pressure–velocity coupling was chosen due to its suitability for buoyancy-affected flows. The fluid was taken to be incompressible, Newtonian and in turbulent flow regime. All the numerical sub-models, discretization and simulation strategies were previously validated against an experimental database [18]. Air properties were set to be temperature-dependent. A time step of 600 s was taken for performing the simulations.

Numerical simulations were performed in a Hybrid Bull machine property of Consorci de Serveis Universitaris de Catalunya (CSUC). Numerical convergence of the model was checked based on the normalized numerical residuals of all computed variables. Heat fluxes and temperatures on all solid surfaces were recorded for analysis and discussion purposes.

4. Results and Discussion

4.1. Velocity and Temperature Profiles inside the Cavity

For all the modeled cases, flow and temperature fields were obtained in order to qualitatively assess the DSF thermal performance. Figures 3 and 4 show the temperature and velocity profiles in a longitudinal mid-section of the flow cavity for different times of the day for the base case simulation. The velocity fields shown in Figure 3 are similar to previously reported velocity fields for upper-crossed lateral ventilation schemes in DSF [19,20], which indicates that the VB has little effect on the inner cavity flow (between the VB and the inner wall). In the outer cavity (between the VB and the outdoor glass) the flow velocities obtained are low for all cases studies.

Temperature contours obtained for the base case (Figure 4) show higher mid-cavity air temperatures than comparable results obtained for the same DSF model without VB [19,20]. This is directly related with the placement of opaque surfaces (VB louvers) in the cavity. These surfaces are the highest temperature zone of the cavity, affecting accordingly the temperature of its surroundings.

The effect of a VB on the indoor solar heat gain was evaluated using a set of numerical simulations of the DSF with/without a VB. For both scenarios, natural and forced ventilation operating conditions were studied. Heat flux through the interior surfaces of a DSF were recorded for a 24-hour period for all studied cases. These heat fluxes were integrated and area-weighted averaged to compute the solar heat gains. For the modeled cases with VB, the base case parameter values shown in Table 3 were used. Figure 5 shows examples of the heat flux recorded and Table 4 presents the results obtained for this set.

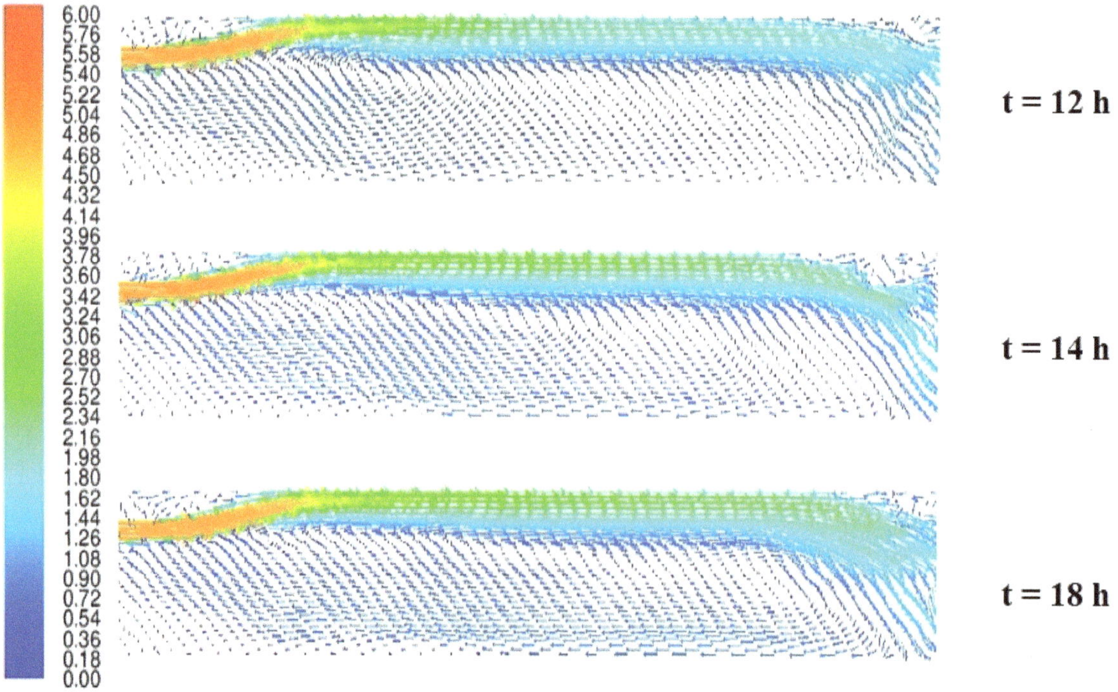

Figure 3. Mid-cavity velocity contours for the base case. Velocity scale in m/s.

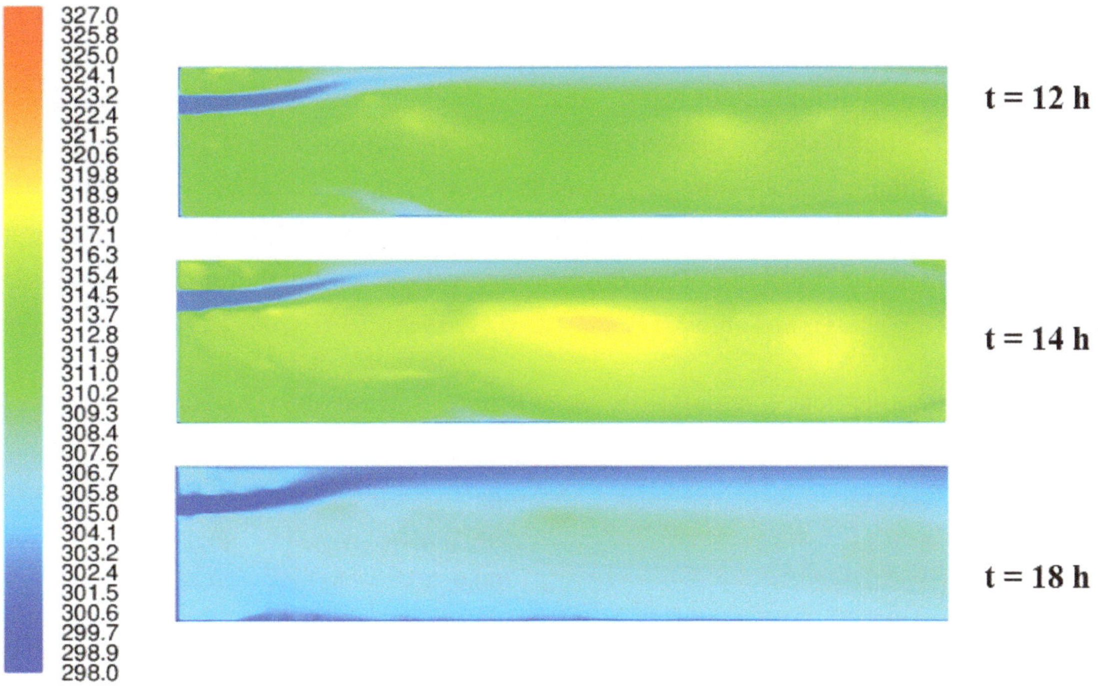

Figure 4. Mid-cavity temperature contours for the base case. Temperature scale in K.4.2. Effect of VB on Solar Heat Gains.

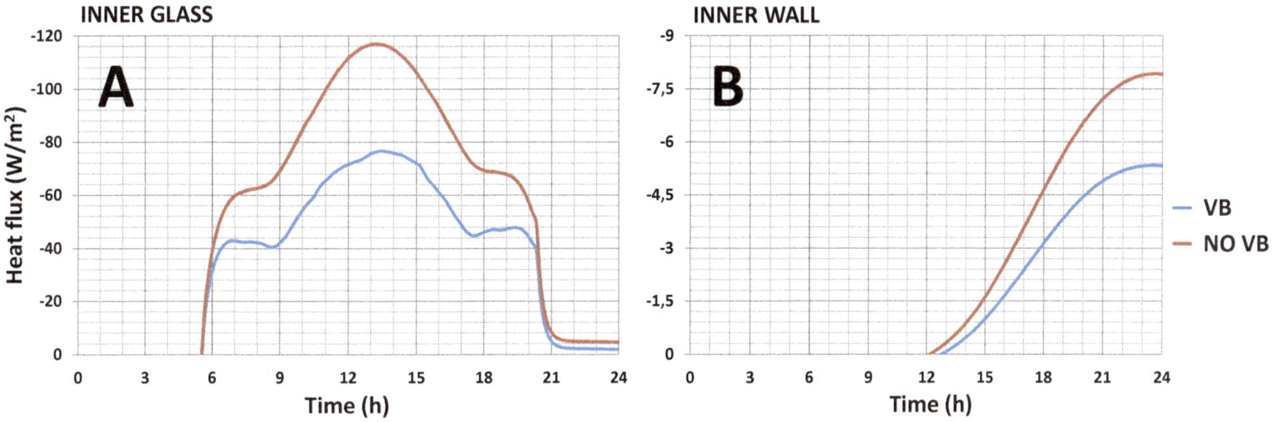

Figure 5. Heat flux through the inner glass (**A**) and inner wall (**B**) for a DSF with forced ventilation with and without VB.

Table 4. Solar heat gains with and without VB.

Solar heat gains (kJ/m^2-day)	Natural ventilation	Forced ventilation
DSF without VB	3173.4	2627.8
DSF with VB	2317.0	1708.1 (base case)
Reduction in solar heat gains	27%	35%

As expected, Table 4 shows that the use of a VB reduces the solar heat gains through the inner surfaces of the DSF. These reductions are not only due to the reflected radiation but also to the heat absorbed by the VB solid surfaces that is dissipated to the ventilation air, which may help explain the larger reductions obtained when forced ventilation is used instead of natural ventilation.

4.3. VB inner Surfaces Absorptivity

Double reflection can play an important role in heat and daylight transmission through a DSF [28]. Reducing the double reflection can be achieved by modifying the VB inner surfaces absorptivity. Gray or pastel paints on aluminum surfaces are a low cost solution that offers an interesting range of absorptivities ($0.2 < \alpha_i < 0.9$) [29]. In order to evaluate the effect of the VB inner surface absorptivity on the thermal performance of the DSF, four absorptivity values within the suggested range were tested. Results obtained are shown in Table 5.

Table 5. Solar heat gains for different VB inner surface absorptivities.

VB inner surface absorptivity	Solar heat gain (kJ/m^2-day)	Reduction (%)
0.18	1708.1	
0.3	1707.2	0.05
0.6	1700.9	0.42
0.9	1695.6	0.72

Results obtained show that increasing the absorptivity of the VB louvers inner (lower) surface reduces the heat flux through the DSF inner layer. Besides reducing double reflection effects, an increased absorptivity reduces the air temperature within the cavity (see Figure 6).

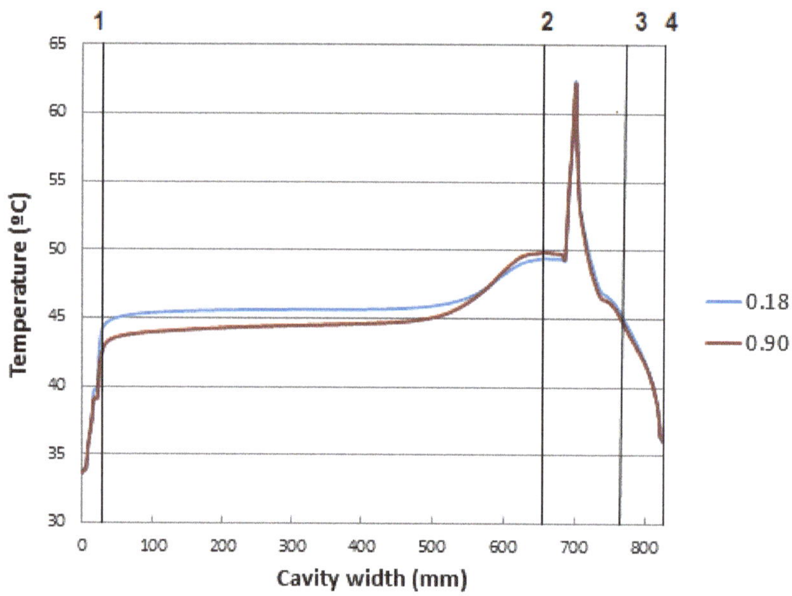

Figure 6. Temperature profile on a mid-cavity line for different VB inner surface absorptivities. (1) Indoors double glazing; (2–3) VB; (4) Outer glazing.

This reduced cavity temperature may improve the DSF thermal performance by reducing the conductive/convective heat transfer towards the building interior.

4.4. VB Outer Surfaces Emissivity

Surface heat rejection is a desired quality for applications where avoiding heat transfer towards the interior is the goal. Increasing the heat rejection capability of a surface by modifying its emissivity is a regular subject of study in the aerospace industry. Dielectric film coatings on polished metals offer a wide range of emissivities [29] that may be of application for an aluminum VB. In order to evaluate the effect of the VB outer surface emissivity on the thermal performance of the DSF, four emissivity values within the suggested range were tested. Results obtained are shown in Table 6.

Table 6. Solar heat gains for different VB outer surface emissivity.

VB outer face emissivity	Solar heat gain (kJ/m^2-day)	Reduction (%)
0.1	1708.1	
0.4	1687.4	1.21
0.6	1667.5	2.38
0.8	1666.6	2.43

An enhanced emissivity allows the VB upper surfaces to dissipate the absorbed heat faster into the cavity air. Figure 7 shows the temperature contours for a mid-cavity plane for different VB outer surface emissivities. Higher surface emissivities translate into lower surface temperatures, and thus lower cavity temperatures, reducing also the conductive/convective heat flux towards the interior.

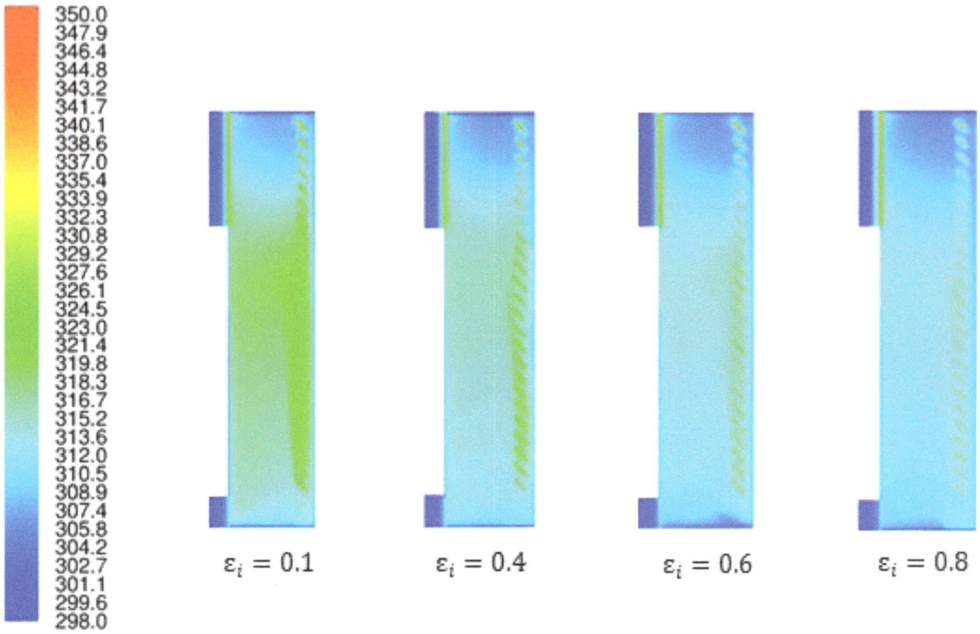

Figure 7. Temperature contours on a mid-cavity plane at 14h for different VB outer surface emissivities. Temperature scale in K.

4.5. VB Louver Distance to Outer Glass

The VB divides the cavity into an outer and an inner cavity. The outer cavity (comprised of between the outer layer and the VB) is a low-velocity to stagnation flow zone, while in the inner cavity the flow presents higher velocities due to the forced ventilation. Therefore is reasonable to expect that the relative size of the outer cavity with respect to the cavity width should have influence on the thermal performance of the DSF. In order to evaluate this effect, four VB positions within the cavity were tested. For this purpose, three new geometrical models and meshes were generated. The results obtained are shown in Table 7.

Table 7. Solar heat gains for different VB positions.

$R = \frac{r}{w}$ (Figure 1)	Solar heat gain (kJ/m²/day)	Reduction (%)
0.0254	1708.1	
0.0215	1707.5	0.03
0.0177	1685.1	1.34
0.0138	1684.3	1.39

As it can be seen in the results shown in Table 7, moving the VB closer to the outer glazing improves the DSF thermal performance. Figure 8 shows the temperature profiles on a mid-cavity line for the extreme positions tested. A small outer cavity (R = 0.0138) translates into higher outer cavity and lower inner cavity temperatures than a DSF with larger outer cavity (R = 0.0254). This favors the heat transfer from the outer cavity to the exterior and reduces the convective/conductive heat fluxes towards the interior of the DGF.

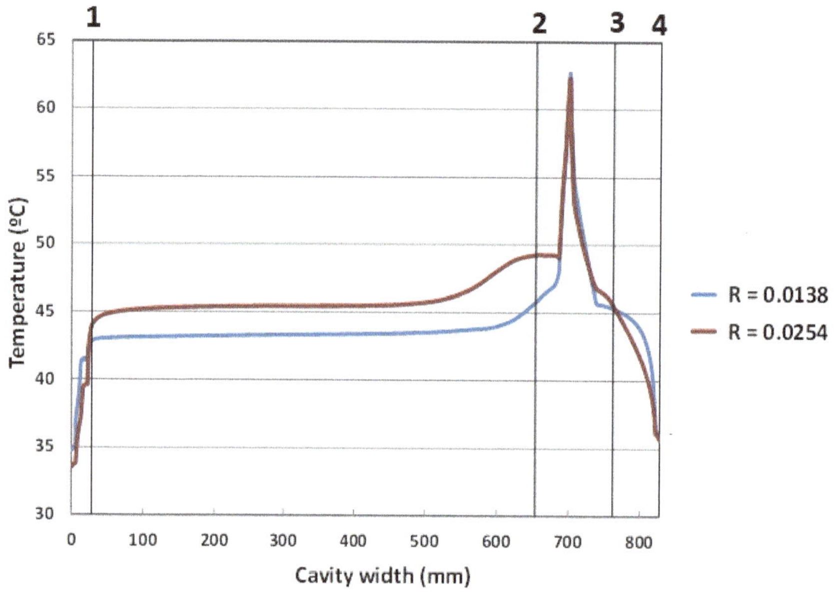

Figure 8. Temperature profile on a mid-cavity line for different VB positions. (1) Indoors double glazing; (2–3) VB; (4) Outer glazing.

4.6. Forced Ventilation Flow Rate

Four different ventilation air flow rates (1.21, 1.88, 2.63 and 3.32 kg/s) were imposed to the computational model in order to evaluate the influence of the forced ventilation in the thermal performance of the DSF with VB. Results obtained are shown in Table 8.

Table 8. Solar heat gains for different ventilation flow rates.

$\dot{m}_r = \frac{\text{Mass flow rate}}{\text{Maximum mass flow rate}}$	Solar heat gain (kJ/m²-day)	Reduction (%)
0.3645	1708.1	
0.5663	1706.1	0.11
0.7922	1696.0	0.70
1.0000	1685.9	1.30

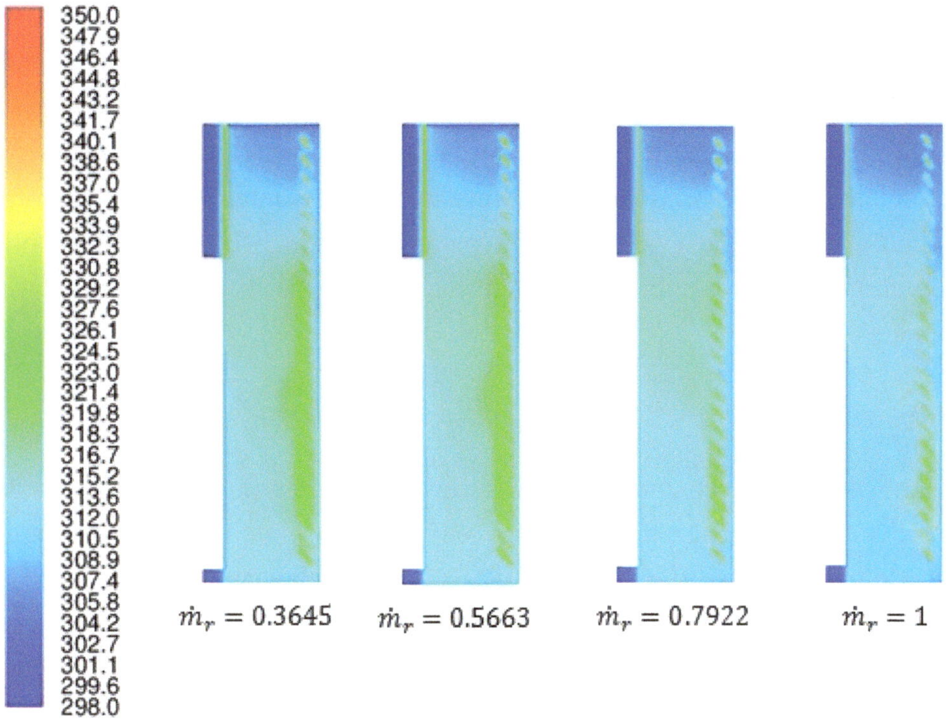

Figure 9. Temperature contours on a mid-cavity plane at 14 h for different ventilation flow rates. Temperature scale in K.

As expected, a higher ventilation flow rate reflects into better DSF thermal performance. It is important to notice that almost tripling the ventilation flow rate only improves the performance of the base case by 1.30%, which is a small solar heat gain reduction considering the higher operational costs associated with higher ventilation flow rates.

5. Conclusions and Future Work

CFD proves to be a useful tool when modeling conductive/convective/radiative heat transfer in ventilated DSF. Numerical simulations were run for several cases and velocity and temperature fields together with heat fluxes through all surfaces were obtained for the studied scenarios. A previously validated modeling strategy [3,18] was used to obtain the presented results, consolidating the idea that CFD can offer tailored solutions for DSF performance optimization.

Heat fluxes through the inner layer of a DSF were evaluated for a DSF with/without VB in natural and forced ventilation operating conditions in order to assess the influence of VB on DSF thermal performance. Results obtained show that a VB can reduce solar heat gains up to 35%. These reductions are not only due to the reflected radiation but also to the heat absorbed by the VB solid surfaces that is dissipated to the ventilation air, which may help explain the larger reductions obtained when forced ventilation is used instead of natural ventilation.

The effects of several parameters as VB's inner surface absorptivity, outer surface emissivity and its distance to the outer glass as well as the forced ventilation flow rate on the thermal performance of a DSF were studied using single-parameter variation simulation sets. It was found that increasing the VB outer surfaces emissivity (*i.e.*, using dielectric film coatings) or VB inner surfaces absorptivity (*i.e.*, using pastel/gray paints) can improve the thermal performance of the DSF due to the enhancement of heat rejection towards outdoors and the reduction of double reflection towards indoors. VB distance to the outer glass can also affect the thermal performance of the façade, as the solar heat gains are lower when the VB is closer to the outer glazing. Increasing the ventilation flow rate can also reduce the DSF solar heat gains, but the thermal benefits obtained by higher flow rates may be overshadowed by the operating costs associated with the ventilation equipment required to achieve these ventilation rates.

Future work should include the evaluation and parametrization of the effects of other shading elements as textile curtain walls or micro-perforated materials in the thermal performance of DSF in Mediterranean climates.

Acknowledgments

The authors acknowledge the economic support received from Grupo JG Consultora de Proyectos S.A., Catedra UPC-JG and Universitat Internacional de Catalunya. Funding from the Spanish Ministry of Economy and Competitiveness (Grant DPI2012-36264) is also appreciated. The authors also acknowledge Consorci de Serveis Universitaris de Catalunya (CSUC) for granting access to their supercomputing facilities for developing this work.

Author Contributions

Eduard Egusquiza and Pere Alavedra conceived and designed the numerical experiments; Jordi Parra performed the numerical experiments; Alfredo Guardo analyzed the data and wrote the paper.

Nomenclature

a	Absorption coefficient	m^{-1}
c_p	Specific heat	$J \cdot kg^{-1} \cdot K^{-1}$
D	Louvers shape factor	
$\boldsymbol{F}_b$	External specific body forces	$kg \cdot m^{-2} \cdot s^{-2}$
$\boldsymbol{F}_{w,\infty}$	Radiation shape factor	
g	Gravity forces	$m \cdot s^{-2}$
h	Heat transfer coefficient	$W \cdot m^{-2} \cdot K^{-1}$
k	Thermal conductivity	$W \cdot m^{-1} \cdot K^{-1}$
L	Characteristic length	m
	Louvers width	m
m_r	Flow rate parameter	
p	Static pressure	Pa
$\dot{q}$	Heat flux	$W \cdot m^{-2}$
r	Louvers distance to outer glass	m
R	Louvers position parameter	
SL	Solar load gain	$J \cdot m^{-2}$
t	Time	s
T	Static temperature	K
u	Velocity	$m \cdot s^{-1}$
w	Cavity width	m

Greek Letters

α	Absorptivity	
β	Louvers tilt angle	
δ	Louvers thickness	m
ζ	Emissivity	
λ	Louvers width	m
μ	Dynamic viscosity	$kg \cdot m^{-1} \cdot s^{-1}$
ρ	Fluid density	$kg \cdot m^{-3}$
σ	Stefan-Boltzmann constant	$W \cdot m^{-}2 \cdot K^{-4}$
$\boldsymbol{\tau}$	Stress tensor	$N \cdot m^{-2}$
Φ	Normalized solar load gain	
χ	Parameter value	

Dimensionless Groups

Bi	Biot number	$[h \cdot L \cdot k_w^{-1}]$
Br	Brinkman number	$[\mu \cdot u^2 \cdot k_f^{-1} \cdot T^{-1}]$
Ec	Eckert number	$[u^2 \cdot C_p^{-1} \cdot T^{-1}]$, $[Br.Pr^{-1}]$
Eu	Euler number	$[\delta p \cdot \rho^{-1} \cdot u^{-2}]$
Fr	Froude number	$[u^2 \cdot L^{-1} \cdot g^{-1}]$
Ma	Mach number	$[u \cdot u_{sound}^{-1}]$
Nu	Nusselt number	$[h \cdot L \cdot k_f^{-1}]$
Pr	Prandtl number	$[c_p \cdot \mu \cdot k_f^{-1}]$
Re	Reynolds number	$[L \cdot u \cdot \rho \cdot \mu^{-1}]$
Sr	Strouhal number	$[L \cdot t^{-1} \cdot u^{-1}]$
Th	Thring number	$[\rho \cdot c_p \cdot u \cdot \zeta^{-1} \cdot \sigma^{-1} \cdot T^{-3}]$

Acronyms

CFD	Computational fluid dynamics
DSF	Double skin façade(s)
VB	Venetian blind(s)

Sub/Super—Index

0	Referred to a reference quantity
^	Referred to a dimensionless quantity
bc	Referred to the base case
case	Referred to the studied simulation case
f	Referred to a fluid
fc	Referred to the free convection case
i	Referred to an inner/indoors position
o	Referred to an outer/outdoors position

rad	Referred to radiation
T	Referred to a turbulent quantity
w	Referred to a wall

Conflicts of Interest

The authors declare no conflict of interest.

References

1. Poirazis, H. *Double Skin Façades for Office Buildings*; Report EBD-R—04/3; Lund Institute of Technology, Lund University: Lund, Sweden, 2004.
2. Gratia, E.; de Herde, A. Optimal operation of a south double-skin façade. *Energy Build.* **2004**, *36*, 41–60.
3. Guardo, A.; Coussirat, M.; Egusquiza, E.; Alavedra, P.; Castilla, R. A CFD approach to evaluate the influence of construction and operation parameters on the performance of active transparent façades in Mediterranean climates. *Energy Build.* **2009**, *41*, 534–542.
4. Phillips, J.; Naylor, D.; Oosthuizen, P.H.; Harrison, S.J. Modeling of the conjugate heat transfer from a window adjacent to a louvered shade. In Proceedings of the 6th International Conference on Advanced Computational Methods in Heat Transfer, Madrid, Spain, 10–14 March 2002.
5. Collins, M.R.; Harrison, S.J.; Oosthuizen, P.H.; Naylor, D. Heat transfer from an isothermal vertical surface with adjacent heated horizontal louvers: Numerical analysis. *ASME J. Heat Transf.* **2002**, *124*, 1072–1077.
6. Collins, M.R.; Harrison, S.J.; Oosthuizen, P.H.; Naylor, D. Heat transfer from an isothermal vertical surface with adjacent heated horizontal louvers: Validation. *ASME J. Heat Transf.* **2002**, *124*, 1078–1087.
7. Athienitis, A.; Santamouris, M. *Thermal Analysis and Design of Passive Solar Buildings*; James & James Science Publishers Ltd.: London, UK, 2002.
8. Collins, M.R.; Harrison, S.J. Calorimetric analysis of the solar and thermal performance of windows with interior louvered blinds. *ASHRAE Trans.* **2004**, *110*, 474–485.
9. Chantrasrisalai, C.; Fisher, D.E. An *in situ* experimental method for the development and validation of slat-type blind models in cooling load calculations. *J. Sol. Energy Eng.* **2006**, *128*, 189–198.
10. Manz, H.; Menti, U. Energy performance of glazings in European climates. *Renew. Energy* **2012**, *37*, 226–232.
11. Tzempelikos, A.; Athienitis, A. The impact of shading design and control on building cooling and lighting demand. *Sol. Energy* **2007**, *81*, 369–382.
12. Tzempelikos, A. The impact of Venetian blind geometry and tilt angle on view, direct light transmission and interior illuminance. *Sol. Energy* **2008**, *82*, 1172–1191.
13. Oh, M.H.; Lee, K.H.; Yoon, J.H. Automated control strategies of inside slat-type blind considering visual comfort and building energy performance. *Energy Build.* **2012**, *55*, 728–737.
14. Glória Gomez, M.; Santos, A.J.; Moret Rodrigues, A. Solar and visible optical properties of glazing systems with venetian blinds: Numerical, experimental and blind control study. *Build. Environ.* **2014**, *71*, 47–59.

15. Manz, H. Total solar energy transmittance of glass double façades with free convection. *Energy Build.* **2004**, *36*, 127–136.
16. Manz, H.; Schaelin, A.; Simmler, H. Airflow patterns and thermal behavior of mechanically ventilated glass double façades. *Build. Environ.* **2004**, *39*, 1023–1033.
17. Sedlák, J.; Mráček, P. Simulation of the double facade in the Brno Metropolitan Library. In Proceedings of the Dynastee 2005. Dynamic Analysis, Simulation and Testing Applied to the Energy and Environmental Performance of Buildings, Athens, Greece, 12–14 October 2005.
18. Coussirat, M.; Guardo, A.; Jou, E.; Egusquiza, E.; Cuerva, E.; Alavedra, P. Performance and influence of numerical sub-models on the CFD simulation of free and forced convection in double-glazed ventilated façades. *Energy Build.* **2008**, *40*, 1781–1789.
19. Guardo, A.; Coussirat, M.; Valero, C.; Egusquiza, E.; Alavedra, P. CFD assessment of the performance of lateral ventilation in double glazed façades in Mediterranean climates. *Energy Build.* **2011**, *43*, 2539–2547.
20. Valentín, D.; Guardo, A.; Egusquiza, E.; Valero, C.; Alavedra, P. Use of Coandă nozzles for double glazed façades forced ventilation. *Energy Build.* **2013**, *62*, 605–614.
21. Valentín, D.; Guardo, A.; Egusquiza, E.; Valero, C.; Alavedra, P. Assessment of the economic and environmental impact of double glazed façade ventilation systems in Mediterranean climates. *Energies* **2013**, *6*, 5069–5087.
22. Baldinelli, G. Double skin façades for warm climate regions: Analysis of a solution with an integrated movable shedding system. *Build. Environ.* **2009**, *44*, 1107–1118.
23. Safer, N.; Woloszyn, M.; Roux, J.J. Three-dimensional simulation with a CFD tool of the airflow phenomena in single floor double-skin facade equipped with a venetian blind. *Sol. Energy* **2005**, *79*, 193–203.
24. Zeng, Z.; Li, X.; Zhu, Y. Modeling ventilation in naturally ventilated double-skin façade with a Venetian blind. *Build. Environ.* **2012**, *57*, 1–6.
25. Código Técnico de la Edificación, Real Decreto 314/2066 del 17 de marzo. Sección S1 5, Apartado 2. Gobierno de España. Available online: http://www.codigotecnico.org/cte/export/sites/default/web/galerias/archivos/DB_SI_19feb2010.pdf (accessed on 19 April 2015). (In Spanish)
26. Gratia, E.; de Herde, A. The most efficient position of shading devices in a double-skin façade. *Energy Build.* **2007**, *39*, 364–373.
27. Mitjà, A. *Atlas de Radiació Solar a Catalunya*; Institut Català d'Energia: Barcelona, Spain, 2001.
28. Chan, Y.C.; Tzempelikos, A. Efficient Venetian blind control strategies considering daylight utilization and glare protection. *Sol. Energy* **2013**, *98*, 241–254.
29. Gilmore, D.G. *Spacecraft Thermal Control Handbook, Volume I: Fundamental Technologies*, 2nd ed.; American Institute of Aeronautics and Astronautics: Reston, CA, USA, 2002; pp. 139–159.

6

Energy Efficiency Comparison between Hydraulic Hybrid and Hybrid Electric Vehicles

Jia-Shiun Chen

Department of Vehicle Engineering, National Taipei University of Technology, 1, Sec. 3, Zhongxiao E. Road, Taipei 10608, Taiwan; E-Mail: chenjs@mail.ntut.edu.tw

Academic Editor: Joeri Van Mierlo

Abstract: Conventional vehicles tend to consume considerable amounts of fuel, which generates exhaust gases and environmental pollution during intermittent driving cycles. Therefore, prospective vehicle designs favor improved exhaust emissions and energy consumption without compromising vehicle performance. Although pure electric vehicles feature high performance and low pollution characteristics, their limitations are their short driving range and high battery costs. Hybrid electric vehicles (HEVs) are comparatively environmentally friendly and energy efficient, but cost substantially more compared with conventional vehicles. Hydraulic hybrid vehicles (HHVs) are mainly operated using engines, or using alternate combinations of engine and hydraulic power sources while vehicles accelerate. When the hydraulic system accumulator is depleted, the conventional engine reengages; concurrently, brake-regenerated power is recycled and reused by employing hydraulic motor–pump modules in circulation patterns to conserve fuel and recycle brake energy. This study adopted MATLAB Simulink to construct complete HHV and HEV models for backward simulations. New European Driving Cycles were used to determine the changes in fuel economy. The output of power components and the state-of-charge of energy could be retrieved. Varying power component models, energy storage component models, and series or parallel configurations were combined into seven different vehicle configurations: the conventional manual transmission vehicle, series hybrid electric vehicle, series hydraulic hybrid vehicle, parallel hybrid electric vehicle, parallel hydraulic hybrid vehicle, purely electric vehicle, and hydraulic-electric hybrid vehicle. The simulation results show that fuel consumption was 21.80% lower in the series hydraulic hybrid vehicle compared to the series hybrid electric vehicle; additionally, fuel consumption was 3.80% lower in the parallel hybrid electric vehicle compared to the parallel hydraulic hybrid vehicle.

Furthermore, the hydraulic–electric hybrid vehicles consumed 11.4% less electricity than the purely electric vehicle did. The simulations indicated that hydraulic-electric hybrid vehicle could provide the best energy cost among all the configurations studied.

Keywords: backward simulation; hybrid electric vehicle; hydraulic hybrid vehicle; New European Driving Cycle

---

1. Introduction

The internal combustion engine (ICE) is a widely used and well-developed technology, however, pollution and energy resource issues are growing concerns. Hybrid vehicles (HVs) are an effective and common solution to these issues. Most manufacturers have developed hybrid electric vehicle (HEV) systems to improve vehicle fuel economy and cost. Energy management and emission control for electric and IC engine power systems have been studied and developed for various hybrid systems [1–4]. In a hybrid electric vehicle, energy management is important because it can substantially affect vehicle performance and component size. Several control strategies have been implemented in HEVs [5]. Rule-based and fuzzy-based control strategies are widely used in light and medium-sized HEVs. Rule-based control is simpler but less flexible than fuzzy-based control. Fuzzy-based control can identify and learn driver behaviors under various driving conditions. Both strategies are applicable for real-time control [6]. Although a dynamic programming (DP) solution can numerically optimize a specific drive-cycle [7,8], stochastic dynamic programming (SDP) can optimize solutions for assumed road-load conditions with known probabilities [9]. Studies of heuristic rule-base methods [10,11] and experiments in fuzzy logic control (FLC) with an intelligent supervisory control strategy suggest that refined ICE speed transitions are also good candidates for implementation in control algorithms. Another configuration of hybrid vehicle is the hydraulic hybrid vehicle (HHV), in which a hydraulic pump replaces the electric motor as the primary or assistant driving power and the accumulator replaces the battery for energy storage. The hydraulic hybrid system provides large power capacity and more suitable for heavy duty vehicle [12,13]. Compared to HEVs, HHVs recapture more energy during braking, and it makes the HHV more suitable for urban driving applications which include frequent stops and starts [14–16]. New energy management strategies have been proposed to improve fuel efficiency [17,18].

Generally, conventional manual or automatic transmissions are used to regulate an engine within a certain range of engine speeds and to output rotational speed and torsional moments according to driver demands. The main disadvantage of these transmissions is their dependence on a gearbox for gearshift actions. During gearshifts, sudden changes in engine speed, which are common in conventional gearboxes, produce discontinuities in gear-shifting. Hydraulic transmission systems provide variable speed ratios between engine and wheels which remove the sudden changes during the gear-shifting. Continuously variable transmissions (CVTs) also are unaffected by gearshift discontinuities during the shifting process, thus providing passengers with a comfortable and uninterrupted ride. The advantage of the CVT is that by minimizing the power loss of an engine, engines can operate at peak efficiency or at constant engine speeds. Moreover, CVTs are structurally simple and easily maintained. However, CVTs cannot withstand excessive frictional forces generated by the torsional moments of engines because of limited belt or chain strengths. Therefore, hydraulic transmission systems have the following advantages over CVTs:

(1) Steady transmission: Hydraulic transmission devices operate by using hydraulic oil, which is nearly incompressible at ordinary pressures and using continuously flowing hydraulic oil to make gearshifts. Because hydraulic oil has shock absorbing capacities, hydraulic buffer devices can be installed in the oil lines to provide steadily operating transmissions.
(2) Light weight and compact size: In contrast to mechanical and electrical transmissions, hydraulic transmissions are substantially lighter and smaller under identical output power conditions. Thus, hydraulic transmissions have low inertia and are highly responsive.
(3) High capacity: Hydraulic transmissions can readily receive large forces and torsional moments.
(4) Readily obtainable infinitely variable speeds: In hydraulic transmissions, infinitely variable speeds can be obtained by regulating fluid flow rates. The adjustment ratio can vary widely from 1:1 to 2000:1. Extremely low speeds are easily achieved.

Regarding vehicle transmission systems, hydraulic transmission systems can achieve similar continuously variable speed functions. In principle, hydraulic pumps convert mechanical energy provided by engines into fluid hydraulic energy. Finally, hydraulic motors convert high pressure hydraulic energy into mechanical energy to drive the vehicles. To obtain continuously variable speed functions, only the hydraulic motor or hydraulic pump valve opening must be adjusted. This system provides continuously variable and broad gear reduction ratios. However, hydraulic motors and pumps that operate at high pressures and small openings are usually inefficient.

Hydraulic systems have a high pressure accumulator for energy storage and hydraulic pumps/motors to transfer power and are often used for heavy duty vehicles in the past. Recently, the applications can be seen in the passenger or light duty vehicles. Sizes of the hydraulic systems are applicable for the stop-start operation. The accumulator has high power density, can be completely charged and discharged for many cycles and constantly switched between charging and discharging states without being concerned about functional degradation or capacity loss. All these features make hydraulic systems very attractive for stop-start operation, especially in city driving.

Many studies have been done for Manual Transmission Vehicles, Series Hybrid Electric Vehicles, Parallel Hybrid Electric Vehicles, and Electric Vehicles. Growing interest in Series Hydraulic Hybrid Vehicles, Parallel Hydraulic Hybrid Vehicles, and Hydraulic-Electric Hybrid Vehicles can also be seen. However, most studies have focused on improving energy management strategies, improving designs for single configuration, or optimizing their components/subsystems efficiencies, such as hydraulic pumps/motors, DC motors, or batteries, *etc.*, for given configurations. Energy control strategies have been specifically tailored to the particular systems or for special events or applications, and the improvement comparisons were made against their own baseline designs. Few studies have compared energy efficiency among different configurations of hydraulic hybrid vehicles and hybrid electric vehicle systems. Therefore, this research uses MATLAB Simulink to construct all seven complete models for backward simulations, evaluates the fuel economy with comprehensive comparison among all the listed systems, and discusses the suitable applications for different powertrain configurations. This research compares the energy efficiency based on New European Driving Cycle (NEDC) through various power component models, connection with energy storage component models, and combinations of series or parallel configurations, and concludes that hydraulic-electric hybrid vehicle (HEHV) have the highest cost efficiency among all the configurations studied.

2. System Modeling

This section introduces various model components. Methods and empirical equations were established for each component model. To compare the differences between the hydraulic hybrid vehicle (HHV) and hybrid electric vehicle (HEV) models, hydraulic subsystem models were established and comprised hydraulic pump, hydraulic motor, and accumulator models. This study also established models of electrical subsystems, including the electric motor, generator, and lithium-ion battery. Electrical subsystem models supply power to series and parallel HEVs (SHEVs and PHEVs), and pure electric vehicles (EVs), whereas hydraulic subsystem models supply power to series and parallel HHVs (SHHVs and PHHVs) and hydraulic–electric hybrid vehicles (HEHVs). Table 1 presents the powertrain configurations and components applied in the research.

Table 1. Vehicle Configurations.

Configuration/Components & Subsystems		Internal Combustion Engine	Electric System	Hydraulic System	Brake Regenerating	Parallel	Series
MT vehicle (manual transmission)		x					
Electric Vehicle	SHEV	x	x		x		x
	PHEV	x	x		x	x	
	EV		x		x		
Hydraulic Vehicle	SHHV	x		x	x		x
	PHHV	x		x	x	x	
	HEHV		x	x	x	x	

Modeling was performed by backward simulation [9]. The required speed-time relationships were entered into the model of driving cycles before running each simulation. In this study, vehicle speed follows the New European Driving Cycle (NEDC).

2.1. Driving Cycle Model

The entire simulation analysis began from the driving cycles based on the NEDC (Figure 1) for standardized testing procedures to provide instantaneous driving forces required to overcome the aerodynamic, rolling, grade, and inertia resistances. The entire driving cycle (total time, 1180 s) including urban driving cycles (UDCs) and extra-urban driving cycles (EUDCs). The first 780 s was the UDC (maximum speed 50 km/h), and the next 400 s was the EUDC (maximum speed 120 km/h).

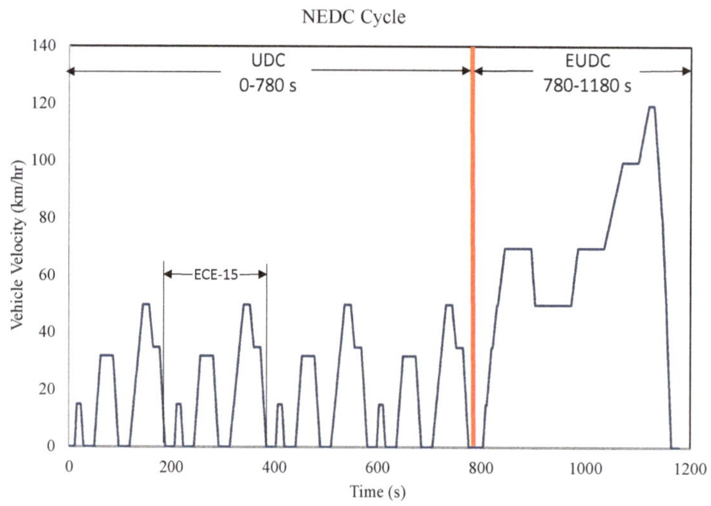

Figure 1. NEDC driving cycle.

2.2. Vehicle Dynamic Model

Figure 2 (red arrows) shows the vehicle driving resistances, including the rolling, aerodynamic, grade, and acceleration resistances. The vehicle driving forces are calculated through the dynamic vehicle model by using the following equations:

$$F_t = R_r + R_a + R_c + R_s \tag{1}$$

$$R_r = R_{r,A} + R_{r,B} = \mu_r \cdot W \tag{2}$$

$$R_a = C_D \cdot \frac{\rho}{2} \cdot A_f \cdot (v - v_w)^2 \tag{3}$$

$$R_c = W sin(\theta) \tag{4}$$

$$R_s = (W + W_e) \cdot {}^{a}/_{g} \tag{5}$$

where F_t is the tractive effort; R_r is the rolling resistance; μ_r is the rolling resistance coefficient, W is the vehicle weight; R_a is the aerodynamic dragging force; C_D is the aerodynamic dragging coefficient; v is vehicle speed; v_w is wind speed; R_c is the grade resistance; W_e is the equilibrium weight of rotational part; a is vehicle acceleration; and R_s is the inertia force. The grade was set to 0 in all fuel economy simulations.

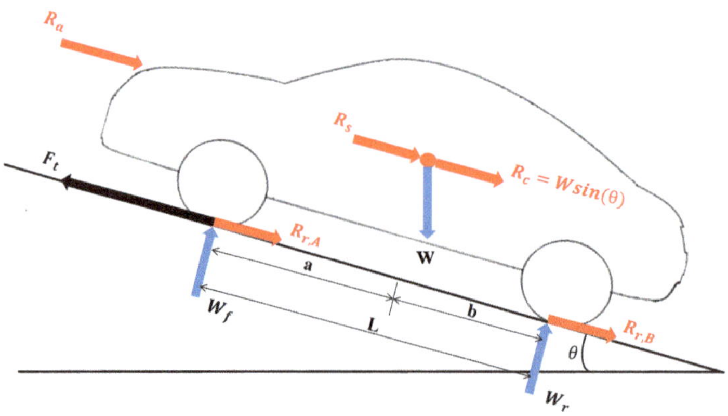

Figure 2. Vehicle dynamic analysis.

2.3. Engine Model

The engine model in this study was a steady state engine model. Vehicle fuel consumption is related to the engine operating conditions. Therefore, the speed and time required for the driving cycles is converted to engine speed and torque. The look-up table method was applied to obtain the brake specific fuel consumptions (BSFC) for the known engine speeds and engine torques and subsequently calculate the vehicle fuel consumptions. Engine specifications for series, parallel, and conventional vehicles were designed by dividing gasoline engines into three displacement volumes of 1.0, 1.3, and 1.8 L, which generate maximal powers of 41, 63, and 95 kW, respectively. Figure 3 presents BSFC diagrams. The three engines were designed to be compatible with each basic configuration. The conventional manual transmission vehicle (MT) is equipped with a 1.8 L engine. The parallel hybrid systems, PHEV and PHHV, use a 1.3 L engine. Since they have an electric motor that assists driving power, their engine power can be lower than that of the MT vehicle. The series hybrid vehicles, which are driven by motors, use a 1.0 L engine. The engine is used to charge the battery. The average tractive power through the driving cycle is much lower than the peak power, therefore, the engine for series hybrid vehicles can be much smaller. The engine data in Figure 3 were retrieved from the ADVISOR 2003 database [19].

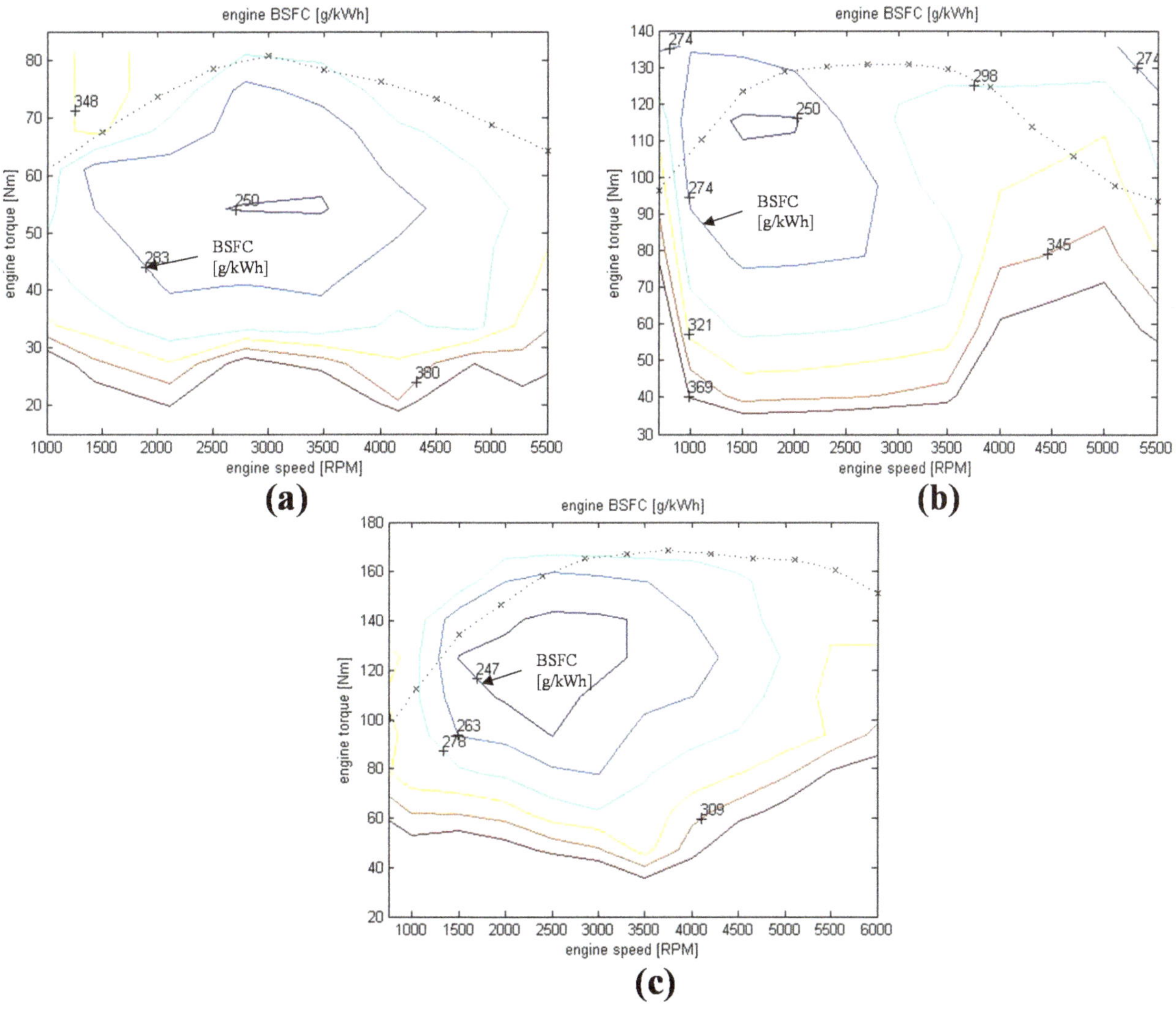

Figure 3. The BSFC contour diagrams for (**a**) 41, (**b**) 63, and (**c**) 95 kW gasoline engines.

2.4. Power Component Model

2.4.1. Electric Motor Model

The electric motors were permanent magnet motors with outputs of 75 and 10 kW, which were used in the series and parallel systems, respectively. This direct current motor model features high torque at low operating speed. The powers required of the motors were calculated from the known required motor torques and motor speeds by using motor efficiency curves (Figure 4). The motor data in Figure 4 were retrieved from the ADVISOR 2003 database [19].

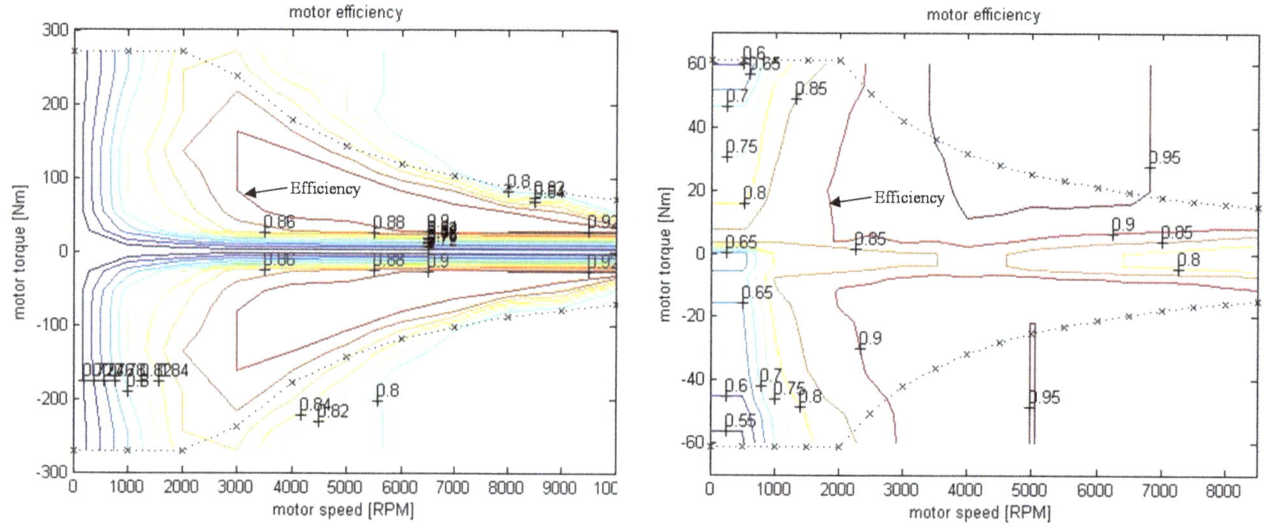

(a) (b)

Figure 4. Efficiency contour diagrams for the (**a**) 75 and (**b**) 10 kW electric motors.

2.4.2. Generator Model

The generator model was used in only the SHEV system. In this model, a 32 kW permanent magnet generator is attached to a 1.0 L gasoline engine. The engine and generator are mainly controlled in a high-efficiency region during operation. The 1.0 L engine has high efficiency region around 3000 to 3500 rpm. The motor efficiency is also high (84% efficiency) in this region. Figure 5 shows the operating point. Therefore, the engine speed and engine torque were maintained at 3500 rpm and 55 Nm, respectively, to obtain the optimal fuel economy. Figure 5 shows the generator efficiency contour diagram. The data is retrieved from the ADVISOR 2003 database [19].

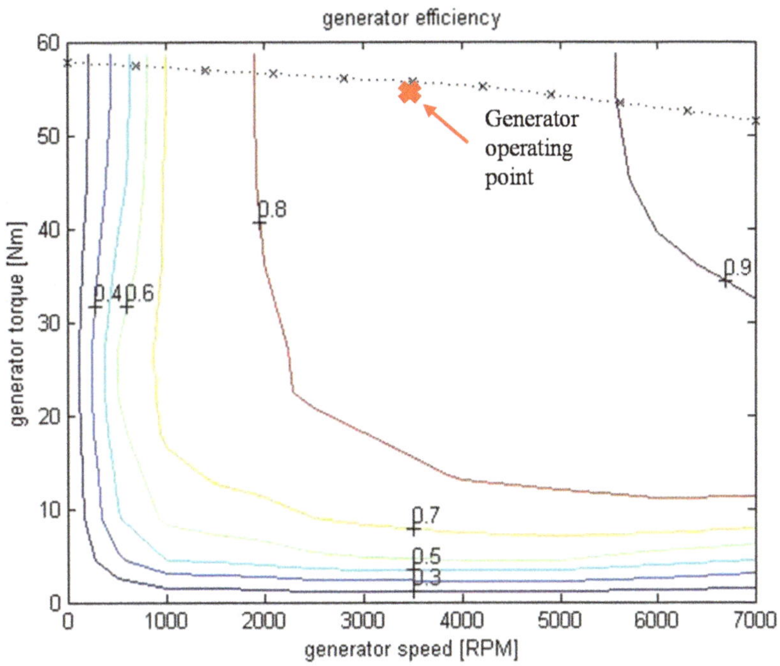

Figure 5. Efficiency contour diagram for the 32-kW generator.

2.4.3. Hydraulic Motor and Hydraulic Pump Models

In hydraulic plunger motor–pumps, plungers reciprocate in limited volumes, vacuuming low-pressure hydraulic oil into the plunger chamber and displacing high-pressure hydraulic oil outside the chamber through plunger compression. During this process, displacements can be regulated from pressure and flow rate variations by adjusting axial (clinoaxis plunger) or swashplate (swashplate plunger) angles, subsequently changing torsional moment and rotational speed relationships. The total efficiency of the hydraulic motor–pump is the product of volumetric and mechanical efficiencies. Operating variables such as pressure difference and volumetric flow rate, hydraulic motor–pump parameters (volumetric displacement), and fluid parameters (fluid viscosity coefficient, density, and bulk modulus) typically influence efficiency. The efficiency of the hydraulic pump and motor varies according to the operating conditions such as pressure and volumetric flow rate [17]. The majority of the pump efficiency map, at a fixed flow rate, is populated with efficiency above 80%. Therefore, an overall efficiency of 80% was used for simulation. The hydraulic motor–pump-related equations are as follows:

The hydraulic motor/pump volumetric fluid flow rate is:

$$Q_{PM} = x_{PM}\omega_{PM}D_{/M}(\eta_{vPM})^{z} \tag{6}$$

The hydraulic motor/pump shaft torque is:

$$T_{PM} = x_{PM}\Delta P_{PM}D_{PM}(\eta_{tPM})^{-z} \tag{7}$$

The hydraulic motor/pump shaft power is:

$$P_{PM} = x_{PM}\Delta P_{PM}D_{PM}\omega_{PM}(\eta_{tPM})^{-z} \tag{8}$$

where Q_{PM} is volumetric fluid flow rate; x_{PM} is the fraction of volume displacement; D_{PM} is the maximum volumetric displacement; ω_{PM} is the shaft angular velocity; ΔP_{PM} is the pressure difference between the hydraulic motor/pump outlet and the inlet; η_{vPM} is volumetric efficiency; η_{tPM} is mechanical efficiency; and Z is mode factor (+1 for pumping mode, −1 for motor mode).

2.5. Energy Storage Component Model

2.5.1. Lithium-Ion Battery Model

Lithium-ion batteries were adopted in this study. The total electric energy varied among PHEVs, SHEVs, and EVs, yielding 1, 2.5, and 25 kWh outputs, respectively. The battery model [20] can be established according to the simple resistor–capacitor circuit shown in Figure 6. The relevant equation is:

$$V_t = V_{oc} - I_{bat} \cdot R_{int} \tag{9}$$

where V_{oc} is the open circuit voltage of battery; R_{int} is the internal resistance of the battery; V_t is the battery terminal voltage; and I_{bat} is the battery output current.

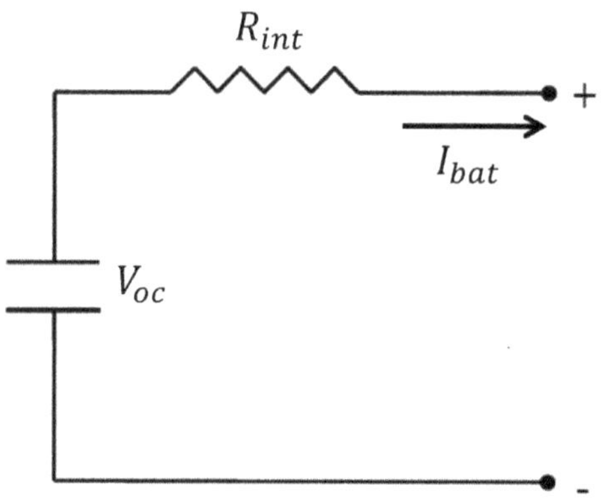

Figure 6. Resistor–capacitor circuit diagram for battery mode.

After measuring terminal voltages and currents, the following equation is used to obtain the battery power outputs (P_{bat}):

$$P_{bat} = I_{bat} \cdot V_{oc} \tag{10}$$

Furthermore, the following equation can be obtained by substituting Equation (9) into Equation (10):

$$I_{bat} = \frac{V_t - ({V_t}^2 - 4 \cdot R_{int} \cdot P_{bat})^{0.5}}{2R_{int}} \tag{11}$$

Typically, battery SOCs are expressed using the battery capacity unit of amp-hour (Ah). Since SOCs vary by charge–discharge current, the following equation is used to obtain battery SOCs:

$$SOC = SOC_{int} - \frac{\int_0^t I_{bat} dt}{Ah} \tag{12}$$

where SOC_{int} is the initial SOC.

2.5.2. Accumulator Model

The energy that the accumulator requires from regenerative braking was calculated using the method proposed in [21]. The equation $E_k = 1/2\ mv^2$ was used to determine the energy of the accumulator.

Assuming regenerative braking commences at v = 60 km/h, 209 kJ of energy can be recycled during each regenerative braking event. Therefore, accumulator volume was set to 18 L, and working pressure was set to 172 to 344 bar. The accumulator model can be established according to the polytropic process of the laws of thermodynamics. The variation process of the gaseous state must be considered to investigate the gaseous pressures and volumes because the accumulator is operated frequently in the HHV. Therefore, the gaseous state changes were considered to be an adiabatic process (rapid changes; n = 1.4) in this study. Temperature variations were not considered. During actual gaseous expansion and compression, the pressure and volume relationship is:

$$PV^n = C \tag{13}$$

$$P_0{V_0}^n = P_1{V_1}^n = P_2{V_2}^n = C \tag{14}$$

Moving boundary work:

$$W_b = \int_1^2 PdV = P_1V_1 \ln\frac{P_1}{P_2} \tag{15}$$

Oil displacements:

$$V_f = V_1' - V_2' = (V_0 - V_1) - (V_0 - V_2) = V_2 - V_1 \tag{16}$$

$$= P_0^{\frac{1}{n}}V_0\left\{\left(\frac{1}{P_2}\right)^{\frac{1}{n}} - \left(\frac{1}{P_1}\right)^{\frac{1}{n}}\right\} \tag{17}$$

$$= \left(\frac{P_0}{P_1}\right)^{\frac{1}{n}}V_0\left\{\left(\frac{P_1}{P_2}\right)^{\frac{1}{n}} - 1\right\} \tag{18}$$

where P_0 is the initial enclosed gas pressure of accumulator; P_1 is the maximum actuation pressure; P_2 is the minimum actuation pressure; V_0 is the accumulator volume; V_1 is the volume with pressure P_1; an V_2 is the volume with pressure P_2; and V_f is the volume of discharge oil from the accumulator.

Accumulator SOCs are typically expressed volumetrically because SOCs vary with volumetric flow rates. Therefore, the following equation can be used to obtain the accumulator SOC:

$$SOC = SOC_{int} - \frac{\int_0^t Qdt}{V_f} \tag{19}$$

2.6. Manual Transmission Vehicle

Figure 7 shows the conventional powertrain system model which comprised the subsystems of driving cycle, vehicle dynamic, transmission, and engine models. In the simulations, gearshifts were operated according to vehicle speeds and were assumed to be smooth and free from clutch slippage.

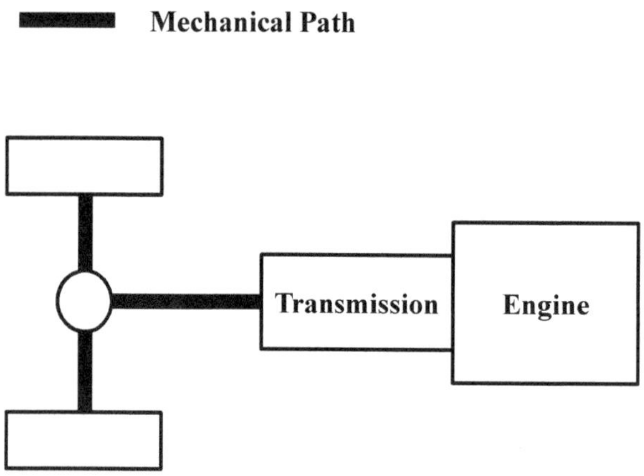

Figure 7. Structural diagram of the manual transmission vehicle.

2.7. Series Hybrid Electric Vehicle

Figure 8 shows the SHEV powertrain system which comprised the driving cycle, vehicle dynamic, transmission system, power components (electric motor and generator), energy storage component (lithium-ion batteries), and engine model subsystems. In the simulations, the ON–OFF states of the engine were used to maintain the state-of-charge (SOC) of the lithium-ion batteries within a range of 0.35–0.65. The SOC range is designed to maintain the system in a high efficiency state, considering the combined efficiencies of both charging and discharging. The charging efficiency is high when SOC is low whereas the discharging efficiency is high when SOC is high [22]. This study chose the middle 30% of SOC, 0.35 to 0.65. The control procedures involved inspecting the engine status, determining the SOC, and determining whether the engine must be switched on to charge the lithium-ion batteries.

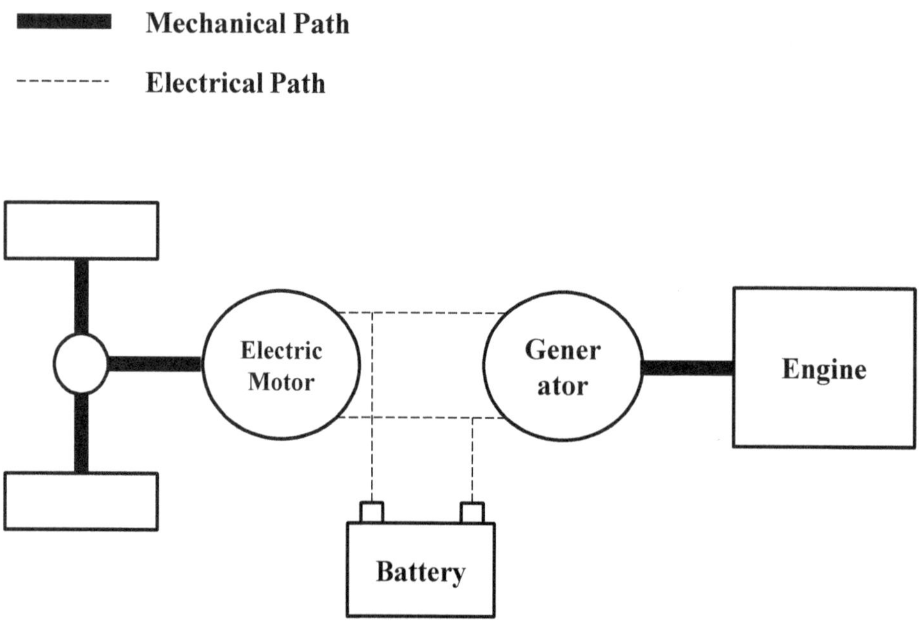

Figure 8. Structural diagram of the SHEV.

2.8. Series Hydraulic Hybrid Vehicle

Figure 9 shows the SHHV system model which comprised the driving cycle, vehicle dynamic, transmission system, power component (hydraulic motor and hydraulic pump), energy storage component (accumulator), and engine model subsystems. Regarding simulation controls, the ON–OFF states of the engine were used to control the SOC of the accumulator within the range of 0–1. The detailed control procedures were identical to those of the SHEV. The primary difference was that the accumulator had a larger SOC range than that of the lithium-ion batteries.

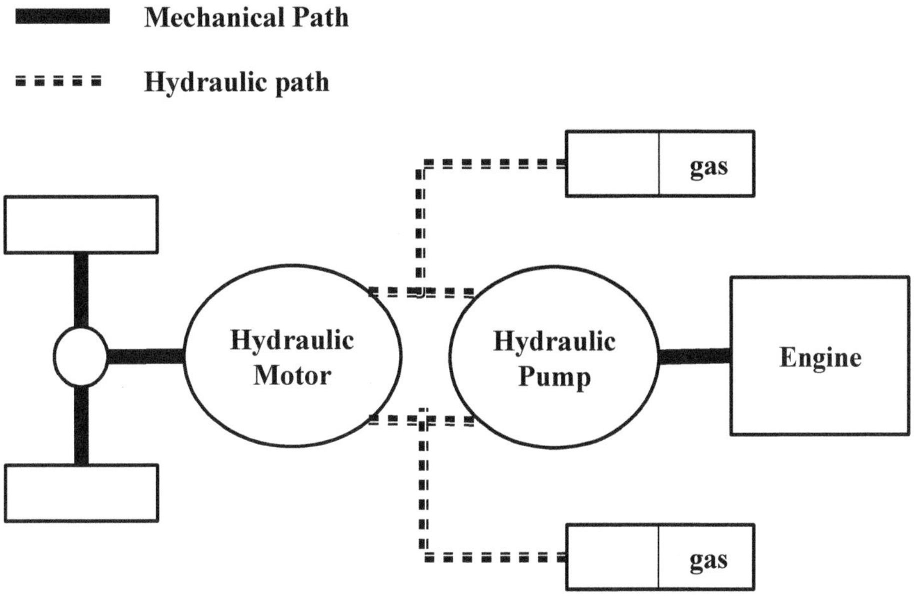

Figure 9. Structural diagram of the SHHV.

2.9. Parallel Hybrid Electric Vehicle

Figure 10 shows the PHEV system model which comprised the driving cycle, vehicle dynamics, transmission system, power component (electric motor), energy storage component (lithium-ion batteries), and engine model subsystems. Additionally, the simulation controls comprised pure electric (Mode A), pure engine (Mode B), hybrid (Mode C), engine charging (Mode D), and regenerative braking (Mode E) modes. In simulation, engine and electric motor torques were distributed according to specific road conditions (deciding among Modes A, B, and C) first. Then, the engine statuses were inspected to decide between the upper and lower SOC thresholds. The system proceeds to (engine) charging mode when battery SOC is below the set threshold; otherwise, the system proceeds according to the operating mode selected in the first step. The SOC operating range is maintained between 0.35 and 0.65.

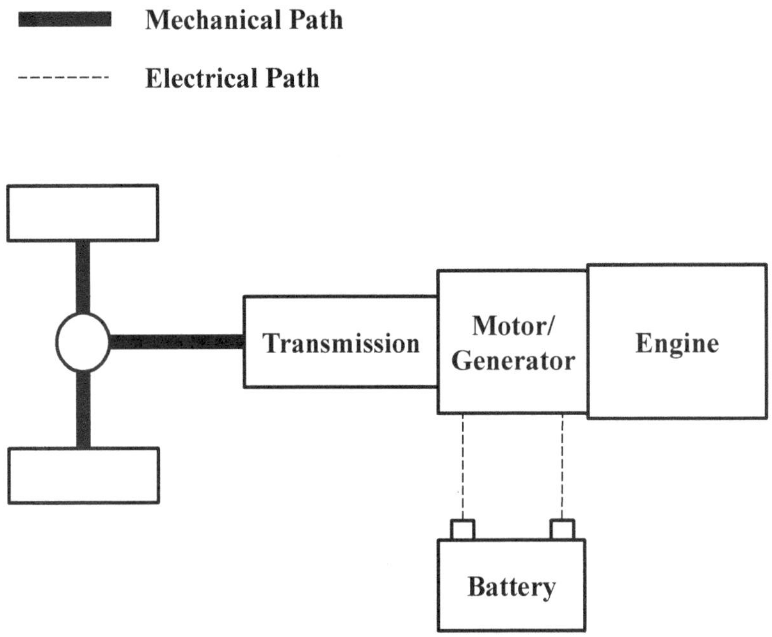

Figure 10. Structural diagram of the PHEV.

2.10. Parallel Hydraulic Hybrid Vehicle

Figure 11 shows the PHHV system which comprised the driving cycle, vehicle dynamic, transmission system, power component (hydraulic motor), energy storage component (accumulator), and engine model subsystems. The start–stop method was applied to improve fuel consumption during vehicle starting and stopping [23]. The hydraulic motor and accumulator generated power for acceleration during vehicle startup. The vehicle was subsequently propelled by the engine as the accumulator depleted, and the engine was the only power source before the vehicle braking. The engine would be shut down when brake was applied. By preventing the engine from operating in an inefficient zone, this control strategy improves fuel economy. During vehicle braking, the hydraulic motor acts as a hydraulic pump that converts kinetic energy regenerated during braking into hydraulic energy and then recycles this energy to the accumulator for the next vehicle startup process. First, the state of vehicle acceleration was determined. When acceleration exceeds 0, the system proceeds to the second step to determine the SOC of the accumulator and to decide whether to operate with the hydraulic motor or in an engine mode. When acceleration equaled to 0, the system operates on engine mode, and when acceleration was less than 0, the system operates on regenerative braking mode, Figure 12.

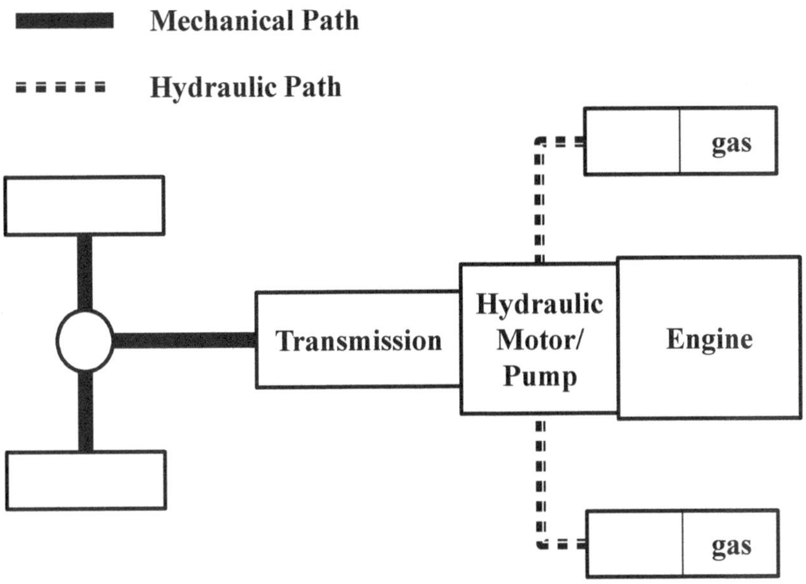

Figure 11. PHHV structural diagram.

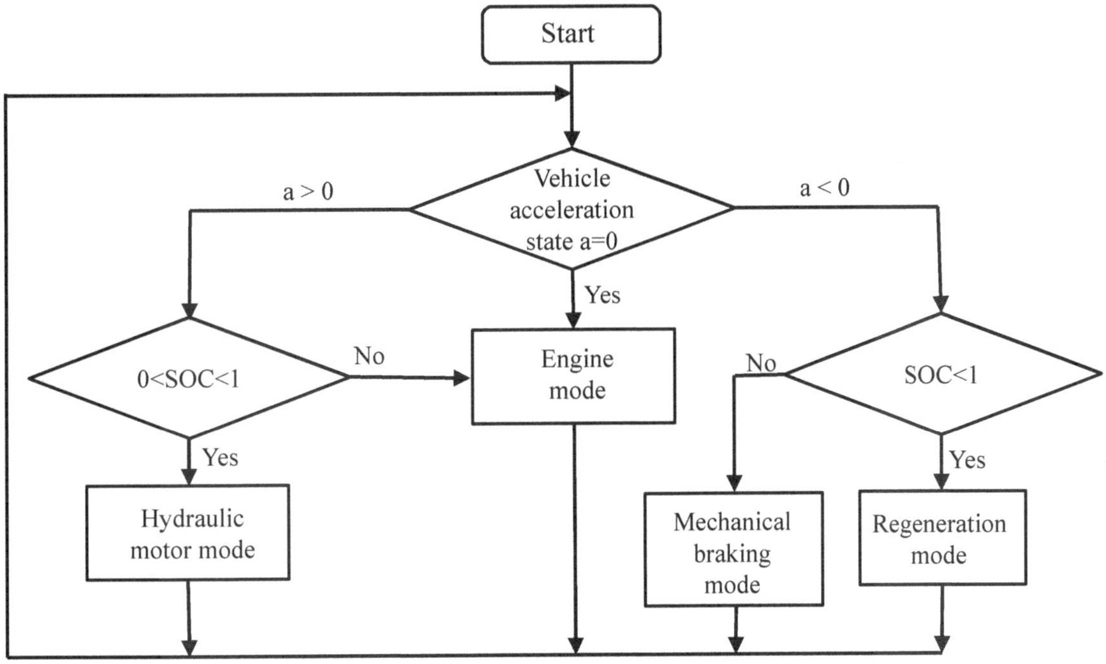

Figure 12. PHHV control flow chart.

2.11. Electric Vehicle

Figure 13 shows EV system which comprised the driving cycle, vehicle dynamics, transmission system, power component model (electric motor), and energy storage component (lithium-ion batteries) model subsystems. The simulations placed restrictions on the maximum motor speed, maximum motor torque, and maximum power of the electric motor. Additionally, the maximum and minimum voltages were configured for the batteries to ensure normal operation.

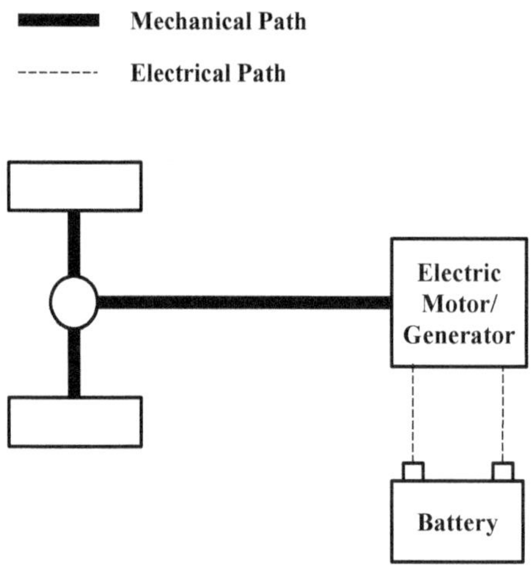

Figure 13. Structural diagram of the EV.

2.12. Hydraulic–Electric Hybrid Vehicle

HEHVs are primarily powered by electrical systems (electric motor and lithium-ion battery submodels) and assisted by hydraulic systems (hydraulic motor–pump and accumulator submodels). These systems reduce electrical energy consumption and extend the driving range (Figure 14).

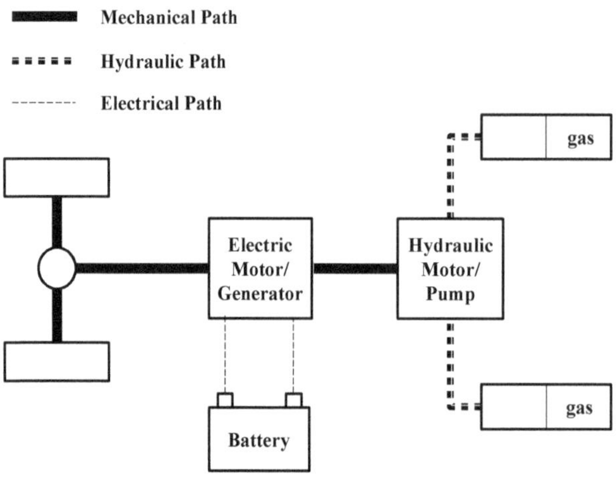

Figure 14. Structural diagram of the HEHV.

In simulations, the controls adopted the start–stop method identical to that applied in the PHHV system. In this study, the HEHV architecture was derived directly from that of the PHHV except the engine was replaced with an electric motor and battery. Therefore, the HEHV used the same control strategy as PHHV did, and the braking energy recycling was done by hydraulic motor instead of electric motor. The hydraulic motor is generally more efficient than the electric motor. In the HEHV, a hydraulic system is mainly used to recycle braking energy.

3. Simulation and Analysis

This section discusses the simulation results for the seven vehicle systems. Energy consumption was analyzed by simulating the NEDCs, and the performance of each component was compared. The required vehicle driving forces and powers for the NEDC were calculated first. Then, according to the performance analyzed for the manual transmission (MT) vehicles, the SHHV, PHHV, SHEV, PHEV, EV, and HEHV performance results were evaluated. Finally, the fuel economy of the SHHV, SHEV, PHHV, PHEV, EV, and HEHV systems was obtained.

3.1. Driving Force and Power Required for the Driving Cycles

Regardless of vehicle type, specifications must be consistent in subsequent performance comparisons. Table 2 shows the common vehicle specifications. For hybrid vehicles, the fuel economy is less sensitive to its vehicle mass since part of the kinematic energy used to accelerate the vehicle can be recovered by regenerative braking. Moreover, the study focused on the fuel efficiency of different powertrain configurations. To minimize the effects of different component masses, this study tentatively disregarded the mass differences of the vehicles in order to compare the actual functional contribution of different powertrain configurations. During simulations, the required wheel driving force and power must be calculated using the driving cycle model (Figure 15). For example, the first, second, and third diagrams show the vehicle speeds, wheel driving force, and work required for the vehicles. The diagrams indicate that the maximal vehicle speed, wheel driving force, and power were 120 km/h, 1787 N, and 35 kW, respectively. Therefore, the power from the engine and electric motor in the series, parallel, and conventional vehicle systems should satisfy these conditions.

Table 2. Common vehicle specifications.

Parameter	Symbol	Value	Unit
Vehicle mass	m	1500	kg
Frontal projected area	A_f	2.26	m^2
Air resistance coefficient	C_D	0.28	—
Rolling resistance coefficient	μ_r	0.008	—
Wheel radius	r	0.315	m
Air density	ρ_{air}	1.225	kg/m^3
Fuel density	ρ_{fuel}	0.74	g/cm^3
Gravitational acceleration	g	9.81	m/s^2

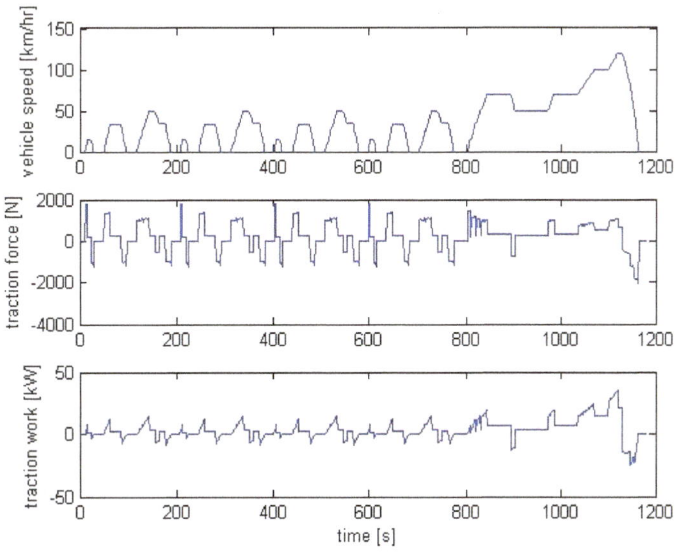

Figure 15. Vehicle speed, wheel driving force, and power.

3.2. Performance Analysis of Conventional Manual Transmission Vehicles

The performance of the MT vehicle was analyzed to facilitate comparison with the HEV and HHV. A model was established according to the specifications of a Toyota Camry with manual transmission and 1.8L gasoline engine. Figure 16 shows the engine operating conditions of an MT vehicle during NEDC. This figure presents, in descending order, the vehicle speed, engine speed, engine torsional moment, engine power, and gearshift conditions. These performance comparison results show that the NEDC reached a constant cruising speed after three stages of acceleration. To overcome these acceleration (including air and rolling) resistances, the engines must generate 8.6, 13.8, and 16.6 kW of power. Additionally, the maximal vehicle speed, engine speed, engine torque, and power during the entire driving cycle were 120 km/h (sixth gear), 2700 rpm, 135 Nm, and 37.5 kW, respectively. Figure 17 shows the engine operating conditions of the MT vehicle during the NEDC, which indicates that optimal fuel consumption (within 247 g/kWh) was obtained at high operating speed. During intermediate and low speed operations, fuel consumption ranged from 247 to 278 g/kWh and from 278 to 323 g/kWh, respectively. To evaluate the accuracy of simulation models, the Camry 2.0E fuel economy data was applied for the comparison between labeled fuel economy data and the MT simulation data. The fuel economies are presented in Table 3. During UDC and EUDC, the fuel economies were 8.88 and 6.59 L/100km, respectively, and average fuel consumption was 7.42 L/100km. The fuel economy of the MT vehicle during the UDC was 23.18% higher than that of Toyota Camry 2.0E. During the EUDC, the fuel economies of these two types of vehicle differed by 1.05%. Regarding average fuel consumption, the two vehicles differed by 12.08% in fuel economy, because the UDC varied more drastically within each driving cycle than EUDC did.

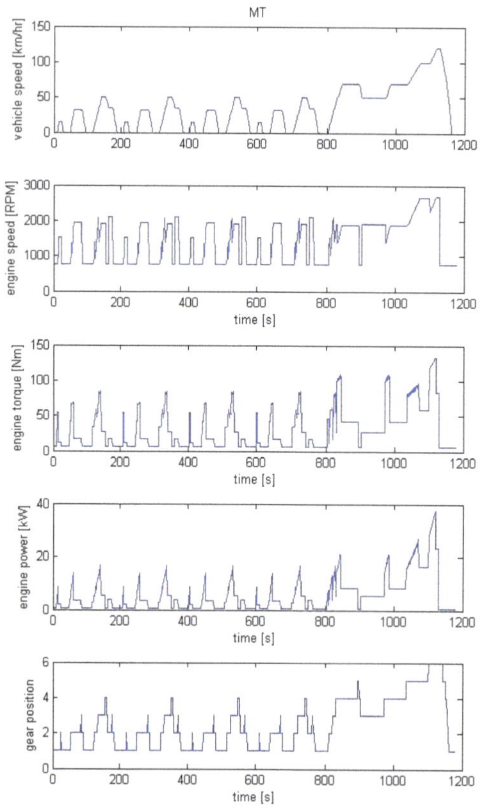

Figure 16. Engine operating conditions of the MT vehicle.

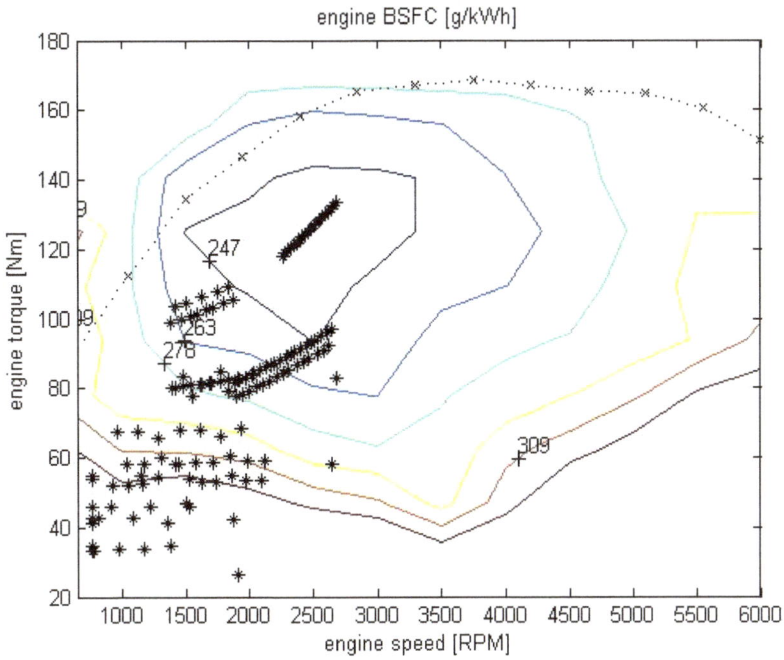

Figure 17. Engine operating points of the MT vehicle in the NEDC.

Table 3. Fuel economies of the Toyota Camry 2.0E and an MT vehicle.

Vehicle	Urban Fuel Consumption (L/100 km)	Extra-Urban Fuel Consumption (L/100 km)	Overall Fuel Consumption (L/100 km)
2014 Toyota Camry 2.0E	11.56	6.66	8.44
MT	8.88	6.59	7.42
Difference (%)	−23.18	−1.05	−12.08

3.3. Performance Comparison between Series Hybrid Electric and Hydraulic Hybrid Vehicles

Figure 18a shows the performance diagrams for each component of the SHEV in the NEDC. The figure shows, in descending order, the vehicle speed, electric motor power, lithium-ion battery current, generator power, and lithium-ion battery SOC conditions. The data indicate that the engine started at 524 s and charged the battery until 733 s. This charging process was repeated at 1082 s, at which time the SOC was below 0.35. The engine started twice during this NEDC, when discharging-charging conditions (positive-negative variations) of the lithium-ion batteries were observed. Regarding the HHV, Figure 18b illustrates the performance diagrams for each component of the SHHV. The figure shows (from top to bottom) the vehicle speed, hydraulic motor power, accumulator pressure volumetric flow rate, pump power, and accumulator SOC. These simulations results revealed that the engine started ten times, which was eight more starts than in the SHEV, because the storage energy density of the accumulator was lower than that when lithium-ion batteries are used.

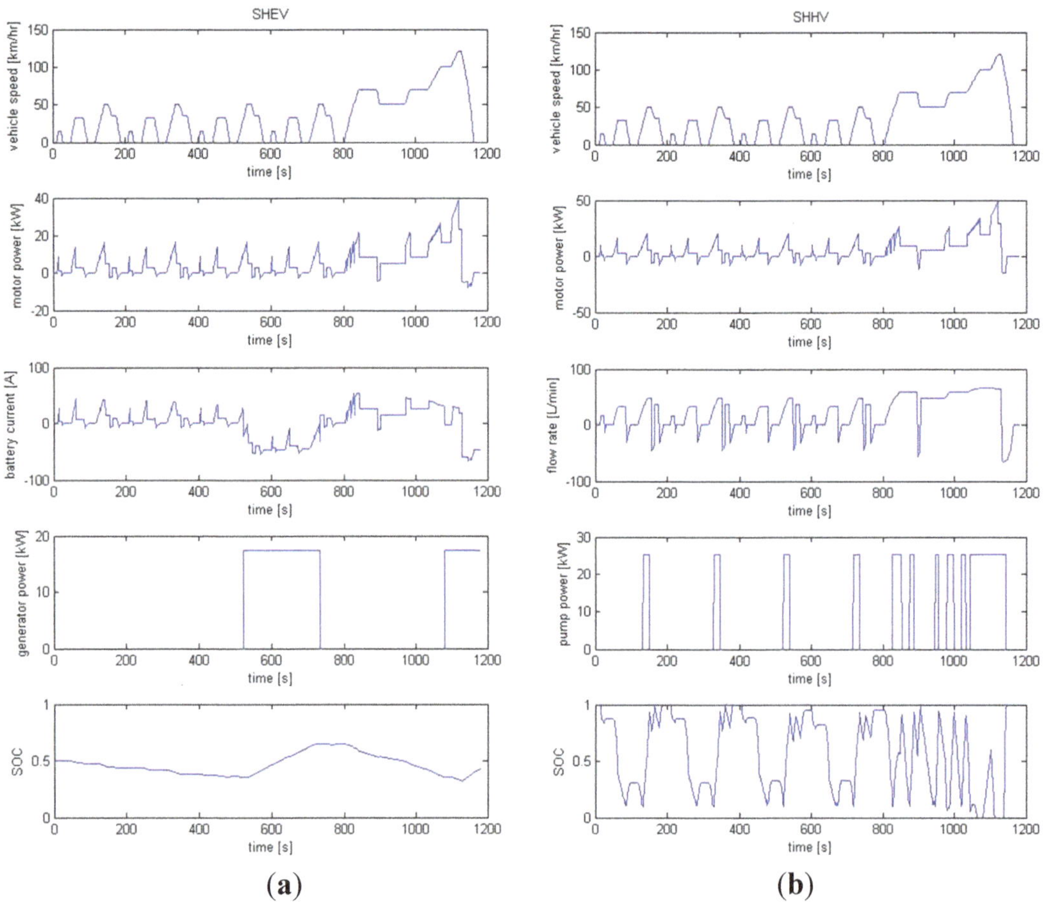

Figure 18. Performance diagrams for each component in the (**a**) SHEV and (**b**) SHHV.

Regarding regenerative braking, the hydraulic motor (pump mode) could absorb high-power vehicle kinetic energy; thus, noticeable high power generation was observed during each vehicle deceleration (braking) session. Table 4 shows that, during UDC, fuel efficiency was 33.03% higher in the SHHV than in the SHEV.

Table 4. Fuel economies of the SHEV and SHHV.

Vehicle Configuration	Urban Fuel Consumption (L/100 km)	Extra-Urban Fuel Consumption (L/100 km)	Average Fuel Consumption (L/100 km)
SHEV	5.62	6.66	6.28
SHHV	3.77	5.57	4.91
Difference (%)	−33.03	−16.37	−21.80

3.4. Performance Comparison between Parallel Hybrid Electric and Hydraulic Hybrid Vehicles

Figure 19a shows the performance diagrams for each component of the PHEV in the NEDC. The figure presents in descending order, the vehicle speed, electric motor power, engine power, and lithium-ion battery SOC order. The second and third diagrams show the coordination between the electric motor and engine. Figure 19b illustrates the performance diagrams for each component of the PHHV in the NEDC. In descending order, the figures show vehicle speed, hydraulic motor power, engine power, and accumulator SOC. These simulation results indicate that the hydraulic motor generated a substantially higher power compared with the electric motor during startup. Therefore, the control strategy used the hydraulic motor for startup and low-speed conditions and hydraulic pumps for energy recycling during deceleration (braking) sessions. Figure 20 illustrates the power distributions of engine and electric motor in pure electric motor, pure engine, engine charging, hybrid, and regenerative braking modes of the PHEV in the NEDC. In the default strategy, a torque of less than 40 Nm indicates a pure electric motor mode (Mode A), which can prevent engine operation in a region of low torque and low fuel efficiency. Between 40 and 85 Nm, the system operates in a pure engine mode (Mode B). Hybrid mode (Mode C) commences when the required torque exceeds 85 Nm, which is the region of optimal engine efficiency. During the engine charging mode (Mode D), the vehicle continues to operate and charge the batteries. Moreover, the engine can operate in the region of low fuel consumption. Figure 21 shows the engine operating point of the hydraulic assist system, which indicates that the engine generally operates in the high efficiency region. Finally, Table 5 shows the fuel economies of PHEV and PHHV. In the UDC, PHHV exhibited 6.32% higher fuel conservation compared with the PHEV. This difference occurred mainly because the control strategy for the PHHV prefers low-speed, startup, and regenerative braking conditions. In the EUDC, a reduced number of regenerative braking reflected 9.26% worse fuel economy in the PHHV compared with the PHEV.

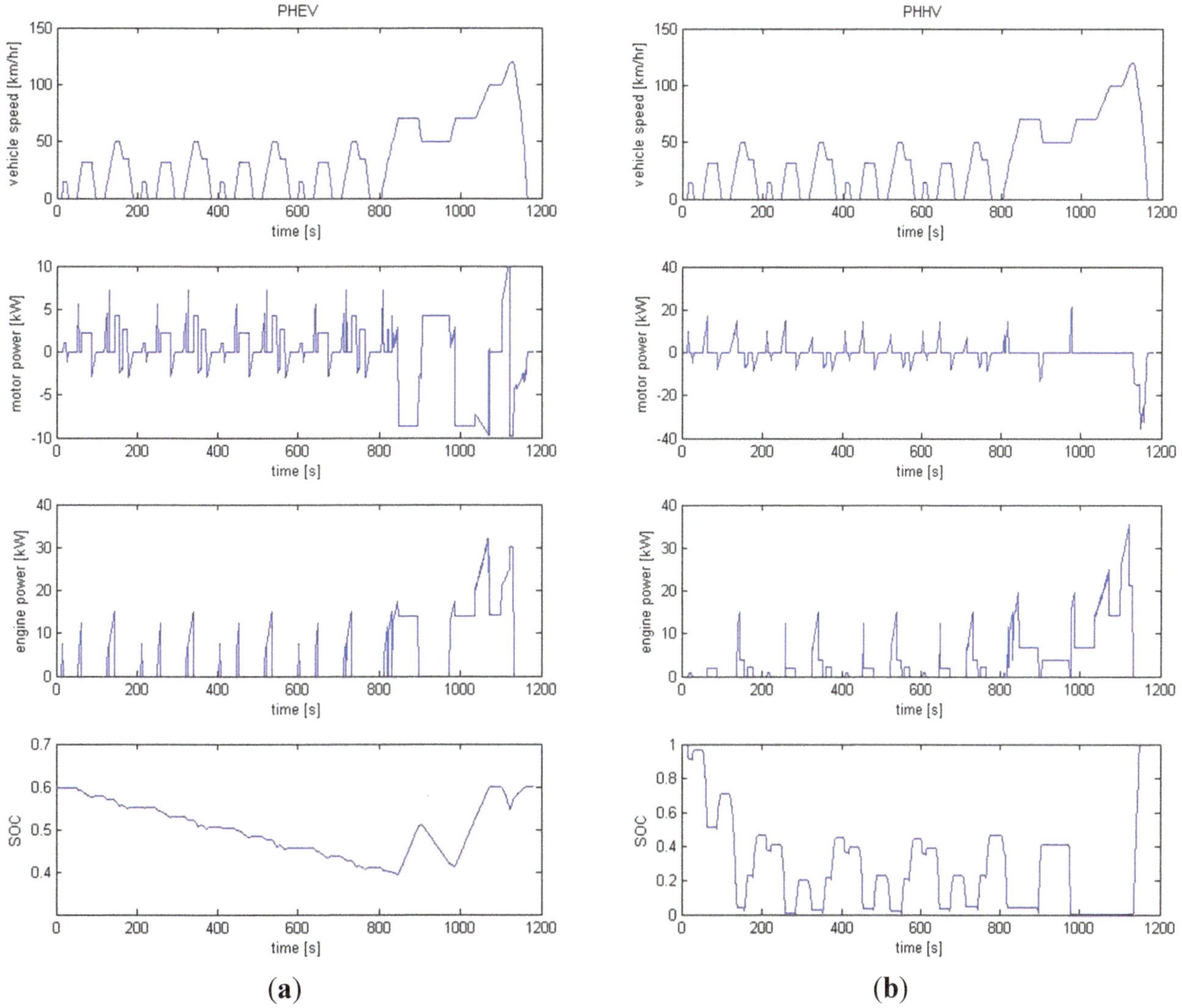

Figure 19. Performance diagrams for each component in the (**a**) PHEV and (**b**) PHHV.

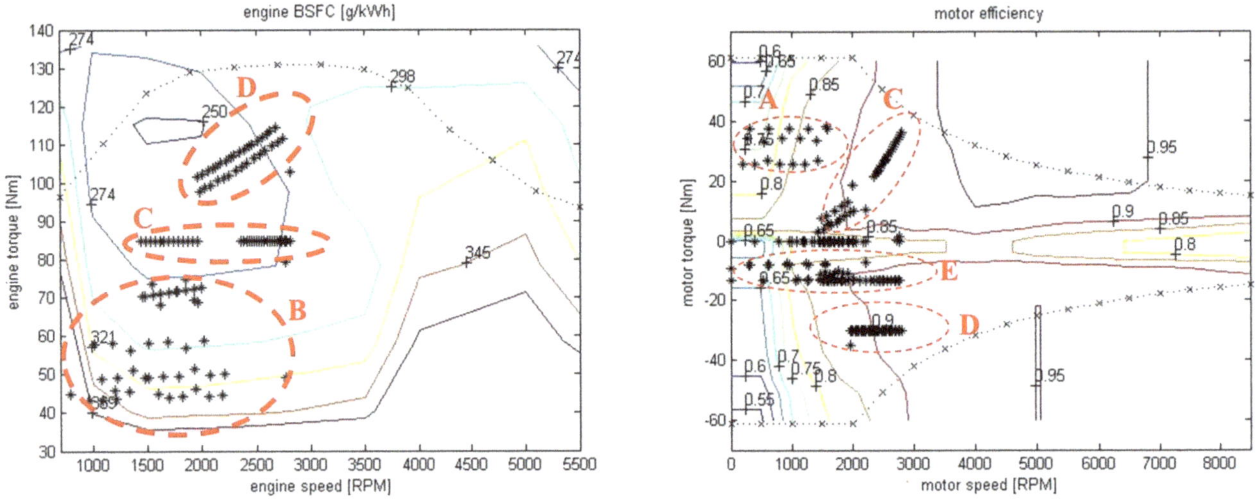

(a) (b)

Figure 20. PHEV operating points of the (**a**) engine and (**b**) electric motor.

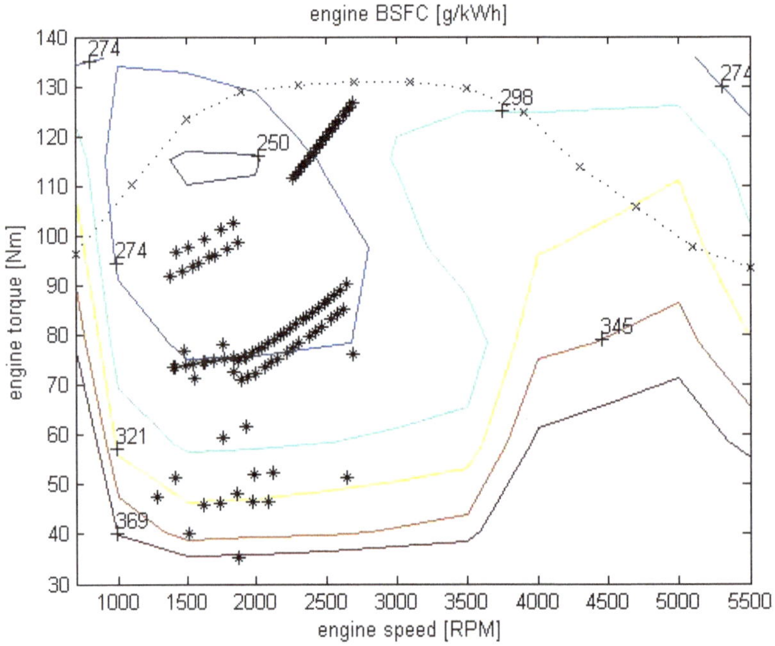

Figure 21. PHHV engine operating points.

Table 5. Fuel economies of the PHEV and PHHV.

Vehicle Configuration	Urban Fuel Consumption (L/100 km)	Extra-Urban Fuel Consumption (L/100 km)	Average Fuel Consumption (L/100 km)
PHEV	4.41	4.68	4.58
PHHV	4.13	5.12	4.76
Difference (%)	−6.32	9.26	3.80

3.5. Performance Comparison between Electric and Hydraulic–Electric Hybrid Vehicles

Figure 22a shows the performance diagrams for each component of the EV in the NEDC. In descending order, the figure presents the vehicle speed, electric motor torque, electric current, power, and lithium-ion battery SOC. Figure 22b illustrates the performance diagrams for each component of the HEHV in the NEDC. In descending order, the figure shows the vehicle speed, electric motor power, hydraulic motor power, lithium-ion battery SOC, and accumulator SOC. According to the electric motor operating points shown in Figure 23, the HEHV was driven by the hydraulic assist system in the low-speed region (electric motor in idling condition); therefore, the electric motor can be replaced by the hydraulic motor to power the vehicle during low speeds. In the high-speed region, the hydraulic assist system disengages. During regenerative braking, kinetic energy is recycled by using the hydraulic assist system (pump mode) instead of the electric motor. Table 6 indicates that, in the UDC, the HEHV consumed less energy than the EV did mainly because the hydraulic assist system prevented the electric motor from operating in the low-efficiency region (72% efficiency) at low speeds. Therefore, electricity consumption was decreased by 23.38%. In the EUDC, the energy consumption of the HEHV and EV did not substantially differ because the hydraulic assist system primarily operated during startup and stopping. Therefore, the hydraulic assist system was less frequently used in the EUDC than in the UDC, consequently conserving only 4.34% of electricity. Regarding total energy consumption, the HEHV provided 11.4% more electricity conservation compared with the EV.

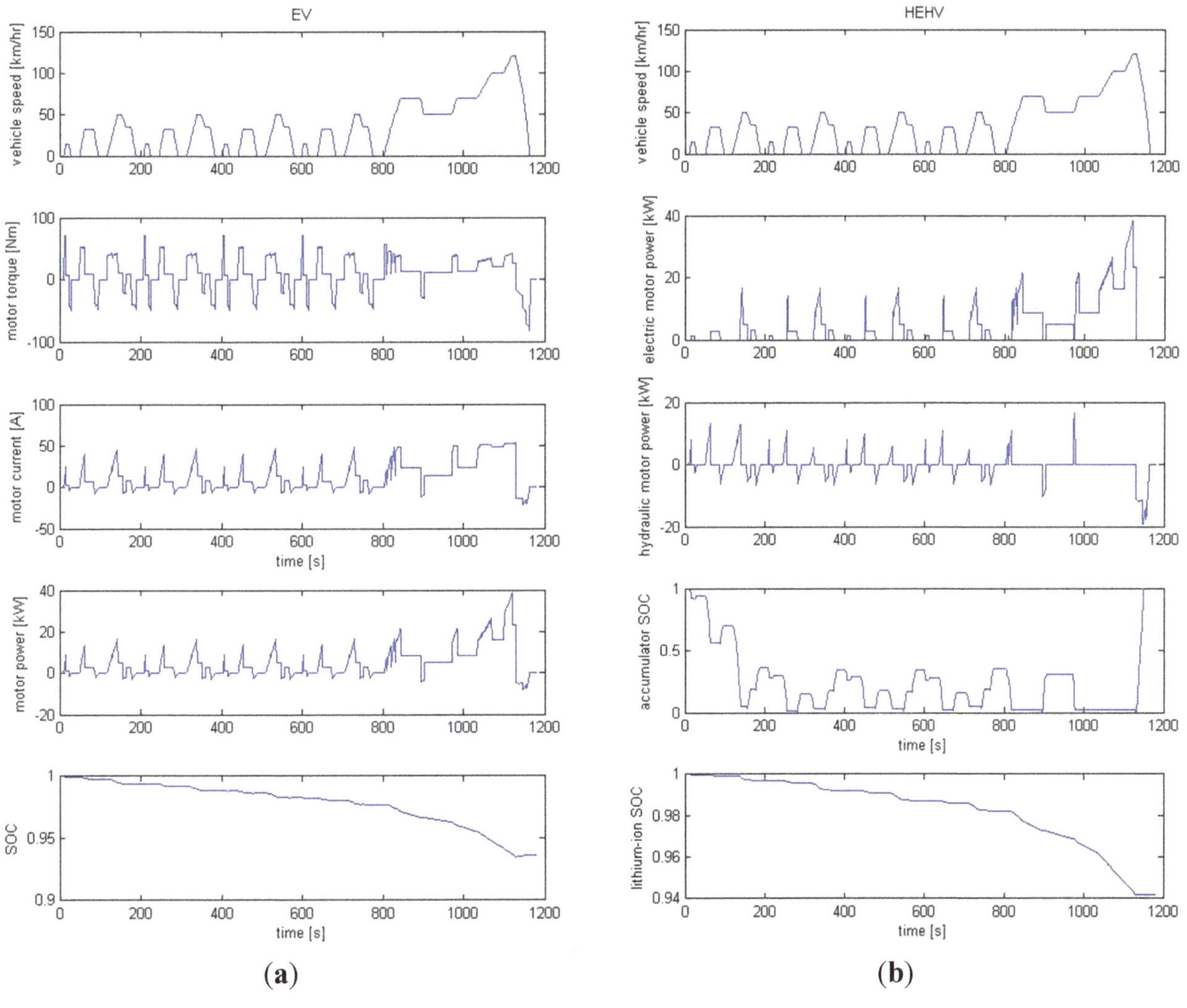

Figure 22. Performance diagrams for each component in the (**a**) EV and (**b**) HEHV.

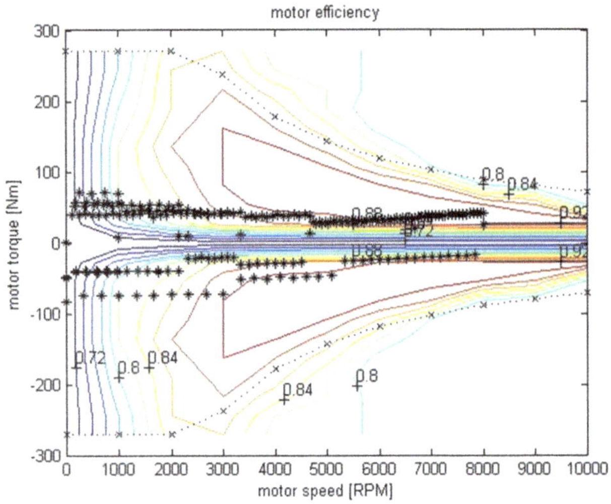

(a)

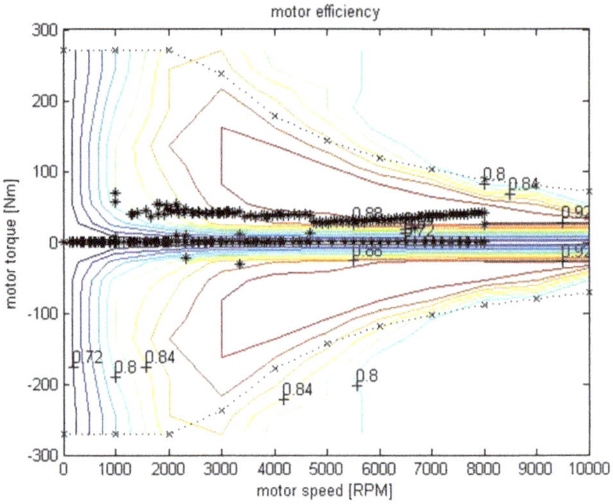

(b)

Figure 23. Electric motor operating points of the (**a**) EV and (**b**) HEHV.

Table 6. Energy consumption comparisons between the EV and HEHV.

Vehicle Configuration	Urban Energy Consumption (MJ)	Extra-Urban Energy Consumption (MJ)	Total Energy Consumption (MJ)
EV	2.079	3.528	5.607
HEHV	1.593	3.375	4.968
Difference (%)	−23.38	−4.34	−11.40

Operating costs were converted to new Taiwan dollars per km (NT$/km) for a consistent comparison of all seven configurations of vehicles. According to the assumptions that the oil and electricity prices are NT$35/L and NT$5/kWh in Taiwan, Table 7 was obtained. The energy consumed per km was also included in Table 7.

Table 7. Energy costs per kilometer for the seven vehicle systems.

Items	Urban		Extra-Urban		Average	
Metric	**NT$/km**	**kJ/km**	**NT$/km**	**kJ/km**	**NT$/km**	**kJ/km**
MT vehicle	3.11	3073	2.31	2279	2.60	2568
SHEV	1.97	1946	2.33	2304	2.20	2174
SHHV	1.32	1303	1.95	1927	1.72	1700
PHEV	1.54	1526	1.64	1620	1.60	1586
PHHV	1.45	1430	1.79	1770	1.67	1646
EV	0.726	523	0.705	507	0.712	513
HEHV	0.556	401	0.674	485	0.631	455

3.6. Comparison of the Mass Effect on Energy Consumption

In all of the above simulations, vehicle mass was set to 1500 kg in order to focus on the energy efficiency of different powertrain systems. In this section, a 100 kg mass increment is added to the six derived powertrain systems to simulate the extra add-on masses from the energy storage medium, e.g., batteries and accumulators, and any energy transformation devices, e.g., motor/generators and pump/motors. Table 8 summarizes the analytical results for the effect of mass on energy consumption, which indicate that, with extra mass increment, HEHV consume relatively less energy.

Table 8. Mass effect on the per kilometer energy consumption.

kJ/km	m = 1500 kg	m = 1600 kg	Diff.
SHEV	2174	2244	70
SHHV	1700	1793	93
PHEV	1586	1630	44
PHHV	1646	1707	61
EV	513	532	19
HEHV	455	469	15

4. Conclusions

To effectively design environmentally friendly, energy efficient and cost beneficial vehicles, this research compares the energy consumption based on New European Driving Cycle using backward simulation in MATLAB/Simulink through various power component models, connection with energy storage component models, and combinations of series or parallel configurations. The study focuses on the energy efficiency comparison among different powertrain configurations. The setting for vehicle mass was similar in all simulation models to minimize the effects of different component masses and to clarify the actual functional contribution of each powertrain system. The operating efficiencies for each component were comprehensively compared. The fuel economy, power component, and energy storage component SOC of the HEV and HHV were obtained through the NEDC simulation. The conclusions of this study are as follows:

(1) In the MT vehicle performance analysis, UDC and EUDC had fuel consumptions of 8.88 and 6.59 L/100 km (average, 7.42 L/100 km), respectively.

(2) In the performance comparison between the SHEV and SHHV, SHHV conserved 33.03% more fuel than did the SHEV in the UDC. The improved efficiency is obtained by the hydraulic motor (pump mode), which has a higher capacity for absorbing the vehicle kinetic energy compared with the electric motor. Therefore, high power outputs by the hydraulic pump were clearly observed during each vehicle deceleration (braking) session.

(3) In the performance comparison between the PHEV and PHHV, the PHHV conserved 6.32% more fuel compared with the PHEV. This is because the PHHV control strategy focused on low-speed, startup, and regenerative braking. In the EUDC, the PHHV had a 9.26% worse fuel economy compared with the PHEV because of the reduced number of regenerative braking sessions.

(4) The performance comparison of the EV and HEHV revealed that the HEHV exhibited 23.38% higher electricity conservation compared with the EV in the UDC primarily because the hydraulic assist system prevented the electric motor from operating in the low-efficiency region (72%) during low-speed operations. In the EUDC, the HEHV conserved only 4.34% electricity. Regarding total energy consumption, the HEHV conserved 11.4% more electricity compared with the EV.

(5) The experimental results revealed that, regardless of whether the UDC or EUDC was used, the HEHV had the best energy-conservation performance (average, NT$0.631/km). Table 9 shows that, for the MT vehicle used in this study, the HEHV presented a 75.7% improvement in economic savings.

Table 9. Percentage improvement in economic savings.

Configuration	Percentage Improvement (%)
MT vehicle used in this study	— (Baseline)
SHEV	15.4
SHHV	33.8
PHEV	38.2
PHHV	35.9
EV	72.6
HEHV	75.7

Nomenclature

C_D	aerodynamic dragging coefficient
D_{PM}	maximum volumetric displacement
EV	electric vehicle
F_t	tractive effort
HEHV	hydraulic-electric hybrid vehicle
I_{bat}	battery output current
PHEV	parallel hybrid electric vehicle
PHHV	parallel hydraulic hybrid vehicle
P_0	initial enclosed gas pressure of accumulator
P_1	maximum actuation pressure
P_2	minimum actuation pressure
Q_{PM}	volumetric fluid flow rate
R_a	aerodynamic dragging force
R_c	grade resistance
R_{int}	internal resistance of the battery
R_r	rolling resistance
R_s	inertia force
SHEV	series hybrid electric vehicle
SHHV	series hydraulic hybrid vehicle
V_0	accumulator volume
V_1	volume with pressure P_1
V_2	volume with pressure P_2
V_f	volume of discharge oil from the accumulator
V_{oc}	open circuit voltage of battery
V_t	battery terminal voltage
W	vehicle weight

W_e	equilibrium weight of rotational part,
Z	mode factor (+1 for pumping mode, −1 for motor mode)
a	vehicle acceleration
m	vehicle mass
x_{PM}	percentage of volume displacement
v	vehicle speed
v_w	wind speed
η_{vPM}	volumetric efficiency
η_{tPM}	mechanical efficiency
μ_r	rolling resistance coefficient
ω_{PM}	haft angular velocity
ΔP_{PM}	pressure difference between the hydraulic motor/pump outlet and the inlet

Acknowledgments

The author would like to thank the Ministry of Science and Technology of the Republic of China for financially supporting this research under Contract No. MOST 103-2221-E-027-024.

Conflicts of Interest

The author declares no conflict of interest.

References

1. Doucette, R.T.; McCulloch, M.D. Modeling the prospects of plug-in hybrid electric vehicles to reduce CO_2 emissions. *Appl. Energy* **2011**, *88*, 2315–2323.
2. Chau, K.T.; Wong, Y.S. Overview of power management in hybrid electric vehicles. *Energy Convers. Manag.* **2002**, *43*, 1953–1968.
3. Wu, B.; Lin, C.C.; Filipi, Z.; Peng, H.; Assanis, D. Optimal power management for a hydraulic hybrid delivery truck. *Veh. Syst. Dyn.* **2004**, *42*, 23–40.
4. Wang, L.; Collins, E.G., Jr.; Li, H. Optimal design and real-time control for energy management in electric vehicles. *IEEE Trans. Veh. Technol.* **2011**, *60*, 1419–1429.
5. Syed, F.U.; Kuang, M.L.; Czubay, J.; Ying, H. Derivation and experimental validation of a power-split hybrid electric vehicle model. *IEEE Trans. Veh. Technol.* **2006**, *55*, 1731–1747.
6. Gao, Y.; Ehsani, M. A torque and speed coupling hybrid drivetrain—Architecture, control, and simulation. *IEEE Trans. Power Electron.* **2006**, *21*, 741–748.
7. Lin, C.C.; Filipi, Z.; Wang, Y.; Louca, L.; Peng, H.; Assanis, D.; Stein, J. Integrated feed-forward hybrid electric vehicle simulation in SIMULINK and its use for power management studies. *SAE Tech. Paper* **2001**, doi:10.4271/2001-01-1334.
8. Lin, C.C.; Peng, H.; Grizzle, J.W.; Liu, J.; Busdiecker, M. Control system development for an advanced-technology medium-duty hybrid electric truck. *SAE Tech. Paper* **2003**, doi:10.4271/2003-01-3369.

9. Lin, C. Modeling and Control Strategy Development for Hybrid Vehicles. Ph.D. Thesis, University of Michigan, Ann Arbor, MI, USA, 2004.
10. Piccolo, A.; Ippolito, L.; Zo Galdi, V.; Vaccaro, A. Optimization of energy flow management in hybrid electric vehicles via genetic algorithms. In Proceedings of the IEEE/ASME International Conference on Advanced Intelligent Mechatronics, Como, Italy, 8–12 July 2001; Volume 1, pp. 434–439.
11. Schouten, N.J.; Salman, M.A.; Kheir, N.A. Fuzzy logic control for parallel hybrid vehicles. *IEEE Trans. Control Syst. Technol.* **2002**, *10*, 460–468.
12. Kim, N.; Rousseau, A. A comparative study of hydraulic hybrid systems for class 6 trucks. *SAE Tech. Paper* **2013**, doi:10.4271/2013-01-1472.
13. Woon, M.; Nakra, S.; Ivanco, A.; Filipi, Z. Series hydraulic hybrid system for a passenger car: Design, integration and packaging study. *SAE Tech. Paper* **2012**, doi:10.4271/2012-01-1031.
14. Ning, X.; Xu, Y.; Wang, Q.; Chen, J. Analysis of hydraulic regenerative braking system for electric vehicle. *Adv. Mater. Res.* **2012**, *490–495*, 195–202.
15. Midgley, W.J.B.; Cebon, D. Comparison of regenerative braking technologies for heavy goods vehicles in urban environments. *Proc. Inst. Mech. Eng. Part D J. Automob. Eng.* **2012**, *226*, 957–970.
16. Zhang, Z.; Chen, J.; Wu, B. The control strategy of optimal brake energy recovery for a parallel hydraulic hybrid vehicle. *Proc. Inst. Mech. Eng. Part D J. Automob. Eng.* **2012**, *226*, 1445–1453.
17. Deppen, T.O.; Alleyne, A.G.; Stelson, K.A.; Meyer, J.J. Optimal energy use in a light weight hydraulic hybrid passenger vehicle. *ASME Trans. J. Dyn. Syst. Meas. Contr.* **2012**, *134*, 041009.
18. Bender, F.A.; Kaszynski, M.; Sawodny, O. Drive cycle prediction and energy management optimization for hybrid hydraulic vehicles. *IEEE Trans. Veh. Technol.* **2013**, *62*, 3581–3592.
19. Advanced Vehicle Simulator. Available online: http://sourceforge.net/projects/adv-vehicle-sim/?source=navbar (accessed on 30 April 2015).
20. Wiegman, H.L.N.; Vandenput, A.J.A. Battery state control techniques for charge sustaining applications. *SAE Tech. Paper* **1998**, doi:10.4271/981129.
21. Toulson, E.R. Evaluation of a hybrid hydraulic launch assist system for use in small road vehicles. In Proceedings of the IEEE International Symposium on Industrial Electronics, Cambridge, UK, 30 June–2 July 2008; pp. 967–972.
22. Ehsani, M.; Gao, Y.; Emadi, A. *Modern Electrci, Hybrid Electric, and Fuel Cell Vehicles*, 2nd ed.; CRC Press: Boca Raton, FL, USA, 2010; pp. 384–385.
23. Liu, G.Q.; Yan, Y.C.; Chen, J.; Na, T.M. Simulation and experimental validation study on the drive performance of a new hydraulic power assist system. In Proceedings of the IEEE Intelligent Vehicles Symposium, Xi'an, China, 3–5 June 2009; pp. 966–970.

7

Co-Production of Olefins, Fuels, and Electricity from Conventional Pipeline Gas and Shale Gas with Near-Zero CO_2 Emissions: Economic Performance

Yaser Khojasteh Salkuyeh and Thomas A. Adams II *

Department of Chemical Engineering, McMaster University, 1280 Main St W, Hamilton, ON L8S 4L7, Canada; E-Mail: khojasy@mcmaster.ca

* Author to whom correspondence should be addressed; E-Mail: tadams@mcmaster.ca

Academic Editor: Enrico Sciubba

Abstract: In this paper, techno-economic analyses of a polygeneration system for the production of olefins, transportation fuels and electricity are performed, considering various process options. Derivative-free optimization algorithms were coupled with Aspen Plus simulation models to determine the optimum product portfolio as a function of a wide variety of market prices. The optimization results show that the proposed plant is capable of producing olefins with the same production costs as traditional petrochemical routes while having effectively zero process CO_2 emissions (including the utilities). This provides an economic and more sustainable alternative to traditional naphtha cracking.

Keywords: polygeneration; gas-to-olefins; gas-to-liquids; methanol-to-olefins; CO_2 capture

---

1. Introduction

The production of olefins from methanol, called methanol-to-olefins (MTO), is a novel process concept that can produce petrochemical feedstocks from alternative fuels [1]. Therefore, it can be an interesting option to meet global demand for petrochemical feedstocks, which is growing relatively steadily [2]. In addition, by incorporation of advanced CO_2 capture systems, the MTO process can also diminish the carbon emissions of petrochemical industries by replacing traditional steam cracking units [3].

The MTO process can be improved by incorporating it into a polygeneration process, in which other products are co-produced. By tightly integrating the different parts of the polygeneration process together, significant profitability and efficiency improvements can result [4–6]. Several studies have been performed on the techno-economic analysis of various polygeneration models. Most of these works focused on the hybrid systems that use multiple feedstocks such as coal and natural gas [7,8], coal and biomass [9], or coal and oven gas [10], for the co-production of transportation fuels, methanol, hydrogen and power. However, none of these works considered the impact of incorporating innovative CO_2 capture technologies such as oxy-fuel combustion and chemical looping combustion processes.

The main objective of this work is techno-economic optimization of a novel polygeneration process that co-produces methanol, liquid transportation fuel (dimethyl ether), and olefins from natural gas. The economic performance of different advanced power generation options are studied in this paper: chemical looping combustion using nickel oxide, iron oxide, oxyfuel combustion, and conventional combustion with a gas turbine. The economic optimization also included investigating the impact of the feedstock composition by considering conventional natural gas as well as different types of shale gas. Details of the impact of gas composition on the plant's configuration and geographic location are described in part I of this work.

In this paper, optimization algorithms were used in an economic analysis to determine the highest profitability for each design variant, or the highest production rate of olefins (both are considered as separate objective functions in separate optimization problems), considering a selected subset of the most important design parameters. Because market conditions are always subject to change, this was repeated for a wide variety of market conditions, ranging from low to high prices for the various products generated in each process. Finally, the most promising variants of the MTO concept were compared to naphtha cracking and ethane cracking processes for olefin production, the closest competing processes, showing promising results. The models used for each of the individual unit operations were either developed in the prior work of our group or other groups.

2. Economic Baseline

The profitability of each process was determined by computing the net present value (NPV) as the primary economic indicator. Capital costs were determined from published sources wherever possible, such as for the reformer section, MeOH synthesis section, power generation section and carbon sequestration units [7,11]. The capital costs of other units which did not have available cost data were estimated using the Aspen Icarus software package. This tool is commonly used to predict the capital cost of common chemical engineering unit operations and is updated regularly. In this work, it was used to predict the capital costs of individual unit operations such as pumps, compressors, and distillation columns which were not included in the sections listed previously. Table 1 lists all economic analysis parameters, market prices and assumptions for calculating the net present value (NPV) of all design cases, with citations to justify each parameter included in that table. Note that the utility and market prices shown are those used only for "base-case" calculations. Since prices vary from place to place and change somewhat unpredictably in the future, a sensitivity analysis which includes the effects of changes in price is discussed in Section 4. All financial parameters, such as interest rates, are chosen based on recommendations of prior work. Other business related expenses such as labor, overhead, laboratory operations, and maintenance are included using estimate formulas common to most analyses of this type. A detailed listing of the costs of each line item for each specific design option can be found in the supplementary material. Although all of these parameters can change from case to case, an extensive sensitivity analysis of all parameters considered is outside the scope of this work. However, even though there is uncertainty in the exact values of the NPV, the parameters are applied equally to all process variants and so meaningful conclusions can be made about the relative comparison of one process option to another.

Table 1. Base-case market prices and economic assumptions.

Feeds and Products	***Market Prices***
Natural gas & Ethane prices, $/MMBtu	3.88, [12]
NiO (commercial grade, 76% wt.), $/kg	20, [13]
NiO Disposal Cost, $/kg	5 [(1)]
NiO life span, h	10,000, [14]
Fe_2O_3 (commercial grade, 96% wt.), $/kg	1.44, [15]
Fe_2O_3 Disposal Cost, $/kg	0.36 [(1)]
Fe_2O_3 life span, h	3000, [16] (page 203)
Electricity, ¢/kWh	5.67, [17]
MeOH price, $/tonne	482, [18]
DME, $/tonne	962, [19]
Propylene, $/tonne	1340, [20]
Ethylene, $/tonne	1424, [21]
Economic Assumptions, [7,22]	***Values***
Plant capacity (shale gas inlet rate)	1111 MW, LHV
Operation time (h/year)	8760
Capacity factor	85%
Chemical engineering plant cost index	574.3, [23]
Plant lifetime (year)	30
Loan lifetime (year)	30
Interest rate on loan	9.5%
Debt percentage	50%
Inflation	2.79%
Federal + state tax rate	40%
Equity return rate	20%
NPV Calculation Elements, [22]	***Values***
Indirect cost	20% Fixed capital cost
Working capital cost	15% Total investment
Operating labor	10% Total production cost
Direct supervisory and clerical labor	10% Operating labour
Utilities	10% Total product cost
Maintenance and repairs	5% Fixed capital investment
Operating supplies	10% Maintenance and repair
Laboratory charges	10% Operating labour
Plant-overhead	50% of cost for operating labour, supervision, and maintenance
Administrative costs	15% of cost for operating labor, supervision, and maintenance
R&D, distribution and selling costs	4% of total product cost

[(1)] Due to lack of published data, assumed to be 25% of the metal-oxide price.

Process Optimization

For each process variant, three design variables were subject to optimization, as noted in Figure 3 of part I of this study: the recycle ratio of unreacted gas sent to the MeOH reactor, the split ratio of MeOH which is sent to the MTO section (stream 2–7), and the split ratio of the MeOH stored as the final product (stream 2–8). Other process variables are adjusted to either meet the process constraints and product specifications, or selected based on suggested values in literatures. The feasibility range of all decision variables is fixed between 1% and 99%. Bounds of 0% and 100% were not used in order to avoid simulation problems in the software associated with zero flows. Instead, if the lower bound of 1% on a variable was reached, the simulation was reconstructed manually using 0% and adjusting the others proportionally (where in some cases the corresponding process section was eliminated), and if it improved the objective function, it was taken as the final result for that instance. A similar technique was used for 99%, which was reconstructed manually using 100%.

Because there were only three decision variables, a coarse-grain method (an exhaustive search over a grid) was used to sample the decision space in three dimensions for each design variant. For example, the effect of the recycle ratio and the ratio of MeOH to the MTO section are shown in Figure 1 for two different MeOH to MTO ratios. It can be seen that unlike the low olefin production ratio (Figure 1a), the optimum MeOH to DME ratio is between 20%–40% in the high olefin production scenario (Figure 1b). In addition, the influence of electricity, MeOH, DME and olefin market price fluctuations as a function of different process variables are depicted in Figure 2.

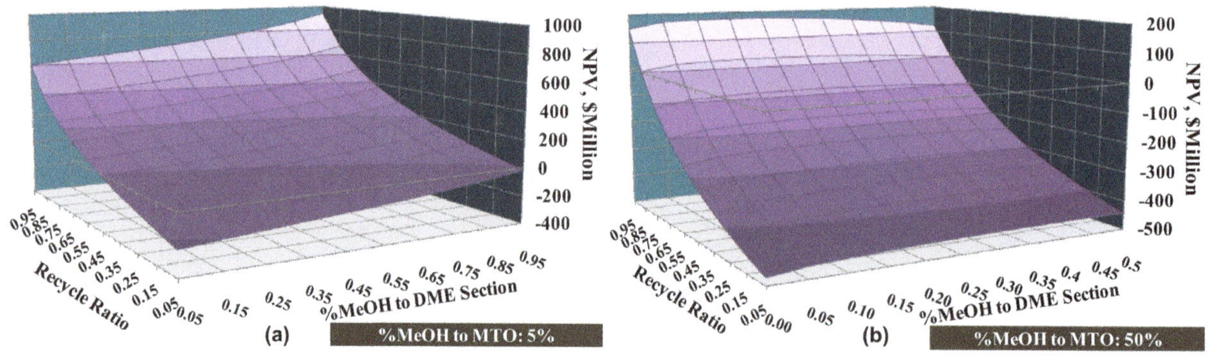

Figure 1. Effect of recycle ratio of unreacted gases to the methanol synthesis reactor and %MeOH to DME section on the NPV of plant at (**a**) low olefin (5%); (**b**) high olefin (50%) production ratios.

The sensitivity analysis shown in Figures 1 and 2 illustrate the impact of the main three decision variables on the NPV of plant at different market prices. Due to the relative long run time required per simulation (approximately one minute), the spacing between sample points was too large to sufficiently locate an optimum in a reasonable amount of time. The built-in Aspen Plus optimization tool was also found to be insufficient and improved the results only trivially compared to the coarse-grain method, even when starting from a variety of different initial conditions.

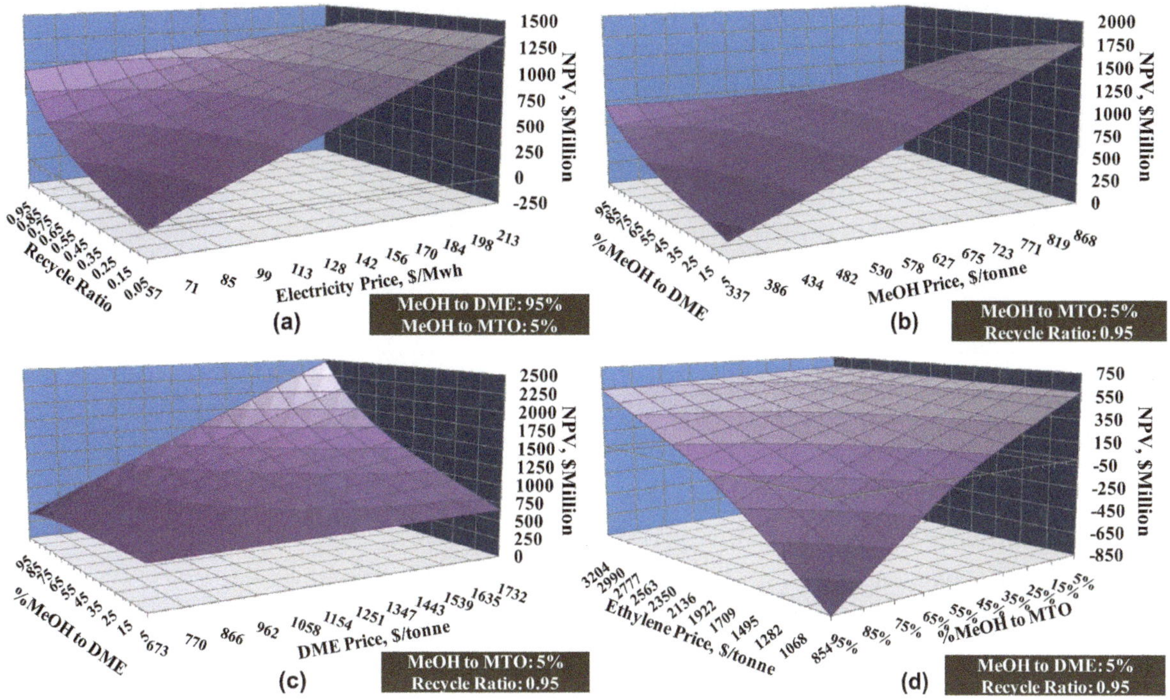

Figure 2. The NPV of plant at different market prices and selected decision variables changes: (**a**) recycle ratio and Electricity price; (**b**) %MeOH to DME and MeOH price; (**c**) %MeOH to DME and DME price; (**d**) %MeOH to MTO and ethylene price.

Instead, three heuristic-based, derivative-free black box optimization algorithms were considered, namely particle swarm optimization, genetic algorithms, and simulated annealing. These are stochastic methods which use different strategies of exploring the decision variable space broadly at first and then later exploring promising subspaces in more and more depth. Eventually, the algorithms terminate once no more improvements can be found, and the best known point is usually locally optimal within tolerances, with a good chance that it is also the global optimal if the problem is not too "bumpy" and the heuristics were chosen correctly. Fortunately, finding the true optimum point is not critical to this work, and since the coarse-grain method samples points across the entire range of the state space, suboptimum results are expected to deviate from the true global optimum by only small amounts. These algorithms were implemented in Matlab and were coupled with the Aspen Plus simulations through the Excel interface. The three algorithms were used on a subset of cases to determine which derivative-free algorithm was the most promising. Standard algorithms have been used for each optimization technique and have not been detailed here for brevity. However, the heuristic parameters chosen for each algorithm are listed and referenced in Table 2. Particle swarm optimization was found to give the best results in all test cases, and so this algorithm was chosen for all remaining cases used in this study (see the supplementary material).

Table 2. Optimization algorithms.

Particle Swarm Optimization Parameters [24]	**Values**
Number of function evaluation (NFE)	200
Number of particles	5
a_1, a_2	2.05
φ	$a_1 + a_2$
Inertia weight (w)	$\frac{2}{\left\|2-\varphi-\sqrt{\varphi^2-4\varphi}\right\|}$
Personal and global learning ratios (c_1, c_2)	a_1w, a_2w
Genetic Algorithm Parameters [25,26]	**Values**
Number of function evaluation	200
Number of particles	10
Crossover ratio	0.9
Mutation ratio	0.6
Mutation rate	1 process variable
Simulate Annealing Parameters [27,28]	**Values**
Number of function evaluation	200
Initial population	5
Number of particle moves in iteration i	3
Initial temperature (T_0)	10
Temperature decrement rate (α)	0.99

Several objective functions were used: maximizing the NPV, maximizing the power production, maximizing the olefin production, and maximizing the olefin production with the added constraint that the NPV must be non-negative (the constraint was implemented by adding a huge penalty to the objective function when violated). These four objective functions were run for each of the three process variants for each of the four shale gas types, and for each of these the particle swarm optimization was rerun a total of 10 times using different initial guesses, to ensure that the exploration was sufficiently broad. Thus a total of 160 optimization runs were performed, requiring about 15 min each for a total of about 40 h of computer processing time on a modern desktop PC. It is worth noting that the results of the maximum NPV objective function were identical to the results of the maximum DME objective function described in Part I of this work.

3. Economic Results

Stream tables for the "Maximum olefin with non-negative NPV constraint" optimization scenario are shown in supplementary material. Furthermore, the NPV maximization results of the chemical looping combustion (both NiO and Fe_2O_3 oxygen carriers), oxy-fuel combustion and post-combustion technologies for Fayetteville shale gas are shown in Table 3. Looking at the results, the highest NPV is obtained at the base case prices (Table 1) when the split ratios and unreacted gas recycle ratio are such that DME production is maximized in the chemical looping and oxyfuel cases. It can also be seen that the NiO-CLC system achieved the highest NPV compared to other options.

As described in Part I of this study, since the energy requirement of the post combustion CO_2 capture is relatively high, the DME ratio cannot be more than 70% without the import of external steam sources.

3.1. Different Feed Compositions

The breakdown of products and NPV results of various optimization scenarios are shown in Table 4 for each type of shale gas and conventional pipeline gas using NiO-CLC as the power generation configuration since it had the highest NPV. In the first scenario, the objective function was to maximize the NPV of the process. It can be seen from the optimization results that most of the output must be DME for all types of shale gases. In the second and fourth scenarios, the objective function was to maximize the olefins production rate. Their difference is that there is an NPV constraint in the fourth scenario, which must always be non-negative. Therefore, this scenario shows that the maximum possible olefins production while still having a profitable plant is between 45%–54%, depending on the gas type. In the third scenario, the objective function was to maximize the power output of our proposed polygeneration plant. The maximum power generation is around 60% of the output in this scenario.

Table 3. Comparison of the power generation options using Fayetteville shale gas.

Power Generation Option	**Chemical Looping**		**Oxy-Fuel Combustion**	**Post Combustion**
	Iron-Oxide	**Nickel-Oxide**		
NPV, $Million	1138	1165	1026	709
Efficiency, %HHV	52.5	52.1	48.2	54.5
%CO_2 capture	100	100	100	90
Product Portfolio (%)				
Net electricity	5.0	4.2	1.0	11.1
MeOH	0.0	0.0	0.0	47.2
DME	95.0	95.8	99.0	41.7
Olefins	0.0	0.0	0.0	0.0
Capital investment, $Million	565	509	535	714

3.2. Olefins Production Cost

The effect of changing the olefin production ratio on its production cost is shown in Figure 3. The olefin production of traditional cracking processes varied between 0.44–1.3 $/kg in 2012, depending on feedstock type and price [29]. It can be seen that the production cost of the proposed novel system is lower than the average production cost of commercial naphtha cracking and ethane cracking plants when the olefin production ratio is around 44% and 33% of output respectively. The calculated production cost of olefins is the total plant annualized cost subtracted by the revenue from other products. Therefore, its value can be negative when the credit from other products is more than the total cost of plant.

Table 4. Comparison of optimization results different shale gases at the same energy input (1111 MW) and with NiO-CLC power generation approach.

Optimization Scenario	Maximum NPV (Identical to Maximum DME Production)					Maximum Olefin				
Gas Type	**Marcellus**	**Fayetteville**	**New Albany**	**Haynesville**	**Conventional Gas**	**Marcellus**	**Fayetteville**	**New Albany**	**Haynesville**	**Conventional Gas**
NPV, $Million	1139	1165	1158	1177	1143	−813	−769	−746	−728	−803
Product portfolio										
%Power	4.5	4.2	4.3	4.2	4.3	14.3	11.7	12.5	12.9	18.5
%MeOH	0.0	0.0	0.0	0.0	0.0	0.0	0.0	0.0	0.0	0.0
%DME	95.5	95.8	95.7	95.8	95.7	0.0	0.0	0.0	0.0	0.0
%Olefins	0.0	0.0	0.0	0.0	0.0	85.7	88.3	87.5	87.1	81.5
Capital, $Million	519	509	503	501	558	694	678	684	676	705
Optimization Scenario	**Maximum Power**					**Maximum Olefin with Non-Negative NPV Constraint**				
Gas Type	**Marcellus**	**Fayetteville**	**New Albany**	**Haynesville**	**Conventional Gas**	**Marcellus**	**Fayetteville**	**New Albany**	**Haynesville**	**Conventional Gas**
NPV, $Million	28	30	38	35	39	0	0	0	0	0
Product portfolio										
%Power	57.9	57.4	56.1	56.6	57.5	15.8	14.8	13.4	13.2	20.1
%MeOH	0.0	0.0	0.0	0.0	0.0	0.0	0.0	0.0	0.0	0.0
%DME	42.1	42.6	43.9	43.4	42.5	43.5	41.1	41.4	40.8	40.7
%Olefins	0.0	0.0	0.0	0.0	0.0	40.7	44.1	45.3	45.9	39.2
Capital, $Million	530	536	539	538	584	681	666	670	661	698

3.3. Sensitivity Analysis- Effect of MeOH and DME Prices

A sensitivity analysis was performed to determine how changes in the MeOH and DME prices affect the optimization results. For each set of MeOH and DME prices, the optimization was repeated using the maximum NPV objective function, resulting in a potentially different optimal configuration of recycle and product ratios.

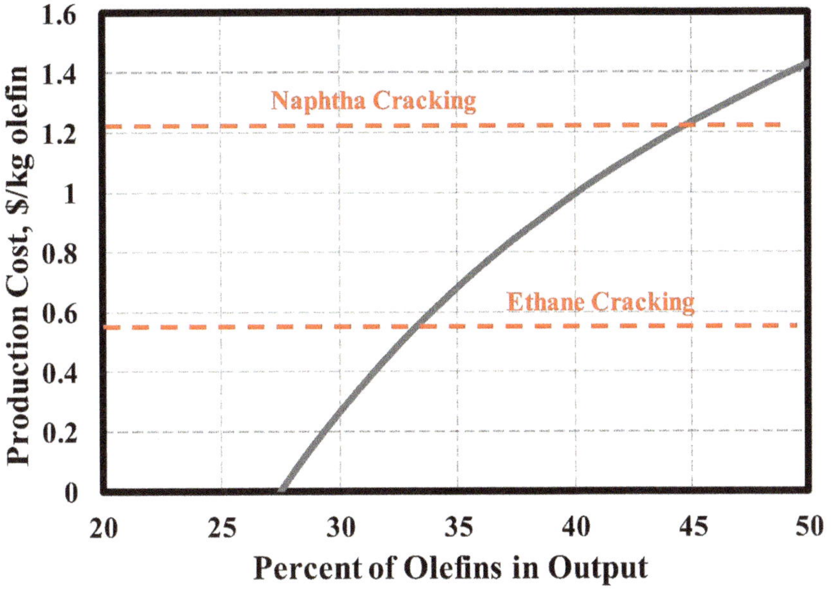

Figure 3. Sensitivity analysis of olefin production cost as a function of olefin production ratio (Fayetteville shale gas).

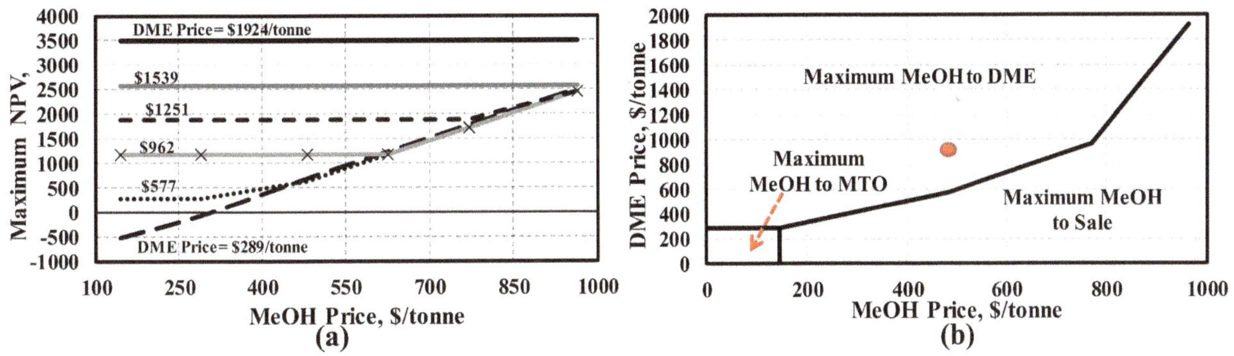

Figure 4. Effect of MeOH and DME price on (**a**) the maximum NPV of process; (**b**) optimal product portfolio (Fayetteville shale gas). The location of the chosen base case market conditions is indicated as a red circle.

As shown in Figure 4a, the design with the maximum NPV is always positive when the MeOH price is \$300/tonne (60% of its base case value) or more. Furthermore, the maximum NPV is always positive when the DME price is \$577/tonne (60% of its base case price) or more. Figure 4b depicts a "price map" of the optimal product portfolios depending on the market prices. For example, maximizing the amount of olefins produced is the most profitable choice when MeOH prices are below about \$150/tonne and DME prices are below about \$300/tonne (the lower-left corner of Figure 4b).

Above those prices, either maximum DME production or maximum MeOH production is the most profitable choice depending on the prices, corresponding to the upper left and lower right regions of the diagram, respectively.

3.4. Sensitivity Analysis—Effect of Olefins Prices

Similarly, the effect of varying the price of olefins (ethylene and propylene) on the maximum NPV is shown in Figure 5. It is assumed that the propylene price changes at the same ratio as the ethylene price varies compared to their base case prices (Table 1). It can be seen that the NPV of the process is always positive regardless the price of olefins and DME products when methanol and electricity prices are at their base case market conditions. The reason is that at low olefins and DME prices, the maximum methanol production option is favoured over the others as shown in Figure 5b. Similarly, maximum DME or maximum olefin production is favoured for relatively high DME or olefin prices (the upper left and lower right regions of Figure 5b). However, when both DME and olefin prices are high, a mixture of DME and olefin products is actually more profitable than maximizing just one.

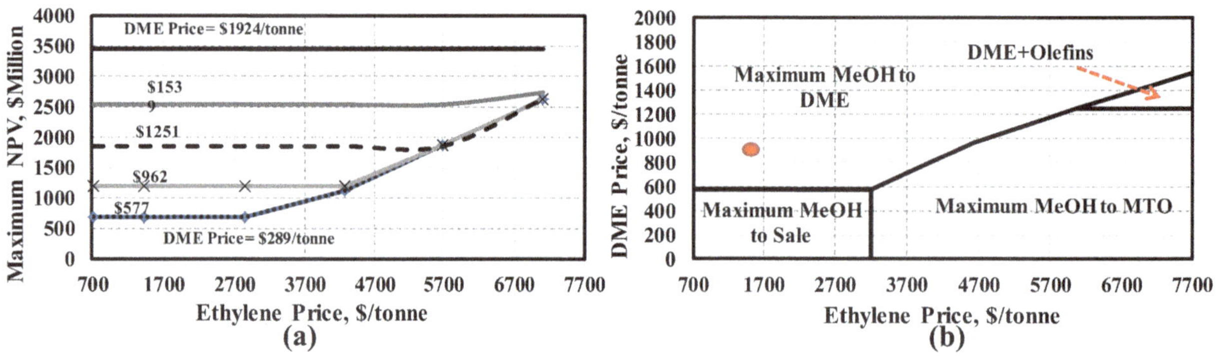

Figure 5. Effect of olefins and DME prices on (**a**) the maximum NPV of process, (**b**) optimal product portfolio (Fayetteville shale gas). The base case prices are marked with a red circle.

4. Conclusions

In this study, a techno-economic analysis of a novel polygeneration process for the co-production of DME, olefins, electricity, and methanol using different qualities of shale gas and also conventional natural gas was performed. The impact of power generation alternatives, gas composition and product portfolio on the thermal efficiency of the plant had been presented in part I of this work.

The economic optimization results, presented in this paper, showed that the CLC technology with nickel oxide was the most profitable choice compared to the other power generation options. The results of the sensitivity analysis showed that in most market conditions, it was usually optimal for only one of methanol, DME, or olefins to be co-generated along with electricity. However, the co-production of DME and olefins was optimal for some cases with very high product prices. At current (base case) market conditions, olefin production was not favored. However, using the polygeneration concept, it is possible to produce a mixture of olefins and electricity with 100% CO_2 capture in current market conditions at costs similar to naphtha or ethane cracking without any CO_2 capture. Thus, the proposed process may be a promising way of displacing traditional olefin production using abundant natural gas sources at similar costs but with significantly reduced environmental impact.

Despite promising results achieved by this proposed polygeneration model, the flexibility and dynamic behaviour of this tightly integrated system has not been investigated and is subject of further research. It has been shown previously that flexible polygeneration systems can have massive improvements in NPV compared to inflexible ones. For example, one study showed that up to 62% increases in net present value were possible if the amount of methanol, synthetic fuels, or electricity produced by that particular example process were allowed to change twice per day in response to changes in market prices with a 100% turndown ratio [30]. Other studies on polygeneration systems have shown that the optimal process design and corresponding product portfolio for inflexible polygeneration plants vary significantly with market prices [7]. These studies also showed that the historical variation in market prices experienced over a several year period is sufficiently large enough such that flexible polygeneration may make economic sense. However, all of these studies assumed that transitions between operating steady states were feasible, instantaneous, and free. It is a significant future challenge to construct dynamic models of all of the relevant process units in sufficient detail such that the complex interaction between tightly integrated process sections can be understood well enough such that information about the time required for transitions, feasible turndown ratios, off-spec products, and their associated costs can be known. Some progress has been made in constructing dynamic models of certain individual process units specifically for the purposes of flexible polygeneration applications, such as for gasification [31], steam reforming [32], water gas shift [33], CO_2 capture [34], and solid oxide fuel cells [35]. However, combining these models into one system requires considerable effort and has not yet been achieved.

Acknowledgments

We gratefully acknowledge financial support by the Ontario Research Fund: Research Excellence and Ontario Graduate Scholarship programs on this project.

Author Contributions

YKS performed research and wrote the manuscript, and TAAII supervised research, gave the technical advice and edited the manuscript.

Nomenclature

Abbreviations

CLC	Chemical looping combustion
DME	Dimethyl ether
HHV	Higher heating value
LHV	Lower heating value
NPV	Net present value
PSO	Particle swarm optimization

Conflicts of Interest

The authors declare no conflict of interest.

References

1. Xiang, D.; Yang, S.; Li, X.; Qian, Y. Life cycle assessment of energy consumption and GHG emissions of olefins production from alternative resources in China. *Energy Convers. Manag.* **2015**, *90*, 12–20.
2. Bok, J.-K.; Lee, H.; Park, S. Robust investment model for long-range capacity expansion of chemical processing networks under uncertain demand forecast scenarios. *Comput. Chem. Eng.* **1998**, *22*, 1037–1049.
3. Ren, T.; Patel, M.K.; Blok, K. Steam cracking and methane to olefins: Energy use, CO_2 emissions and production costs. *Energy* **2008**, *33*, 817–833.
4. Gao, L.; Li, H.; Chen, B.; Jin, H.; Lin, R.; Hong, H. Proposal of a natural gas-based polygeneration system for power and methanol production. *Energy* **2008**, *33*, 206–212.
5. Zhang, X.; Gundersen, T.; Roussanaly, S.; Brunsvold, A.L.; Zhang, S. Carbon chain analysis on a coal IGCC—CCS system with flexible multi-products. *Fuel Process. Technol.* **2012**, *108*, 146–153.
6. Floudas, C.A.; Elia, J.A.; Baliban, R.C. Hybrid and single feedstock energy processes for liquid transportation fuels: A critical review. *Comput. Chem. Eng.* **2012**, *41*, 24–51.
7. Adams, T.A., II; Barton, P.I. Combining coal gasification and natural gas reforming for efficient polygeneration. *Fuel Process. Technol.* **2011**, *92*, 639–655.
8. Adams, T.A., II; Barton, P.I. Combining coal gasification, natural gas reforming, and solid oxide fuel cells for efficient polygeneration with CO_2 capture and sequestration. *Fuel Process. Technol.* **2011**, *92*, 2105–2115.
9. Cormos, C.-C. Assessment of flexible energy vectors poly-generation based on coal and biomass/solid wastes co-gasification with carbon capture. *Int. J. Hydrogen Energy* **2013**, *38*, 7855–7866.
10. Man, Y.; Yang, S.; Zhang, J.; Qian, Y. Conceptual design of coke-oven gas assisted coal to olefins process for high energy efficiency and low CO_2 emission. *Appl. Energ.* **2014**, *133*, 197–205.
11. Haslbeck, J.L.; Kuehn, N.J.; Lewis, E.G.; Pinkerton, L.L.; Simpson, J.; Turner, M.J.; Varghese, E.; Woods, M.C. *Cost and Performance Baseline for Fossil Energy Plants*; Volume 1: Bituminous Coal and Natural Gas to Electricity; Department of Energy: Pittsburgh, PA, USA, 2010.
12. Henry Hub Spot, Average Wholesale Price. Available online: http://www.eia.gov/electricity/wholesale/ (accessed on 1 August 2014).
13. Nickel Monoxide, CAS No.: 1313–99–1. Available online: http://www.alibaba.com/product-gs/747545915/Factory_direct_sales_with_reasonable_price.html (accessed on 1 August 2014).
14. Adanez, J.; Abad, A.; Garcia-Labiano, F.; Gayan, P.; de Diego, L.F. Progress in chemical-looping combustion and reforming technologies. *Prog. Energy Combust. Sci.* **2012**, *38*, 215–282.
15. Factory price of Iron Oxide (Fe_2O_3), CAS No.: 1309–37–1. Available online: http://www.alibaba.com/product-detail/factory-price-of-iron-oxide-fe2o3-_814503142.html?s=p (accessed on 1 August 2014).
16. Fan, L.S. *Chemical Looping Systems For Fossil Energy Conversions*. John Wiley & Sons, Inc.: Hoboken, NJ, USA, 2010.

17. Electricity Wholesale Price, Average 2014. Available online: http://www.eia.gov/electricity/wholesale/ (acessed on 10 November 2014).
18. Methanol Price, North America, October 2014. Available online: https://www.methanex.com/our-business/pricing (accessed on 26 October 2014).
19. DiMethyl Ether. Available online: http://www.alibaba.com/product-detail/DiMethyl-Ether_1659793013.html (accessed on 1 June 2014).
20. Platts Global Propylene Price Index, August 2014. Available online: http://www.platts.com/news-feature/2014/petrochemicals/pgpi/propylene (accessed on 1 November 2014).
21. Platts Global Ethylene Price Index, August 2014. Available online: http://www.platts.com/news-feature/2014/petrochemicals/pgpi/ethylene (accessed on 1 November 2014).
22. Peters, M.S.; Timmerhaus, K.D. *Plant Design and Economics for Chemical Engineers*, 4th ed.; McGraw-Hill: New York, NY, USA, 1991.
23. CEPCI. Chemical engineering plant cost index. *Chem. Eng.* **2014**, *121*, 80.
24. Parsopoulos, K.E.; Vrahatis, M.N. *Particle Swarm Optimization and Intelligence: Advances and Applications*; IGI Global: Hershey, PA, USA, 2010.
25. Gen, M.; Cheng, R. *Genetic Algorithms and Engineering Optimization*; John Wiley & Sons: New York, NY, USA, 2000.
26. Goldberg, D.E. *Genetic Algorithms in Search, Optimization, and Machine Learning*; Addison-Wesley: Boston, MA, USA, 1989.
27. Diwekar, U.M. *Introduction to Applied Optimization*. Springer: New York, NY, USA, 2008.
28. Bertsimas, D.; Tsitsiklis, J. Simulated annealing. *Stat. Sci.* **1993**, *8*, 10–15.
29. Roberts, T. Ethylene–Good Today, Better Tomorrow–A Year Later. Available online: http://www.lyondellbasell.com/NR/rdonlyres/92CE3892-B543-46FB-9660-00E9AAE8889D/0/GoldmanSachsChemicalIntensity_March27.pdf (accessed on 1 November 2014).
30. Chen, Y.; Adams, T.A., II; Barton, P.I. Optimal design and operation of flexible energy polygeneration systems. *Ind. Eng. Chem. Res.* **2011**, *50*, 4553–4566.
31. Kasule, J.S.; Turton, R.; Bhattacharyya, D.; Zitney, S.E. One-dimensional dynamic modeling of a single-stage downward-firing entrained-flow coal gasifier. *Energy Fuels* **2014**, *28*, 4949–4957.
32. Ghouse, J.H.; Adams, T.A., II. A multi-scale dynamic two-dimensional heterogeneous model for catalytic steam methane reforming reactors. *Int. J. Hydrogen Energy* **2013**, *38*, 9984–9999.
33. Adams, T.A., II; Barton, P.I. A dynamic two-dimensional heterogeneous model for water gas shift reactors. *Int. J. Hydrogen Energy* **2009**, *34*, 8877–8891.
34. Harun, N.; Nittaya, T.; Douglas, P.L.; Croiset, E.; Ricardez-Sandoval, L.A. Dynamic simulation of MEA absorption process for CO_2 capture from power plants. *Int. J. Greenh. Gas Control* **2012**, *10*, 295–309.
35. Harun, N.F.; Tucker, D.; Adams, T.A., II. Fuel composition transients in fuel cell turbine hybrid for polygeneration applications. *J. Fuel Cell Sci. Technol.* **2014**, *11*, doi:10.1115/FuelCell2014-6509.

8

Innovative Nuclear Energy Systems: State-of-the-Art Survey on Evaluation and Aggregation Judgment Measures Applied to Performance Comparison

Vladimir Kuznetsov [1,*], Galina Fesenko [1], Aleksandra Schwenk-Ferrero [2,*], Andrei Andrianov [3] and Ilya Kuptsov [3]

[1] International Atomic Energy Agency, Wagramer Strasse 5, P.O. Box 100, Vienna A-1400, Austria; E-Mail: G.Fesenko@iaea.org

[2] Karlsruher Institut für Technologie (KIT) Institut für Kern- und Energietechnik, Hermann-von-Helmholtz-Platz 1, Eggenstein-Leopoldshafen 76344, Germany

[3] National Research Nuclear University MEPhI (Moscow Engineering Physics Institute), Kashirskoe sh. 31, Moscow 115409, Russia; E-Mails: Andrianov@iate.obninsk.ru (A.A.); ISKuptsov@mephi.ru (I.K.)

* Authors to whom correspondence should be addressed;
E-Mails: V.Kuznetsov@iaea.org (V.K.); Aleksandra.Schwenk-Ferrero@kit.edu (A.S.-F.)

Academic Editor: Erich Schneider

Abstract: This paper summarizes the experience gained in the application of multi-criteria decision making and uncertainty treatment methods to a comparative assessment of nuclear energy systems and related nuclear fuel cycles. These judgment measures provide a means for comprehensive evaluation according to different conflicting criteria, such as costs, benefits and risks, which are inevitably associated with the deployment of advanced technologies. Major findings and recommendations elaborated in international and national projects and studies are reviewed and discussed. A careful analysis is performed for multi-criteria comparative assessment of nuclear energy systems and nuclear fuel cycles on the basis of various evaluation and screening results. The purpose of this paper is to discuss the lessons learned, to share the identified solutions, and indicate promising future directions.

Keywords: multi-criteria decision making; nuclear energy systems; nuclear fuel cycles; multi-objective optimization; uncertainty treatment

1. Introduction

The future Nuclear Energy System (NES) should meet the following requirements: to be resource-sufficient producing low amounts of waste in the long run, to be cost effective, to maintain the necessary level of safety and reliability and to ensure effective resistance to nuclear weapon proliferation. While conducting an integrated analysis of the future NESs and related nuclear fuel cycles (NFC) performance, special consideration must be given to the full range of system factors and constraints.

In general, experts identify five significant problem areas associated with nuclear energy development: nuclear safety, risks of unauthorized proliferation of fissile materials and technologies, radioactive waste management, economic efficiency and limited natural fuel resources, which are deterrents to large-scale development. A conflict is evident, not only between various areas, but also for some measures within a certain area which may lead, for instance, to a reduction of long-term risks while increasing short-term ones. However, at present there are no universal, reliable and generally accepted quantitative criteria and methods for planning and assessing NES long-term development and principles for selecting promising technological solutions. Therefore neither a strategy combining NES and NFC deployment balanced on various costs, benefits and risks nor strategic and tactical tradeoff decisions in each related nuclear engineering area is yet possible.

Thus, the assessment of efficiency and sustainability of NES together with an optimization of the system performance is evidently of a multi-criteria nature. Besides conflicting criteria (which means that improving the value of one criterion may lead to a decrease in the values of another) a significant uncertainty of the judgment resulting from the assessment must be taken into account, making uncertainty analysis an inevitable step in order to create a sound basis for decisions. Uncertainty examination indicates the stability of results when making multi-criteria decision comparative assessments.

A lack of commonly applied proven methodologies for decision-making using multi-criteria techniques in the nuclear engineering field and, in particular, in the area of the NES (and NFC) performance assessments, makes the procedure of formulating a coordinated vision of a preferable technological and institutional solutions, either common to several countries or specific to a particular country, rather complicated.

Multi-criteria decision making (MCDM) techniques allow searching for compromises among the conflicting factors that determine the NES and NFC efficiency and sustainability by calculating corresponding trade-off rates and make it possible to carry out comparative analysis of alternatives and selection by ranking, and sorting of the best options.

This paper is organized as follows: Section 2 gives an overview on the experience of MCDM and the uncertainty treatment methods applied to a comparative assessment of NESs and NFCs capability (or potential) performance, Section 3 summarizes and discusses the major findings and recommendations elaborated in different projects and studies, finally in Section 4 lessons learned from existing studies are discussed and promising directions for the future are indicated.

2. Studies and Projects on Multi-Criteria Assessment of NESs and NFCs

The sustainability of NES operating commercially today is often questioned by the public and some decision makers, because of concerns related to safety, nuclear waste disposal and proliferation aspects. To address these concerns and ensure the sustainable development of nuclear energy international and national nuclear agencies and organizations regularly initiate studies and projects oriented to the topic of NES and NFC sustainability and efficiency assessments. Some representative illustrations of such endeavors are presented below, which at the same time do not pretend to cover all possible studies and projects having been realized by the nuclear community on that issue.

2.1. The IAEA/INPRO Related Studies and Projects

The International Atomic Energy Agency (IAEA) is the world's centre for cooperation in the nuclear field to promote the safe, secure and peaceful application of nuclear energy.

2.1.1. IAEA Nuclear Technology Basic Principles

Any integrated analysis on the NES and NFC options related to efficiency and sustainability assessments strives to be pluralistic, by coherently accepting different priorities, intentions, values, and norms, and simultaneously relying on a consensus about the established hierarchy of requirements and basic principles. A basic principle is a statement of a general goal that is to be achieved by nuclear technology and provides broad guidance for its deployment. An appropriate set of basic principles was proposed by the IAEA, which may be divided into three categories: beneficial, responsible, and sustainable [1]. The formulated eight basic principles are intended to provide a holistic approach to the use of nuclear technology to fulfill its potential, help to sustainably meet growing global energy needs.

The beneficial use category includes two basic principles: benefits and transparency, hence the nuclear energy should provide benefits that outweigh the associated costs and risks, and be based on the open and transparent communication of all relevant information. The sustainable use category includes two basic principles: resource efficiency and continuous improvement, wherein resource use and advances in technology and engineering to continuously pursued to improve safety, security, economics, proliferation resistance, and reduce environmental impact. Finally, the responsible use category encompasses four basic principles: the protection of people and the environment, security, non-proliferation, and long-term commitment in compliance with internationally recognized standards. Moreover, the IAEA proposed general nuclear and specific objectives for nuclear power, NFC and radioactive waste management and decommissioning to be achieved at different implementation stages. To adapt these general basic principles for a specific area the experts apply these concepts to elaborate problem- and practical-oriented principles.

2.1.2. INPRO Methodology

A methodology for assessing NES capabilities to meet national sustainability requirements was developed in the International Project on Innovative Nuclear Reactors and Fuel Cycles (INPRO) initiated by the IAEA in 2000 [2]. INPRO developed a set of basic principles, user requirements (UR) and criteria together, with an assessment method, which taken together, comprise the INPRO methodology for the evaluation of NESs, including those with innovative components. INPRO proposed a set of basic principles on which NESs should be based to meet the overall target of sustainable energy supply [3]. To achieve sustainable NES all basic principles shall be met in the area of economics, infrastructure, waste management, proliferation resistance, physical protection, environment and safety.

INPRO hierarchy of requirements to NES design for nuclear energy system assessment (NESA) is illustrated in Figures 1 and 2. At the top-level 14 basic principles are defined. The second level called 'user requirements' contains 52 URs which specify the conditions that should be met to achieve meeting the corresponding basic principle. In this way URs define a means of achieving the top-level goals. Finally, a set of metrics with 125 criteria with indicators and acceptance limits enable a judgment of NES potential. Each INPRO criterion consists of an 'indicator' and an 'acceptance limit'. Indicators may be based on single parameter, on an aggregated variable, on a status statement.

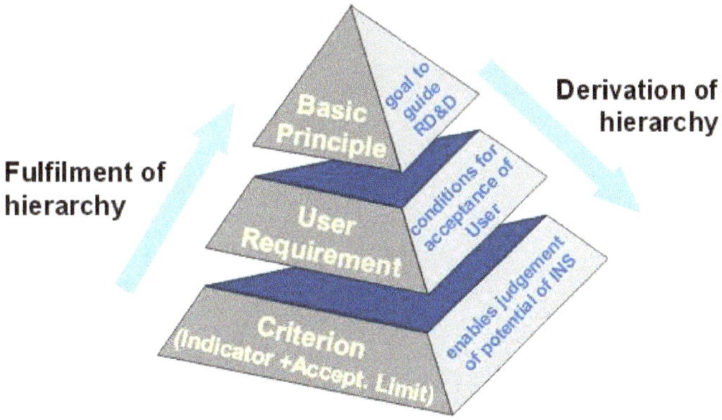

Figure 1. INPRO hierarchy of demands.

Two types of indicators of INPRO criteria are distinguished, numerical and logical. A numerical indicator may be based on a measured or a calculated value that reflects a property of an NES. Examples might be the estimated probability of a major release of radionuclides to the containment obtained from a probabilistic safety analysis or the number of intact safety barriers maintained after a severe accident. A logical indicator is usually associated with some necessary feature of an NES and usually is presented in form of a question.

In the INPRO areas (economics, safety, waste management, *etc.*) some indicators may be applicable to the overall NES, others are valid only for specific components (such as the reactor) or for specific nuclear technologies (e.g., light water reactors), they may relate to the functionality of a system or component or set out measures for implementation or methods ofanalyses. In addition, some indicators utilize 'evaluation parameters'. These parameters were introduced to assist the INPRO assessor in determining whether the acceptance limit for an indicator has been met. In some specific cases these evaluation parameters have their own acceptance limits, in which case they could be called sub-indicators. An example of evaluation parameters could be the use of parameters, such as some combination of design simplification, improved materials, increased operating margins, increased use of passive safety features and systems, increased redundancy, as an indicator for increasing the robustness of an NES component relative to an existing design as a means of enhancing defense-in-depth. In addition to the classification of indicators, another type of indicator, a so-called 'key indicator', may be defined in specific or in all INPRO areas, depending on the preferences of technology developers and Member States.

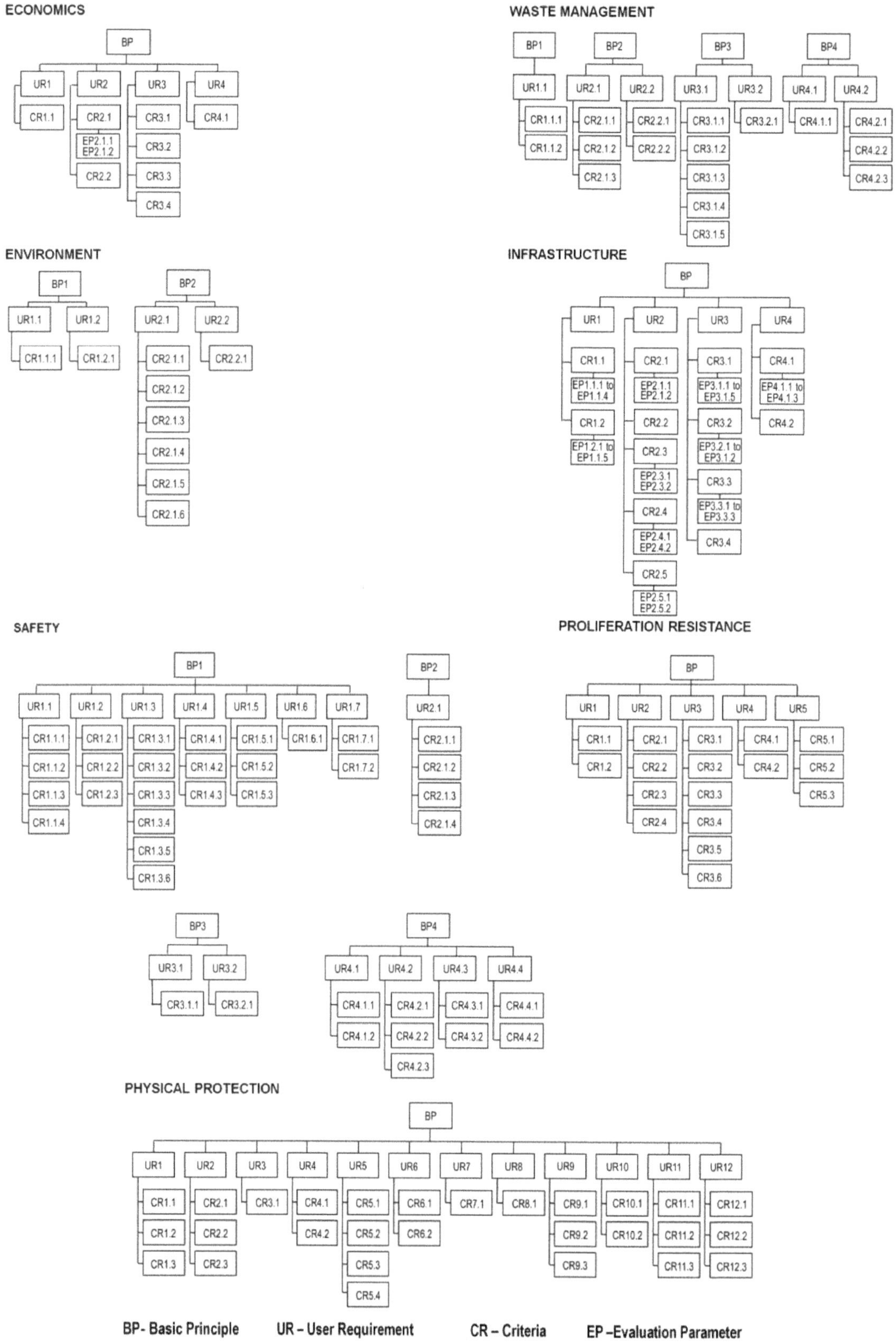

Figure 2. The INPRO hierarchies for each area.

The acceptance limit of an INPRO criterion is a target, either qualitative or quantitative, against which the value of an indicator can be compared by the INPRO assessor leading to a judgment of acceptability (pass/fail, good/bad, better/poorer). In correspondence to the two types of indicators there are also two types of acceptance limits, numerical (for quantitative targets) and logical (for qualitative targets). Typically, a logical acceptance limit is a positive (yes) or negative (no) answer to a question raised by the indicator.

In addition, a 'desired target value' would usually be defined for a key indicator (KI) chosen to represent the ultimate KI value that could be achieved through development. The concept of desired target values can be further extended by defining a so called 'relative benefit index', and a 'relative risk index' for each KI. The risk may include the uncertainty in the desired target value of the KI determined to be achievable for a specific NES, but would also reflect the development effort required and the maturity level of the concept. Thus, a concept may be advanced that has a good possibility of achieving a very substantial improvement in the value of a KI, but at the same time the concept may require the development of specific technical features that are at an early stage of development, and may require significant investment of funds, staff, *etc.*

However, feedback from experts which have used the INPRO methodology for a comparison of different NES options indicates that the current INPRO methodology as described above does not seem to be well suited for comparing different NESs. In order to perform comparative NES assessments advanced methods should be proposed for judgments aggregation.

In elaborating national and international recommendations for the NES development there is need for a structured and objective evaluation of options including: (1) screening an NES (one or more than one), selected by Member State on a national, regional and/or global basis, to evaluate whether it is compatible with the objective of sustainable energy development; (2) comparison of different NESs or components thereof, e.g., to find a preferred or optimum NES tailored to the needs of a Member State; or to make a capability comparison on a global basis; (3) identifying necessary improvements which, in turn, will lead to research, development and demonstration to upgrade the performance. In performing an INPRO assessment, a reference energy scenario or scenarios driven by a projected total demand for nuclear energy as a function of time must be considered and the pace of NES (power plants with related NFC facilities) deployment must be modeled.

2.1.3. The GAINS Analytical Framework

The INPRO Collaborative Project "Global Architecture of Innovative Nuclear Energy Systems Based on Thermal and Fast Reactors Including a Closed Fuel Cycle" (GAINS) [4] developed an analytical framework for assessing transition scenarios from existing to future sustainable NESs.

The analytical framework (hereafter, the GAINS framework) is a part of the integrated services provided by IAEA to Member States which consider either an initial development or the expansion of their nuclear energy programs. The framework was understood as a common methodological approach based on the same assumptions and boundary conditions. The major framework components include:

- Scenarios for long term nuclear power evolution based on IAEA Member States' high and low estimates for nuclear power demand until 2050, and trend forecasts to 2100 based on projections of international energy organizations;

- A so-called 'heterogeneous' global model to capture countries' different policies regarding the back end of the NFC;
- Metrics and tools to assess the sustainability of dynamic scenarios for a NES deployment, including a set of KIs and EPs;
- An international database including characteristics of existing and future innovative nuclear reactors and associated NFCs for material flow analysis, which expands upon other IAEA databases and takes into account different preferences of Member States;
- Findings from analyses of scenarios of a transition from present nuclear reactors and NFCs to future NES architectures with innovative technological solutions.

The INPRO methodology was designed as a tool for assessing the capabilities of a national NES to meet requirements of sustainability, whereas the GAINS framework is aimed at comparing options and possible scenarios at the national, regional, and global levels. Accordingly, the GAINS framework relates to INPRO methodology primarily through the concept of 'key indicators' introduced in INPRO methodology reports.

The GAINS framework measures the transition from an existing to a future sustainable NES by the degree to which the selected targets (e.g., minimized waste, minimized amounts of direct use materials in storage, or minimized natural resource depletion) are approached in particular evolution scenarios. The KIs and EPs are compared to determine the more promising options for achieving the selected targets.

The set of KIs and EPs for the GAINS assessments has been selected for global scenarios with different NES architectures after considering more than a hundred candidate indicators comprising all areas in the INPRO evaluation methodology. GAINS project participants sought to reduce the number of KIs to a minimum to facilitate implementation of a scenario-based approach. As shown in Table 1 these KIs/EPs quantify nuclear power production by reactor types, resource availability, discharged fuel mass, amount of generated radioactive waste, NFC services, costs and investment of a 'global' NES. This set of GAINS KIs and EPs, originally developed for global architectures, can be adapted for a more localized application.

2.1.4. Nuclear Reactor Technology Assessment for Near Term Deployment

Another IAEA guidance effort addresses 'newcomer' Member States which have embarked recently on initiatives to establish or reinvigorate nuclear power programs. The major challenge in this enterprise is the process associated with nuclear reactor technology assessment (RTA) for near term deployment. RTA permits the evaluation, selection and deployment of the best technology to meet the objectives of the nuclear power programme when choosing among various variable reactor designs.

The IAEA report [5] demonstrates the basic elements of the approach and technology that is recommended for use in performing RTA. Table 2 lists the key elements and numbers of features and sub-features that have been chosen as the base categories for the RTA within the context of the ad-joining major tasks of nuclear programme implementation, delivering the core technical evaluations for the project feasibility study, the bid invitation, the bid evaluation and contracting, and the reactor deployment phases. The report [5] describes the use of quantitative measures and metrics to evaluate the worth of design features so that the assessment team can develop a consistent approach for assigning the benefit derived from each required element. The study recommends for application in decision making processes the Kepner–Tregoe analysis and the multi-attribute utility theory techniques, and describes in detail how these approaches can be employed to improve current practices in selecting the best technology for application.

Table 1. The GAINS key indicators and evaluation parameters.

No	Key Indicators and Evaluation Parameters		INPRO Assessment Areas					
	Color coding indicative of relative uncertainty level in estimating specific quantitative values for future NES (can vary based on a particular scenario)	Low / Medium-Low / Medium-high / High	Resource Sustainability	Waste Management and Environmental Stressors	Safety	Proliferation Resistance and Physical Protection	Economics	Infrastructure
	Power Production							
KI-1	Nuclear power production capacity by reactor type							X
EP-1.1	(a) Commissioning and (b) decommissioning rates			X				X
	Nuclear Material Resources							
KI-2	Average net energy produced per unit mass of natural uranium		X	X				
EP-2.1	Cumulative demand of natural nuclear material, *i.e.*, (a) natural uranium and (b) thorium		X	X				
KI-3	Direct use material inventories per unit energy generated		X			X		X
	Discharged Fuel							
KI-4	Discharged fuel inventories per unit energy generated			X				X
	Radioactive Waste Minor Actinides							
KI-5	Radioactive waste inventories per unit energy generated			X				X
EP-5.1	(a) radiotoxicity (b) decay heat of waste, including discharged fuel destined for disposal			X				X
EP-5.2	Minor actinide inventories per unit energy generated			X				X
	Fuel Cycle Services							
KI-6	(a) Uranium enriched and (b) fuel reprocessing capacity, both normalized per unit of nuclear power production capacity					X		X
KI-7	Annual quantities of fuel and waste material transported between groups			X		X		X
EP-7.1	Category of nuclear material transported between groups					X		
	System Safety							
KI-8	Annual collective risk per unit energy generation				X			
	Cost and Investment							
KI-9	Levelized unit of electricity cost						X	
EP-9.1	Overnight cost for Nth-of-a-kind reactor unit: (a) total and (b) specific (per unit capacity)						X	
KI-10	Estimated R&D investment in Nth-of-a-kind deployment						X	X
EP-10.1	Additional functions or benefits						X	

2.2. The NEA Related Studies and Projects

The Nuclear Energy Agency (NEA) is an intergovernmental organization which brings together 31 countries from North America, Europe and the Asia-Pacific region in a small, non-political forum with a mission to assist the member countries in maintaining and further developing nuclear energy for peaceful purposes. In particular NEA provides the authoritative assessments and forges common understandings on key issues in energy and sustainable development area.

Table 2. RTA key elements, features and sub-features.

Key Elements	A Number of Key Features	A Number of Key Sub-Features
Site specific considerations	7	7
Grid integration	5	5
Nuclear plant safety	18	47
Technical characteristics and performance	10	59
Nuclear fuel and NFC performance	13	13
Radiation protection	9	9
Environmental impact	4	4
Safeguards	1	1
Plant and site security	6	6
Owner's scope of supply	7	7
Supplier/technology holder issues	18	18
Project schedule capability	6	6
Technology transfer and technical support	7	9
Economics	5	26

2.2.1. Trends in the Nuclear Fuel Cycle

Several studies pertain to an evaluation of the performance and to screening of advanced NES, employing only simplified comparative assessment techniques. Reference [6] provides a description of the developments and trends in the NFC examining their potential to improve the competitiveness and sustainability in the medium and long term and presents criteria and indicators for the future innovative NESs assessments (see Table 3). The study refers to two MCDM techniques: the multi-criteria method and the life cycle analysis without, however, applying them. The search was based on the available results taken over from the life cycle analysis performed in the past. As far as the multi-criteria analysis is concerned the expert group activities were focused on a definition and an analysis of the applicability of criteria (indicators) and the descriptive evaluation of the different NFC options along with developments according to these criteria. Experts have not attempted to quantify and use both criteria and indicators in a concrete selection and prioritization process. The aggregation judgment process would require the development of an adequate life cycle analyses framework, which was beyond the scope of the study. It can be stated that the objective was to provide a basis for a multi-criteria analysis without at aggregating the indicators for a decision making.

2.2.2. Advanced Nuclear Fuel Cycles and Radioactive Waste Management

The broad range and flexibility of advanced NFC under future development offer a possibility to design economic NES addressing efficiently both natural resource and waste management issues. The impact of various advanced NFC on the uranium consumption rate and on the radioactive waste management was analyzed in [7] relative to present industrial practices and current technologies.

Well established codes and tools (such as DARWIN, CESAR, MARNIE, TOUGH2, PORFLOW, see [7] for further details) were used to make numerical analyses of repository long term performance and to design the hypothetical repositories to be located in granite, salt and boom clay formations.

Table 3. Criteria and indicators.

	Criterion	Indicator	Measure
Economy	Economic competitiveness	Levelised NFC cost	USD/kWh (100%)
		Raw materials(U_3O_8 or Th)	USD/t (%)
		Separation work	USD/SWU (%)
		Conversion	USD/kgHM (%)
		Fabrication	USD/kgHM (%)
		Storage	USD/kgHM (%)
		Reprocessing	USD/kgHM (%)
		Transports	USD/kgHM (%)
		Encapsulation and conditioning	USD/kgHM (%)
		Disposal	USD/kgHM (%)
	Financial expenditure	Total cost	USD
	Technology availability	Research (govt.)	USD
		Development (non-govt.)	USD
		I: basic R&D **II**: laboratory/process **III**: pre-industrial **IV**: industrial	Years
Environment and Public Health	Use of non-renewable resources	Energy recovered per kg U	TWh/kg U
	Energy intensity	Ratio of necessary energy input to obtained output	%
	Transportation	Range of ton-kilometers	t×km/kWh
		Energy intensity ratio	%
	Land occupation	Land area used	km^2/kWh
	Greenhouse gas emission	Amount of GHG emitted	t CO_2 eq./kWh
	Amount of waste	Total volume	m^3/kWh
		Volume at each life cycle step	m^3/kWh
		α-emitters	kg/kWh; Bq/kWh
		γ-emitters	kg/kWh; Bq/kWh
	Confinement time of waste	α-emitters	Years
		γ-emitters	Years
	Radiological impacts	Collective operational dose	manSv/kWh
	Human health effects: acute fatalities	Operational accidents	# of immediate fatalities per accident
Society	Human resources, work opportunities	Change in work opportunity	person × year/kWh
		Work opportunity	person × year/kWh
	Broad economic effects	Autonomy of resources	
		Induced industrial production	
	Social aspects	Acceptance and risk aversion	
	Proliferation resistance		

Three groups of fuel cycles were considered. The first group encompassed NFCs based on current industrial technologies and their possible extensions, namely the current once-through NFC in pressurized water reactors and NFC options with only one plutonium recycling path. The second group includes partly-closed NFC, multi-recycling plutonium, while neptunium has always been transferred to waste. The NFC options differ however with regard to the treatment of americium and curium. A main feature of the third group is the deployment of advanced NES which have necessary potential and flexibility for continuous transuranics (TRU) (plutonium and minor actinides) recycling until their fission.

Cost estimations complete the study. The results show that advanced NFC with P&T of TRUs offer various possibilities for strategic choices regarding the efficient use of uranium resource and the optimization of waste repository sites and capacities, while keeping almost constant both: the radiological impact on the repositories and the financial impact on the complete NFC.

The metrics selected for assessment were: uranium consumption, TRUs loss/transfer to waste activity after 1000 years, decay heat after 50 years, decay heat after 200 years, high level waste (HLW) volume including spent nuclear fuel (SNF), maximum dose for repositories in granite, clay and tuff respectively, NFC cost and the total electricity generation cost. The study does not propose any multi-criteria method. Judgments are performed in a qualitative way. Final results (quantified metrics in form of indicators) depicting the NFC performance were synthesized as shown in Figure 3.

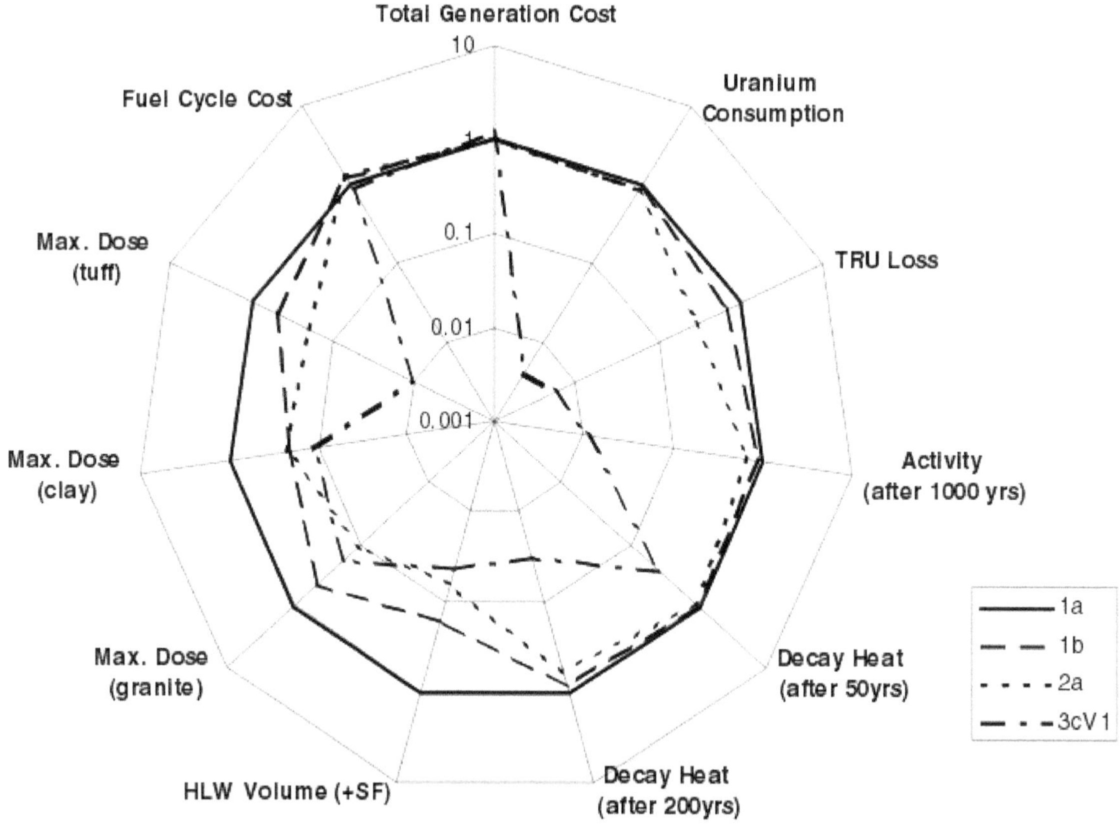

Figure 3. Comparison of eleven representative indicator values for various NFCs (**1a**: once-through NFC; **1b**: full LWR park, SNF reprocessed and Pu re-used once; **2a**: full LWR park, SNF reprocessed and multiple re-uses of Pu; **3cV1**: full fast reactor park and fully closed NFC).

Under the guidance of NEA Nuclear Science Committee and the mandate of Working Party on Scientific Issues of the Fuel Cycle comparative analyses of NFCs in view of resource and nuclear waste management were performed. Expert Groups assessed the impact of innovative NES including P&T on geological repository performance for different host formations [8,9].

2.2.3. Potential Benefits and Impacts of Advanced Fuel Cycles with Partitioning and Transmutation

In order to help promote closer collaboration across the fields of nuclear science and geological disposal the NEA Task Force on Potential Benefits and Impacts of Advanced Fuel Cycles with Partitioning and Transmutation carried out a comparative analyses of studies conducted in several international laboratories on the impact of advanced NESs with P&T on geological repository performance [10].

The outcomes of the analysis can be used to guide the development of appropriate P&T strategies in favorable combination with geological disposal design. The indicators selected for assessment purpose like peak dose rate, radiotoxicity, decay heat, waste form, volume and mass address mainly the repository performance evolution under normal and disturbed scenarios. The role of P&T in the management of uncertainty, which is an essential feature of safety case for geological disposal, was discussed as well as miscellaneous aspects such as proliferation and costs. No method was recommended for judgment aggregation.

2.2.4. Generation IV International Forum (GIF) Methodology

GIF offers a large R&D cooperation framework for preparing the development of six selected future NESs, called Generation IV (Gen IV) systems. Selected Gen IV designs are: Molten Salt Reactor, Super Critical Water Reactor, Gas-Fast Reactor System, Lead-cooled Fast reactor, Sodium-cooled Fast Reactor and High Temperature Reactor. These innovative reactor systems will have enhanced safety, competitiveness, proliferation resistance while using uranium in more efficient way and allowing the optimization of nuclear waste management, thus complying with the concept of sustainable development. A detailed evaluation of NESs may be performed by means of GIF methodology. The results allow the formulation of necessary R&D supporting NES future deployment.

GIF methodology focused mainly on advanced reactor technologies, rather than the overall NFCs, even though NFCs metrics are used in the assessments and does not give any guidelines regarding methods for integrated comparative assessments.

The GIF cost estimation methodology allows two approaches: a top-down method based on scaling and detailed information from similar reactor systems and the conventional bottom-up estimating techniques. To calculate aggregated costs following high level figures of merit are [11] applied:

- Costs to research, develop, and demonstrate the reactor system;
- Capital at risk;
- Annual non-fuel operation costs;
- Annual NFC costs;
- Levelized unit of energy cost and its components.

The GIF proliferation resistance and physical protection methodology is a systematic approach to evaluating vulnerabilities in design concepts. For comparison the IAEA/INPRO methodology for nonproliferation provides 'rules of good practice', therefore both methods together could provide users with overall approach to ensuring robust future design [12]. GIF Proliferation Resistance and Physical Protection measures are:

- *Proliferation technical difficulty*: The inherent difficulty, arising from the need for technical sophistication and materials handling capabilities, required to overcome the multiple barriers to proliferation;
- *Proliferation cost*: The economic and staffing investment required to overcome the multiple technical barriers to proliferation, including the use of existing or new facilities;
- *Proliferation time*: The minimum time required to overcome the multiple barriers to proliferation (*i.e.*, the total time planned by the Host State for the project);
- *Fissile material type*: A categorization of material based on the degree to which its characteristics affect its utility for use in nuclear explosives;
- *Detection probability*: The cumulative probability of detecting a proliferation segment or pathway;
- *Detection resource efficiency*: The efficiency in the use of staffing, equipment, and funding to apply international safeguards to the NES;
- *Probability of adversary success*: The probability that an adversary will successfully complete the actions described by a pathway and it will generate a consequence;
- *Consequences*—The effects resulting from the successful completion of the adversary's action described by a pathway.

The GIF-IV integrated safety assessment methodology consists of five distinct analytical tools or 'elements':

- Qualitative safety features review;
- Phenomena identification and ranking table;
- Objective provision tree;
- Deterministic and phenomenological analyses;
- Probabilistic safety analysis.

It is intended that each element be used to answer specific safety-related questions with different degrees of detail and at different stages of design maturity.

2.3. The U.S. DOE Related Studies and Projects

2.3.1. The Future of Nuclear Power and Nuclear Fuel Cycle

Two Massachusetts Institute of Technology (MIT) studies on the future of nuclear power and NFC were performed in 2003 and in 2011, respectively. The first interdisciplinary study assessed what is required to retain nuclear power as a significant option for reducing greenhouse gas emissions and meeting growing needs for electricity supply [13,14]. Three representative NFCs were considered and evaluated: (1) conventional light water reactor (LWR) operating in a once-through NFC; (2) thermal reactors with reprocessing in a partly-closed NFC with limited recycle; (3) fast reactors with reprocessing in a balanced fully-closed NFC where the fast reactors used to balance LWRs. The NFCs were rated using evaluation criteria including economics, waste management, nonproliferation, and reactor and NFC safety.

The second MIT Study on The Future of the Nuclear Fuel Cycle (2011) performed by MIT and other experts, considered relatively few NFC options for evaluation and was specifically focused on NFC dynamics and transition issues, using designs that would not require fast reactor technologies [15]. Both studies analyze performance characteristics of different NFC options and compare them to the current U.S. open NFC (containing only LWRs). These studies analyze data characterizing NFC material flows and requirements of NFC services peculiar to nuclear power economics, waste management efficiency, nuclear proliferation risks, nuclear safety which after aggregation in each performance area could be used in comparative assessments. Nonetheless these reports contain no recommendations regarding application of judgments aggregation method and tools for comparative assessments. All the conclusions are made based on expert opinions.

2.3.2. Nuclear Fuel Cycle Evaluation and Screening

In 2011, the U.S. D.O.E. Office of Nuclear Energy launched a study on the evaluation and screening of different NFC options, relevant to the U.S. national situation [16]. The study considers the entire NFC, from mining to disposal, including both once-through and partly or fully closed NFC with recycling of Pu or TRU recovered from used fuel. The goal is to identify a relatively small number of promising NFC options with the potential for achieving substantial improvements as compared to the current NFC. The results of this study are intended to strengthen the basis for prioritization of the R&D activities.

Nine evaluation criteria were specified, representing economic, environmental, safety, non-proliferation, security and sustainability goals to identify promising NFC options and measure improvements as compared to the current NFC in the U.S. The first six criteria related to the potential for benefit and the last three reflecting the challenges for developing and deploying a new NFC.

The 40 so-called 'evaluation groups' (EG) of NFC options were collected with similar physics-based performance. The set of NFC options was to be as comprehensive as possible with respect to potential NFC performance. An approach based on the fundamental characteristics of NFC rather than the specific NFC implementation technologies allowed creating a comprehensive set of options which included once-through and recycle NFC; thermal, intermediate and fast neutron reactors; critical and sub-critical reactors (ADS); uranium and/or thorium for fuel along with other distinguishing NFC features (Table 4 and Figure 4). Multi-attribute utility analysis with single-attribute utility step shaped functions for each metrics, proper tradeoff factors and criteria weights (determined using swing technique) were used to evaluate and compare alternative NFCs for multiple criteria simultaneously.

Table 4. NFCs in the Evaluation Groups.

EG	Short Description Indicative of NFCs in the Evaluation Group (EG)	NFC type
	Once-through	
EG01	Once-through using enriched-U fuel in thermal critical reactors	
EG02	Once-through using enriched-U fuel to high burnup in thermal or fast critical reactors	
EG03	Once-through using natural-U fuel in thermal critical reactors	
EG04	Once-through using natural-U fuel to very high burnup in fast critical reactors	***
EG05	Once-through using enriched-U/Th fuel in thermal or fast critical reactors	
EG06	Once-through using Th fuel to very high burnup in thermal ADS	**
EG07	Once-through using natural-U fuel to very high burnup in thermal or fast ADS	**
EG08	Once-through using Th fuel to very high burnup in fast ADS	**
	Limited Recycle	
EG09	Limited recycle of U/TRU with new natural-U fuel to very high burnup in fast critical reactors	**
EG10	Limited recycle of ^{233}U/Th with new Th fuel in fast and/or thermal critical reactors	***
EG11	Limited recycle of ^{233}U/Th with new enriched-U/Th fuel in fast or thermal critical reactors	
EG12	Limited recycle of U/Pu with new natural-U fuel in fast and/or thermal critical reactors	
EG13	Limited recycle of U/Pu with new enriched-U fuel in thermal critical reactors	
EG14	Limited recycle of U/Pu with new natural-U fuel in both fast and thermal critical reactors	***
EG15	Limited recycle of U/Pu with new enriched-U fuel in both fast and thermal critical reactors	
EG16	Limited recycle of U/Pu with new enriched-U fuel in thermal critical reactors and fast ADS	
EG17	Limited recycle of Pu/Th with new enriched-U/Th fuel in thermal critical reactors	
EG18	Limited recycle of 233U/Th with new enriched-U/Th fuel in thermal critical reactors	
	Continuous Recycle	
EG19	Continuous recycle of U/Pu with new natural-U fuel in thermal critical reactors	
EG20	Continuous recycle of U/TRU with new natural-U fuel in thermal critical reactors	
EG21	Continuous recycle of U/Pu with new enriched-U fuel in thermal critical reactors	
EG22	Continuous recycle of U/TRU with new enriched-U fuel in thermal critical reactors	
EG23	Continuous recycle of U/Pu with new natural-U fuel in fast critical reactors	*
EG24	Continuous recycle of U/TRU with new natural-U fuel in fast critical reactors	*
EG25	Continuous recycle of ^{233}U/Th with new enriched-U/Th fuel in thermal critical reactors	
EG26	Continuous recycle of ^{233}U/Th with new Th fuel in thermal critical reactors	**
EG27	Continuous recycle of ^{233}U/Th with new enriched-U/Th fuel in fast critical reactors	
EG28	Continuous recycle of ^{233}U/Th with new Th fuel in fast critical reactors	**
EG29	Continuous recycle of U/Pu with new natural-U fuel in both fast and thermal critical reactors	*
EG30	Continuous recycle of U/TRU with new natural-U fuel in both fast and thermal critical reactors	*
EG31	Continuous recycle of U/Pu with new enriched-U fuel in both fast and thermal critical reactors	
EG32	Continuous recycle of U/TRU with new enriched-U fuel in both fast and thermal critical reactors	
EG33	Continuous recycle of U/Pu with new natural-U fuel in both fast ADS and thermal critical reactors	**
EG34	Continuous recycle of U/TRU with new natural-U fuel in both fast ADS and thermal critical reactors	**
EG35	Continuous recycle of U/Pu with new enriched-U fuel in both thermal critical reactors and fast ADS	
EG36	Continuous recycle of U/TRU with new enriched-U fuel in both thermal critical reactors and fast ADS	
EG37	Continuous recycle of ^{233}U/Th with new enriched-U/Th fuel in both fast and thermal critical reactors	**
EG38	Continuous recycle of ^{233}U/Th with new Th fuel in both fast and thermal critical reactors	**
EG39	Continuous recycle of ^{233}U/Th with new enriched-U fuel in both thermal critical reactors and fast ADS	
EG40	Continuous recycle of ^{233}U/Th with new Th fuel in fast ADS and thermal critical reactors	**

ADS: subcritical reactors, ^{233}U/Th indicates recycle of uranium that is predominantly ^{233}U with thorium; (*): Most Promising NFCs; (**): Additional Potentially Promising NFCs; (***): Other Potentially Promising NFCs.

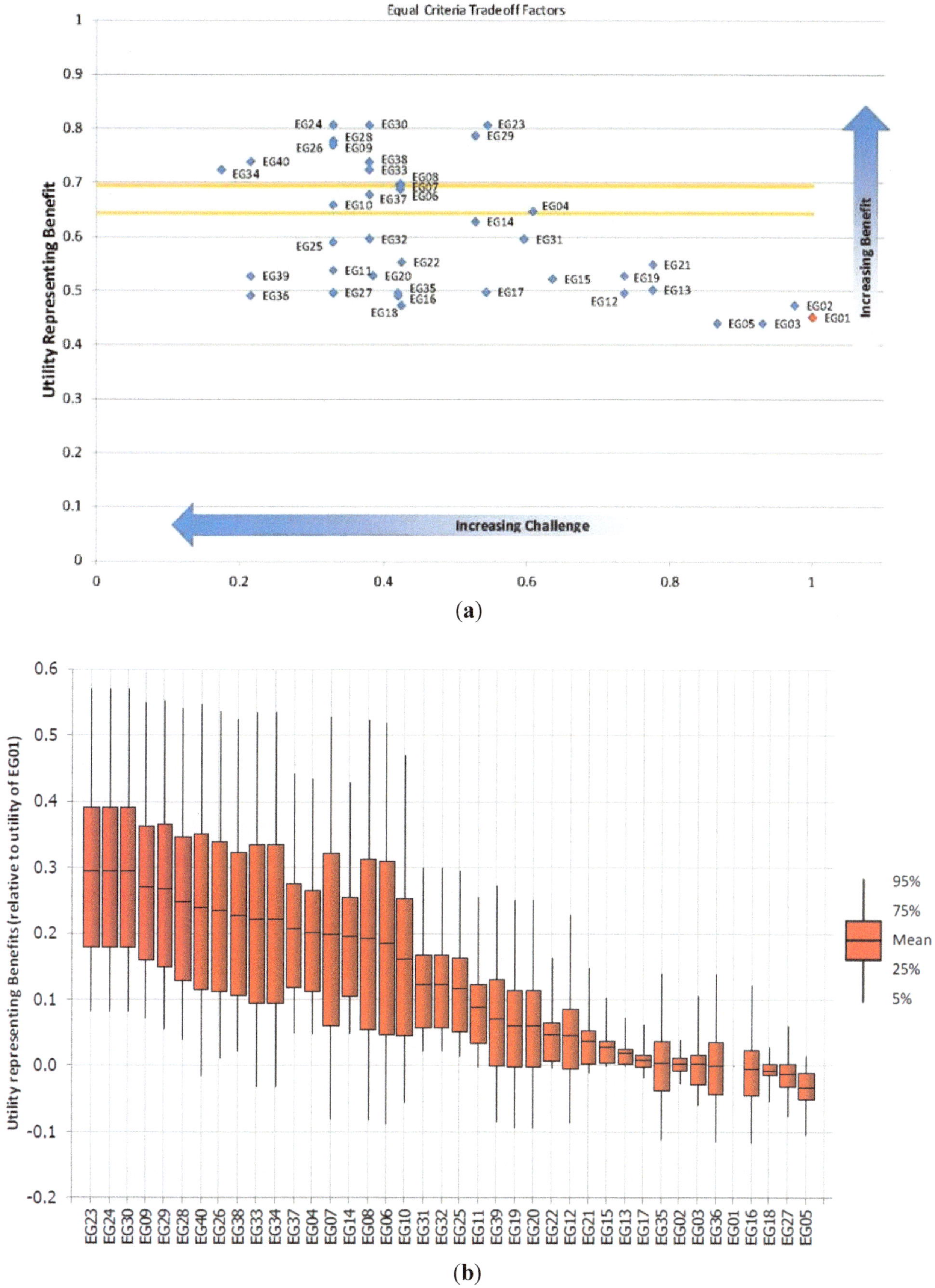

Figure 4. Considered NFCs and evaluation results. (**a**) Benefits *vs.* challenge; (**b**) Utility representing challenge.

2.4. Initiative Studies

2.4.1. Research Articles and Initiative Studies

An exhaustive set of research articles demonstrating results of initiative studies may be found in the journals and web-sites which give an overview on the experience gained with the MCDM methods and tools application in the nuclear engineering area (see, for example, [17–24]). Some of them will be discussed in the following section. The examples analyzed in these studies show that the application of MCDM methods may provide added value to conventional analysis techniques of possible options and increase a degree of clarity (better understanding). Moreover MCDM yields reasonable stable, well-interpreted and decision-making oriented results clarifying the pros and cons of the considered alternatives on quantitatively methodologically proven and well-elaborated foundations.

In a single paper it is impossible to reflect all available studies related to implementation of MCDM tools for nuclear engineering area and, in particular, for NESs and NFCs efficiency and sustainability comparative assessments which are becoming more popular day after day. For these reasons we limit our consideration to some general comments related to these studies.

2.4.2. Performance Indicators

The extensive set of different quantitative performance indicators (we use a 'performance indicator' term to combine in one notion such terms as 'figure of merit', 'key indicator', 'criteria', 'efficiency', *etc*., which are used in different studies to characterize performance of systems) are used in the studies related to NFCs and NESs efficiency and sustainability assessments which encompass economics, waste management efficiency, nuclear proliferation risks, nuclear safety aspects and cross-cutting issues. Some of these indicators which may be objectively evaluated characterize reactor technology performance, NFC material flows and requirements on NFC services, other one can be evaluated only qualitatively and are based on expert judgments. Different ways were proposed for aggregation of corresponding performance indicators into integrated performance indexes in order to assess the efficiency and sustainability associated with the corresponding options.

The lack of an agreement regarding the most representative set of performance indicators which could be generally used implies the need to continue activities on identification of the most informative performance indicators characterizing each of the problem areas. In dependence on the analyzed problem different types of performance indicators may be used such as: qualitative in form of descriptors and quantitative (binary, discrete, continuous, limited and unlimited, *etc*.). The type of indicators chosen determines the selection of appropriate MCDM methods for performing assessment.

Below some examples of the most popular performance indicators used in different studies:

- *Economics area*: net present value, discounted cost, internal rate of return, levelized cost, discounted payback period, overnight capital costs, *etc*.;
- *Non-proliferation area*: different aggregation of material flows (plutonium risk exposure, total amount of fissile materials in the NFC, *etc*.), fissile materials production capabilities, attractiveness of nuclear materials and facilities, time needed to produce a certain quantity of fissile materials, along with others;

- *Waste management area*: radiotoxicity, radioactivity, SNF and HLW amount and volume, peak dose rate, decay heat, time needed to reach a certain level of radiotoxicity of waste, *etc.*;
- *Resources area*: total uranium (thorium, plutonium) consumption, time needed to consume uranium reserves, and so on;
- *Nuclear safety*: events probabilities and consequences, collective radioactive exposure, *etc.*
- *Cross-cutting issues*: maturity of technologies, R&D needs and costs, infrastructural capabilities and others.

While forming a set of performance indicators for the NES and NFC efficiency and sustainability assessments, the authors intend to meet particular requirements, in general not every evaluation parameter may be considered and selected as a performance indicator. A generalized set of requirements should primarily obey completeness, informativeness, non-redundancy, independence, and decomposability rules. Let us briefly discuss these requirements.

1. *Completeness*. All parameters included in the set of performance indicators should provide adequate NES and NFC assessments, *i.e.*, the set should consist of all indicators that characterize the main aspects of both the problem area and the NES and NFC options under evaluation.
2. *Informativeness*. All performance indicators should be clearly understandable to both experts and decision-makers and facilitate the development and adoption of effective solutions. In parallel performance indicators should be analyzed to characterize the main aspects of developed scenarios and provide an opportunity to obtain their unequivocal quantitative assessments.
3. *Non-redundancy*. The set of performance indicators should be non-redundant to avoid duplication of information during the analysis of a problem in order not to complicate the evaluation process. At the same time the set should include all indicators necessary for qualitative assessment of the NES and NFC options.
4. *Independence*. The set of performance indicators should include only those indicators that neither overlap, nor are correlated with other indicators from the set.
5. *Decomposability*. If there are a large number of performance indicators, it is reasonable to divide them into smaller groups for convenience in simultaneous management. A set of performance indicators in this case may be built in the form of hierarchical structure (criteria or objective tree) reflecting hierarchical subordination level. In some cases, it may be necessary to implement several MCDM methods. Therefore the criteria tree construction requires additional work of expert team. It is obvious that in order to aggregate the values of indicators belonging to higher hierarchical level it is necessary to evaluate the indicators values at the lower hierarchical levels. The methods for calculating the values of indicators at the lowest hierarchical level are nowadays well developed.

2.4.3. Judgments Aggregation Procedures

The majority of the studies analyzed usually refer to multi-criteria techniques or methods, without applying them. Sometimes the results of multi-criteria studies performed by other groups are used or the studies demonstrate the implementation of one of the most known multi-criteria methods widely applied for supporting decision-making process in different subject areas. These approaches will be briefly described in the following section. Often experts limit themselves to a definition and analysis of the applicability of used performance indicators along with their evaluation for the different NESs and NFCs options. The aim is to provide a basis for a multi-criteria analysis without making an attempt to aggregate the indicators in a decision making process in the next step.

Integration of individual performance indicators into an aggregated indicator (also known as metrics, aggregated index, composite indicator) is carried out either by formulation of the universal decision rule, where the aggregated indicator is defined a priori, or by interactive (dialog) procedures, where the aggregated indicator will be defined a posteriori. Different types of aggregated indicators are addressed below:

- Arithmetical $F_a = \sum_{j=1}^{k} \mu_j \cdot (z_j / z_{jb})$;
- Geometrical $F_g = \prod_{j=1}^{k} (z_j / z_{jb})^{\mu_j}$;
- Quadratic $F_q = \sum_{j=1}^{k} \mu_j \cdot (z_j / z_{jb})^2$;
- Harmonic $F_h = 1 \Big/ \left(\sum_{j=1}^{k} \mu_j \cdot (z_j / z_{jb}) \right)$;
- General $F_o = \sum_{j=1}^{k} \mu_j \cdot \varphi(z_j / z_{jb})$, where φ is the differentiable function and z_{jb} is the reference value.

Here the k-tuples $z = (z_1, \ldots, z_j, \ldots, z_k)$ denote performance indicators, $z_i = (z_{i1}, \ldots, z_{ij}, \ldots z_{ik})$ – values of performance indicators for an alternative *i*, μ_j – weight of a performance indicator *j*, $\sum_{j=1}^{k} \mu_j = 1$.

All studies based on MCDM methods must formally reproduce all necessary stages of the decision support process: problem specification, formulation of alternatives, performance indicators identification, criteria assessment, selection and implementation of a MCDM method, uncertainty and sensitivity analysis, final conclusions and recommendations. Special attention has to be given to describe the most formal stages of the decision support process related to a specific mathematical method adopted like justification of a MCDM method applied, specification of the method parameters and assumptions, results representation, uncertainty and sensitivity analysis.

Given on one side a wide range of problems with multi-criteria character in the area on efficiency and sustainability assessments for NESs and NFCs and multiple objectives for consideration (at the levels of whole NES, a particular NFC or reactor technology, respectively), and the variety of existing judgment aggregation methods (with/without uncertainties, group/individual, discrete/continuous) on the other side it becomes evident that the application conditions for multi-criteria decision support methods and tools should be examined in view of NESs and NFCs comparative assessments.

3. Multiple-Criteria Decision-Making Framework

Multiple criteria decision making (MCDM) methods are a support tool aimed to help decision makers who are faced with numerous sometimes conflicting assessments to highlight conflicts and perform proper trading off during the decision making process [25,26]. Multi-Criteria Decision Analysis (MCDA) and Multi-Objective Decision Making (MODM) constitute the main components of MCDM. There are different classifications of MCDM problems and methods. The major distinction between MCDM problems is based on whether the solutions are defined explicitly or implicitly.

MCDA problems consist of a finite number of alternatives, explicitly known at the beginning of the decision support process. Each alternative is represented by its performance on multiple criteria. The problem may be defined as searching for the best alternative from the decision maker point of view or finding a set of acceptable trade-off among alternatives.

In general, MCDA will be applied to the following problem: given a set of M alternatives and N criteria for their assessment, where each of the alternatives has been evaluated by either by experts or through objective calculations. A decision rule must be found on the base of the experts' preferences, which will allow ranking all the alternatives according to their values and identifying the best among them.

In the MODM problems, the alternatives are not explicitly known. An alternative (solution) can be found by solving a mathematical optimization problem. The number of alternatives may be either infinite or not countable (when some variables are continuous) or typically very large, if countable (when all variables are discrete).

Both of groups of methods may be applied to assess the sustainability and effectiveness of NESs and NFCs to quantitatively evaluate the most vulnerable areas of the considered options on one hand and to identify the cost-effective risk mitigation measures on the other hand in order to assess the effect from the implementation of these measures on risks minimization associated with the corresponding vulnerabilities. Although tasks, that can be solved using these groups of methods, are different their combined use seems appropriate to identify the most suitable and balanced NESs and NFCs deployment ways despite various controversial factors.

3.1. MCDA Application to Comparative Assessment of NESs and NFCs

3.1.1. The MCDA Paradigm

MCDA is a process which can be divided in the following main phases (Figure 5). At first, the problem should be formulated and structured. All parties interested in the analysis (decision-maker, experts, stakeholders, *etc.*) should find common attitude problem interpretation and understanding. This includes elaborating sets of alternatives, criteria, suitable constraints, involved uncertainties, *etc.*; and identifying goals and preferences, including factors and possible solutions as key points for further discussion and analysis.

The second phase implies construction of a model and its use. The basic MCDA characteristic is the formalization of all preferences involved in the analysis. Based on these preferences, decisions should be made by comparison of refined and elaborated sets of alternatives in a systematic and transparent manner.

Finally, based on the assessments performed and results obtained, including of the outcome of sensitivity and uncertainty analysis, a certain decision on a more preferable solution possible, otherwise it is necessary to turn back to one of the previous stages of MCDA process.

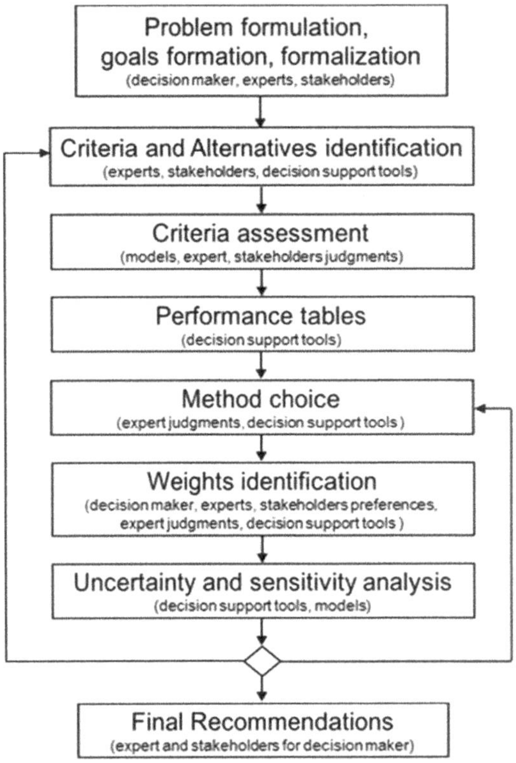

Figure 5. Flowchart of MCDA methods application.

Thus, properly organized studies based on the multi-criteria analysis represent a complex process not only formally operating with a set of mathematical methods and various analytical tools, but also leading to a comprehensive understanding of the problem and its elaboration.

MCDA does not provide a 'right solution'; in this regard it is appropriate to talk about a compromise or a trade-off solution, paying special attention to an analysis of the solution robustness to various methods used. It can be recommended to use an approach consisting of several MCDA methods application which can help all parties involved in the analysis process to understand, recognize and analyze the problem more thoroughly.

Many MCDA methods exist. Most of them were implemented in universal decision support software systems, which enable group analysis of decisions and uncertainty treatment. Choosing a unique solution or further reduction of a set of reasonable options may be performed by means of computer-aided decision-making tools based on different MCDA methods [25]; such as 1000Minds, Analytica, Criterium DecisionPlus, DecideIT, Decision Lab, Decision Lens, Expert Choice, Hiview3, Logical Decisions, MakeItRational, TreeAge Pro, *etc.* Usually the decision support software implements one or mostly two MCDA methods.

3.1.2. The Main MCDA Steps

The decision support process starts with identification of preferences for a given group of decision-makers, experts and the stakeholders (interested in a certain decision), passes through the following phases: problem formulation, specification of alternatives, criteria identification, criteria assessment, selection of MCDA method, uncertainty and sensitivity analysis, final conclusions and recommendations.

The 'problem formulation' step should provide a clear statement of the problem, reflecting both its current state and a vision of the preferred result. Consensus should be reached on problem formulation in order to avoid possible further misunderstandings during the decision support process. At this step it is necessary to specify clearly the requirements, conditions and restrictions that must be satisfied by any possible solution of the problem. The goal formulation is crucial issue specifying directions in a solution finding process.

The subsequent step 'specification of alternatives' provides an identification of possible ways to achieve the goals (a solution, alternative, *etc.*). Each alternative must meet all identified requirements, conditions and restrictions. The number of alternatives may be finite or infinite.

The 'criteria identification' step includes a formulation of quantitative parameters concretizing the goals against which each alternative is measured. Each goal must match one or more criteria. The criteria can be either qualitative or quantitative and, in some MCDA methods they must obey a certain set of requirements (not to be redundant, to be independent, *etc.*).

The 'criteria assessment' step should provide an evaluation of alternatives according to the proposed criteria and the development of the so-called performance table, which is an input data for any MCDA method. It is worth to mention that the assessments may be expressed quantitatively or qualitatively, they may be objective or subjective.

The 'selection of MCDA method' step should be accomplished with a selection of the most suitable method chosen according to the nature of the problem being solved. On the basis of the method chosen an assessment of the alternatives will be made.

The 'uncertainty and sensitivity analysis' step should provide justification for the solutions obtained, whichever method was used or whatever alternative was chosen. Such analysis helps to check stability and robustness of the solution under given uncertainties under considered criteria values and preferences, as well as to identify suitable ways to restructure the problem, including revision of the considered criteria, alternatives, and methods used.

The 'final conclusions and recommendations' step formulates recommendations to decision-makers regarding the more suitable trade-off solutions which were identified during implementation of the decision support process. If an outcome does not satisfy the decision-maker preferences, analysts should return to the previous steps.

In summary, given a wide range of problems having multi-criteria character in the area of NESs and NFCs efficiency and sustainability assessments with different possible range of assessment (e.g., complete NES level, a particular NFC level or reactor technology level, *etc.*), the MCDA technique is recommended to be applied for comparative analysis and assessment as the most suitable among alternatives. Implementation of the MCDA methods may provide useful new information for ranking of the considered options with a set of performance indicators taking into account experts and the decision-maker preferences on hand.

3.2. Review of Relevant MCDA Methods

3.2.1. The Most Commonly Used MCDA Methods

A large number of MCDA techniques have been developed able to deal with different kinds of problems. The assessments presented in different application studies to the nuclear engineering field were made using the following well-known MCDA methods: Multi-Attribute Value Theory (MAVT), Multi-Attribute Utility Theory (MAUT), Technique for Order Preference by Similarity to the Ideal Solution (TOPSIS), ELimination Et Choix Traduisant la REalité (ELECTRE), Preference Ranking Organization METHod for Enrichment Evaluations (PROMETHEE),Analytic Hierarchy Process (AHP) and Simple Scoring Model (SSM) [27–31] (see Figure 6).

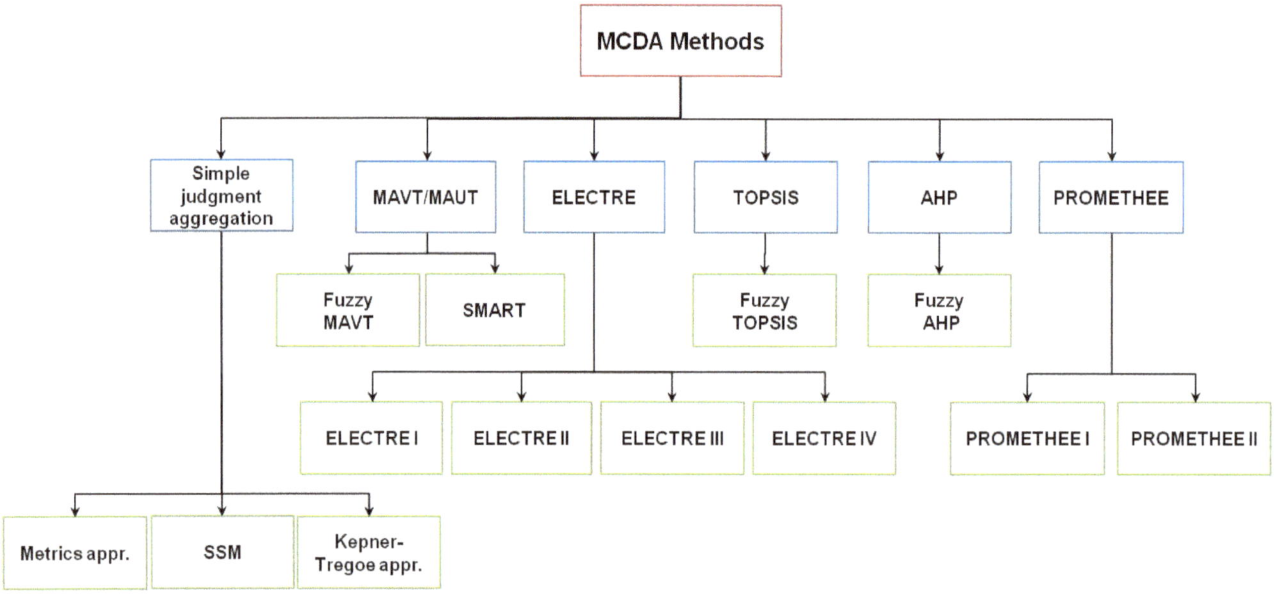

Figure 6. The most commonly used MCDA methods for NESs and NFCs assessments.

In the framework of the classical deterministic MCDA methods, the criteria values and weights are real undistributed (*i.e.*, non-random) numbers. In such methods, uncertainties are examined by means of the sensitivity analysis, generally applying it to changes in the values of weights.

The approach consisting of using several different MCDA methods may facilitate thorough understanding, recognizing and analysis the problem, providing an additional sensitivity analysis of the obtained ranking results to the adopted methods that increase the confidence level to the study. Application of a wide landscape of different methods may have a significant influence on subsequent decision-making since it helps a decision-makers understand and analyze the problem more thoroughly, achieving a consistency in judgments and estimates with stable and robust ranking results. Although the ranks of alternatives may vary for different MCDA methods, the role such analysis plays is important, as it may demonstrate whether different methods may provide non-contradictory results.

3.2.2. The Features of the MCDA Methods

The Multi-Attribute Value Theory (MAVT) is a quantitative comparison method used to combine different measures of costs, risks and benefits into a high-level aggregated performance index—multi-attribute value function accounting for the expert and decision-maker preferences. Aggregated single-attribute value functions can be used, when quantitative information about each alternative is known. A single-attribute value function is created for each indicator. These value functions transform diverse indicators evaluated in their 'natural' scale to one common, dimensionless scale or score (from 0 to 1) in accordance with experts' and decision-maker's judgments. These scores are used in further calculations. The criteria are weighted according to their importance. To identify the preferred alternative, experts should multiply each normalized alternative's scores by corresponding weighting factors for all alternative's criteria, which reflect the experts' and decision-maker's preferences. The total scores (values of multi-attribute value function) indicate the ranks of the alternatives. The preferred alternative will have the highest total score that is the highest rank.

The Multi-Attribute Utility Theory (MAUT) is closely related to MAVT, but it is based upon the expected utility theory, therefore MAUT extends MAVT by using probabilities and expectations in order to deal with uncertainties. A criterion value uncertainty is represented in MAUT by a random variable with given probability density function. The overall utility function for each alternative can be consequently considered in as a random variable. The ranking of alternatives within MAUT is based on the comparison of expected utilities: one alternative exceeds the other if the mathematical expected value of a utility function for this alternative is greater than that the corresponding expected values of the other ones. Despite widespread use of the expected utility function treated as a universal tool to measure the preferences under uncertainties, its application to the choice of alternative ranking is not widely accepted. The MAVT and MAUT methods have been applied to many decision-making problems requiring multi-criteria comparative assessment of nuclear reactors, related NFCs, NESs and technology development areas.

The Technique for Order Preference by Similarity to the Ideal Solution (TOPSIS) is a method of compensatory aggregation that compares a set of alternatives by identifying weights for each criterion, normalizing scores for each criterion and calculating the geometric distance between each alternative and the ideal and anti-ideal alternatives, which are the best and worst scores in each criterion, correspondingly. Thus, TOPSIS is based on a concept that the chosen alternative should have the shortest distance from the most desirable (ideal, or positive ideal) solution and the longest distance from the less desirable (anti-ideal, or negative ideal) solution. Ideal solution is a solution which has the best level for all indicators considered. Negative ideal solution is a solution which has the worst indicator values. TOPSIS selects the solution that is the closest to the ideal solution and farthest from negative ideal solution.

ELimination Et Choix Traduisant la REalité—ELimination and Choice Expressing Reality (ELECTRE) is an outranking method of decision making assuming the construction of outranking relations aimed at comparing each pair of alternatives; and an exploitation procedure that elaborates on the recommendations depending on the type of problem being addressed (choosing, ranking or sorting) by means of two distinct sets of parameters: the importance coefficients and the veto thresholds.

The Preference Ranking Organization METHod for Enrichment Evaluations (PROMETHEE) belongs to the so called outranking methods which imply forming an ordered relation of a given set of alternatives. Outranking methods are based on a pairwise comparison of alternatives for each criterion under consideration, with subsequent integration of the obtained preferences according to a chosen algorithm. In the PROMETHEE method it is required to choose preference function defined in the range from 0 to 1, with specified indifference and preference thresholds.

The Analytic Hierarchy Process (AHP) is a structured technique for organizing and analyzing complex decisions by means of decomposition of the decision problem into a hierarchy of more easily comprehended sub-problems. Once the hierarchy is built, the decision makers systematically evaluate its various elements by comparing them to one another two at a time, with respect to their impact on an element above them in the hierarchy. In making the comparisons, the decision makers can use concrete data about the elements or judgments about the elements' relative meaning and importance. The AHP converts these evaluations to numerical values that can be processed and compared over the entire range of the problem. A numerical weight or priority is derived for each element of the hierarchy, allowing diverse and often incommensurable elements to be compared to one another in a consistent way. In the final step of the process, numerical priorities are calculated for each of the decision alternatives. These numbers represent the alternatives' relative ability to achieve the decision goal.

3.3. Illustrations of Judgments Aggregations for the NESs and NFCs Comparative Assessments

A wide variety of problems can be solved using MCDM methods:

- Identification of optimal or more preferable parameters of reactor technologies comprising a related NFC and NES (type, technical and structural characteristics, *etc.*) Which are promising taking into account preferences of different decision-makers;
- Identification of costs, risks and benefits as well as the corresponding trade-off rates associated with the deployment of certain reactor technologies, related NFC and NES from the system viewpoint;
- Identification of cost-effective risk mitigation measures for reducing risks associated with the deployment of certain reactor technologies, related NFC and NES (technological innovations, diversification, *etc.*);
- Decisions about R&D resource allocation problem supported by an identification of a limited set of reactor technologies worth R&D budgets alignment to provide sustainable NES deployment.

Some application cases of judgments aggregation methods to the NESs and NFCs comparative assessments are discussed below.

3.3.1. Aggregated Indicator (Metrics)

This quite popular approach for judgment aggregation in nuclear engineering is based on using standards of measurement (metrics) by which efficiency, performance, progress, or quality of process or product can be assessed. Aggregated indicator method has been applied for comparative assessments of the nuclear material attractiveness.

Reference [32], for instance, delineates a set of metrics called figures-of-merit (FOM) that are intended to measure the attractiveness or preferences for a range of nuclear materials across a span of credible nuclear adversaries:

$$FOM = 1 - \log\left(M\left[\frac{1}{800} + \frac{h}{4500}\right] + \frac{M}{50}\left[\frac{D}{500}\right]^{\frac{1}{\log 2}} \right) \tag{1}$$

where M is the bare critical mass in kg, h is the heat content in W/kg, and D is the dose rate of 0.2·M evaluated at 1 m from the surface in rad/h.

Another approach, presented in [33] by Saito attempts to account for both spontaneous neutron emission and decay heat using a single attractiveness function. The concept of 'attractiveness' is based on analysis of the physical characteristics of nuclear fuel. Attractiveness may be used for categorization of different nuclear materials hazard. The fuel attractiveness criterion is the attractiveness function accounting for the potential power of a hypothetical explosive device and the technical difficulties arising from its production. The potential yield is determined by the α-Rossi coefficient (ratio of super-criticality to prompt neutron lifetime). The technological difficulties may be represented by the so-called protective barriers, including decay heat (DH) and spontaneously fissile neutron source (NS):

$$ATTR = \frac{\dfrac{\alpha}{\alpha^{239}}}{\dfrac{DH}{DH^{238}} + \dfrac{NS}{NS^{238}}} \tag{2}$$

3.3.2. Multi-Attribute Value/Utility Theory

The MAVT/MAUT methods are widely adopted for various decisions with multi-criteria objectives encountered in nuclear engineering applications. An approach based on the MAVT/MAUT was proposed by Charlton [22] for comparative analysis of diverse NFC for proliferation resistance.

This approach allows one to assess effectiveness of safeguards implementation at front- and at back-end NFC facilities and supports a choice technologies based partly on their effectiveness to deter the proliferation of nuclear materials. The method uses a series of attributes to determine a proliferation resistance measure for each step in a process flow sheet. Each attribute has a weighting that determines its relative importance in the overall assessment and is furnished with associated utility function which. This utility function is derived from both expert knowledge and physical characteristics that relate changes in the value of an attribute to its contribution to the integrated indicator. This approach can be used to estimate risks due to theft by an insider (host), or a threat from outsider (terrorist groups). A method was developed that aggregates proliferation resistance values for each NFC process into an overall nuclear security measure. The nuclear security (NS) measure for the system is determined by:

$$NS = \frac{\sum_{i=1}^{I} m_i\,\Delta t_i PR_i}{\sum_{i=1}^{I} m_i\,\Delta t_i} \tag{3}$$

where m_i is the significant quantity (SQs) of material in the process i, Δt_i is the time for which the material stays in the process i, and the coefficient PR_i determines the proliferation resistance value of the process i. Thus the total nuclear security measure represents the time- and mass-weighted average of the proliferation resistance measure. The mass values as defined by IAEA are: 8 kg for plutonium, 25 kg for high-enrichment uranium, 75 kg for low-enrichment uranium, 25 kg for neptunium-237 and americium (as an element), and 20,000 kg for thorium (as an element). The proliferation resistance for the process i is calculated using following expression:

$$PR_i = \sum_{j=1}^{J} w_j u_j \ (x_{ij}) \tag{4}$$

where w_j is the weight for the attribute j, u_j is the utility function for the attribute j and x_{ij} is the utility function argument for the attribute j in the process i (see Table 5 for attributes, their values and weighting factors).

Table 5. Measures, attributes and weighing factors.

Measures	*j*	Attributes	Weights
Attractiveness	1	Attractiveness level under the US DOE classification (from IB to IVE)	0.10
	2	Plutonium heat release (W/kg)	0.05
	3	Fraction of even plutonium isotopes	0.06
Concentration	4	Concentration (SQ/MT)	0.10
Handling of material	5	Dose rate (rem/h at a distance of 1 м)	0.08
	6	Size/weight	0.06
Type of accounting and control system	7	Measurement frequency	0.09
	8	Measurement accuracy (SQ/year)	0.10
	9	Purity of material	0.03
	10	Share of process operations using accounting systems	0.05
Accessibility	11	Probability of unidentified movements	0.07
	12	Physical barriers	0.10
	13	Inventory (SQ)	0.05
	14	Type of fuel charge	0.06

Fourteen utility functions for selected metrics were constructed as attractiveness level, heating rate from Pu, weight fraction of even Pu isotopes, concentration of fissile material in the process step, radiation dose rates size/weight of a single unit in the process (e.g., a fuel assembly), frequency of measurement of material inventory in the facility, measurement uncertainty in SQs/yr, separability (more separable more conductive for production of weapons), percentage of processing steps that use item accounting, probability of unidentified movement, physical barriers and inventory of fissile material in a facility and finally fuel load type.

3.3.3. Analytic Hierarchy Process

The AHP method has been extensively used for analyzing the issues related to national, energy and economic security. AHP was applied by Silvennonien, Krakowski and others in the nuclear engineering area, to perform comparative analysis of the advanced NFCs for planning of weapons material disposition and other problem. This parametric systems study utilizes a globally aggregated, long-term (~100 years) nuclear-energy model that interprets scenario consequences in terms of material inventories, energy costs, and relative proliferation risks associated with the civilian NFC. Results of pairwise comparison of various plutonium forms assessed in [34] are shown in Table 6. These results are used in comparative analysis of different NFC options in view of reduction capabilities of worldwide plutonium inventories by using conventional reactors and advance fuels.

Table 6. Pairwise comparisons of various plutonium categories.

Plutonium type	Plutonium in a Reactor	Plutonium in SNF	Separated Plutonium	Plutonium at a Reprocessing Facility	Weight
Plutonium in a reactor	1/1	1/4	1/8	1/6	0.05
Plutonium in irradiated nuclear fuel		1/1	4/8	4/6	0.10
Separated plutonium			1/1	8/6	0.55
Plutonium at a reprocessing facility				1/1	0.30

Presented examples are just an illustration reflecting an example of the application of MCDA tools for different judgment measures in nuclear engineering. Obviously the application example set might be expanded. This is however considered here beyond the scope since cases discussed are representative and reflect main common tendencies: diversity of performance indicators used, as a rule only a single MCDA method is applied, sensitivity analysis is performed in a limited scale; uncertainties of subjective and objective nature are not taken into account. These studies lack recommendations regarding applicability of the corresponding methods.

4. Lesson Learned and Promising Directions of the Future Studies

4.1. New Projects and Endeavors

New projects arise which endeavor is to expand the application areas of advanced judgment aggregation tools with uncertainty treatment. The objective is to cope with the abovementioned deficiencies by developing an adequate framework, guidance and tools for comparative evaluation of nuclear technologies and their deployment with respect to the associated costs, benefits and risks.

4.1.1. Strategies for Sustainable Energy Development

Currently any long-term energy supply strategy is strongly focused on sustainable development. Therefore sustainable development indicators are used in conjunction with the MCDA methods. MCDA techniques allow aggregating single indicators and formulating comprehensive judgments about prospective development trends [35–37].

Indicators of sustainable development in a sector of energy production should substantiate the following facts: (1) in selecting energy carriers and relevant technologies for producing, delivering, and using energy products and services, it is required to consider economic, social, and environmental implications; (2) policymakers need methods for measuring and assessing the current and future energy use impact on human health and environment; (3) while using these tools, it is necessary to determine the degree of public energy consumption stability and identify the trends for increasing the degree of energy development sustainability. Actually, such quantitative indices called ‘sustainable energy development indicators’ intended to show to what extent an energy system complies with the three main components of the sustainable development concept, *i.e.*, economic, social, and environmental aspects.

Currently, it has become frequent practice to aggregate single sustainable energy development indicators into a composite index (integral metric) which allows assessing the quality and stability of energy system options. This aggregation is implemented with the use of the MCDA methods, among which the MAVT/MAUT family methods are most popular. A multi-attribute utility function, used as an integral metric, makes it possible to synthesize all the sustainable development aspects into a single aggregated composite index.

4.1.2. Supporting Studies on the Ethics of Nuclear Technology

The ethics of nuclear technology is a multidisciplinary endeavor to examine the problems associated with nuclear technology through ethical frameworks and paradigms [38]. The ethics of nuclear technology provides the analysis of ethical issues related to nuclear technology, so as to overcome nuclear dilemmas and structure nuclear debate.

In general, any professional ethics (such as bioethics, information ethics, *etc.*) provide for the problematization of challenging areas to identify possible trade-offs that balance the risks and benefits associated with conflicting factors. Thus, ethics may be considered a guide for practical decision-making that determines the most effective ways to optimize values. The effective optimization of values, if achieved in one area, does not guarantee its presence in other areas. Thus, advances in nuclear engineering may be incompatible with advances in ecology or economics. Ethics is focused on a general (holistic) optimization, and the search for a compromise between conflicting values.

The studies on the ethics of nuclear technology are a response to the aggravation of nuclear dilemmas and debates stimulated by the shortcomings of nuclear technology and provide a conceptual framework to overcome nuclear challenges. In the last decade, nuclear related contradictions loomed so large as to jeopardize not only the present status but the very future of nuclear technologies. The need for curbing associated challenges is stronger than ever, adding accordingly to the importance of comprehensive quantitative analysis of the nuclear technology ethics related problems, to which no alternative has presented itself yet.

In this regard, widening an application of advanced tools for judgments aggregation for supporting studies on the ethics of nuclear technology may provide solid basis for assessing the proposed ways of enhancing the efficiency and sustainability of technological and institutional measures being analyzed through the ethical frameworks and paradigms.

Contrariwise, consideration of nuclear related issues through the ethical frameworks and paradigms may also be useful for the further development and modification of corresponding comparative assessment and evaluation methodologies, to detail issues related to structuring and managing relevant information, organizing and performing assessments providing thereby conceptual background for such studies.

4.1.3. The IAEA/INPRO KIND Project

The new INPRO collaborative project on key indicators for innovative nuclear energy systems (KIND) has the objective to develop guidance and tools for comparative evaluation of the status, prospects, benefits and risks associated with development of innovative nuclear technologies for a more distant future. The urgent needs for such an activity is that the strong development of innovative technologies and the necessity to support future sustainable NESs deployment. Some IAEA Member States ask for support to evaluate the status, prospects, benefits and risks associated with development of particular technologies in comparison to other available, in order to prioritize/adjust within national programmes on innovative nuclear technology development the allotment of financing and some other resources. To achieve the KIND objective the project will:

- Develop a limited number of KIs for the status, prospects, benefits and risks associated with innovative technologies and NESs on their basis;
- Adapt advanced methods of expert judgment aggregation within the assessments involving also quantifiable data, to enable effective comparative evaluation of the innovative NESs based on the defined set of KIs;
- Using the developed KIs and the adapted methods of expert judgment aggregation, perform case studies on assessment of selected innovative nuclear technologies (NESs on their basis);
- Apply the adapted methods of expert judgment aggregation and uncertainty analysis to comparative evaluation of the transition scenarios to future sustainable NESs based on a set of KIs developed in the GAINS project.

In the framework of the KIND project a number of numerical examples were analyzed oriented to demonstration of MCDA method application for the support of decision making process in the NESs and NFCs efficiency and sustainability assessment. Expedience of application of some well-known MCDA methods for comparative assessment of NESs is presented in [39], for a set of KIs assessed for comparative evaluation of NES evolution scenarios in the INPRO GAINS project. This strengthens the applicability evidence to similar problems.

4.2. Advanced Methods and Tools

4.2.1. Multi-Objective Decision Making

Nuclear energy systems have the potential to contribute to growing energy demands in a sustainable way while playing an important role in economic and social development. However, it is extremely difficult to achieve a common view of an optimal pathway for their deployment and to determine *a priori* the most promising options since the selection of different alternatives for the expansion or deployment of new reactor systems where the environmental criterion is added to commonly used economic, technical and social criteria requires the adoption of powerful decision supporting techniques [40,41].

Therefore to make suitable decisions on the future policy concerning the NES and NFC deployment structure, it is necessary to allow discourse of unbiased experts, free of preferences on the specific modeling assumptions, and able to organize systematic research taking into account all possible NES and NFC options in context of the multi-objective character of the problem by using multi-objective optimization energy models and the MODM based advanced tools.

There is a growing understanding that the traditionally practiced single decision-making approach is no longer able to handle these problems. Optimization of NFC and NES structures has a multi-objective nature with various conflicting objectives characterizing different aspects associated with resource consumption, economics, proliferation risks and waste management. The energy sector and planning affect the interests and resources of multiple decision-maker groups. For this reason the development and application of state-of-the-art MODM tools for comparative NES and NFC deployment scenario assessment and optimization appears necessary because of their potential for highlighting conflicts and finding compromises in the decision making process.

Various MODM methods can be adopted for solving the multi-objective optimization problem: *a priori* or *a posteriori* methods, adaptive methods, methods based on the preliminary construction of the Pareto (efficient, non-dominated) set approximation. The alternatives in the MODM problems, are not explicitly known, thus an alternative (solution) is found by solving a mathematical model. The number of alternatives can be either infinite or not countable (when some variables are continuous) or if countable, typically very large (when all variables are discrete).

The family of MODM-based methods consists of the criteria constraint method, linear convolution method (linear programming case), combinatorial optimization problems, genetic algorithms (NSGA-II, MOCHC, *etc.*), reasonable goals method, *etc.*

Essential to this family is the concept of a set of non-dominated (effective) solutions (Pareto set). The Pareto set is informally defined as a set in which the value of any one among the specific optimality criteria may only be improved by degrading at least one of the remaining criteria. Thus, any solution belonging to the Pareto set will not be improved by all the specific optimality criteria simultaneously. It is a difficult separate activity to form the Pareto set, and the information about the Pareto boundary may form the basis of multiple criteria interactive systems of decision making support. A decision maker consulted by experts will ultimately have to take a solution belonging to the Pareto set.

In the framework of numerous MCDM techniques any dominating alternatives must be identified and excluded before proceeding further. The concept of dominance is a relative concept however and much attention should be paid to the context in which it is used, but the identification of dominating and non-dominating alternatives provides additional information useful in the working out the final decision.

Formation of the Pareto set in multiple criteria optimization of dynamic systems necessitates repeated modeling of the system performance with different values of parameters varied. Modeling of a dynamic system is a problem per se and its solution requires the utilization of software systems. Pareto set in a multiple criteria case can be visualized and enables involving interactive decision maps, radar diagrams, and profiles of alternatives in the decision process.

The MODM methods able to solve the multi-objective optimization problem which are based on pre-approximation of the Pareto set have been recently gaining in popularity. The advantage is that using these methods a decision maker will find a trade-off solution on the Pareto front meeting his preferences. However, the main disadvantage of such methods, deterring their extensive application, is their high calculation complexity.

The interactive decision maps technique was developed for the exploration of multi-criteria decision problems with a more than two criteria. The interactive decision map is a graph displaying several overlapped differently colored slices of the envelope. Each slice contains reasonable values of the first and the second criteria for a certain value of the third criterion (in case when three criteria are considered). Values of the first and the second criteria are displayed on the vertical and horizontal axes. The value of the third criterion is related to the color of the slice and is given on the color scale in a slider of scroll-bar displayed on one side of the figure. Pictures of this kind are known as decision maps. The frontier of a slice displays efficient tradeoffs between two criteria for a given value of the third. Values of other criteria (fourth and fifth, if exist) can be changed by moving the additional sliders of the scroll-bars [42].

An interactive decision map is a tool allowing experts to explore different areas in the feasible alternatives space and to determine the closest alternatives to a given point (goal) in **n**-dimensional space (where n is the number of criteria). This goal is not an 'ideal' point, as used in the methods of ideal point family, but it is a 'reasonable' goal which may in principle be reached by the system. The approach does not require weights identification and may be classified as an interactive approach as illustrated in Figure 7 (taken from [43]) generated for evaluation of the attractiveness of highly enriched uranium production scenarios at a clandestine enrichment facility using centrifuge enrichment technology. The trade-off on NES and NFC structures obtained by means of such methods deliver balanced solutions in spite of conflicting indicators and satisfy the cost-effective requirement. These options are laying on the trade-off surface, which is a surface of non-dominated alternatives and allows identifying additional cost caused by an implementation of complex measures taken to increase the NESs efficiency due to reduction of uranium consumption, proliferation risks and SNF accumulation *etc.* (according to a specified objectives). The analysis of different areas on the trade-off surface supports a hypothesis about the potential of each considered NESs to achieve a specified goal by changing the deployment structure and offers a possibility to evaluate corresponding trade-off rates using indicators [44].

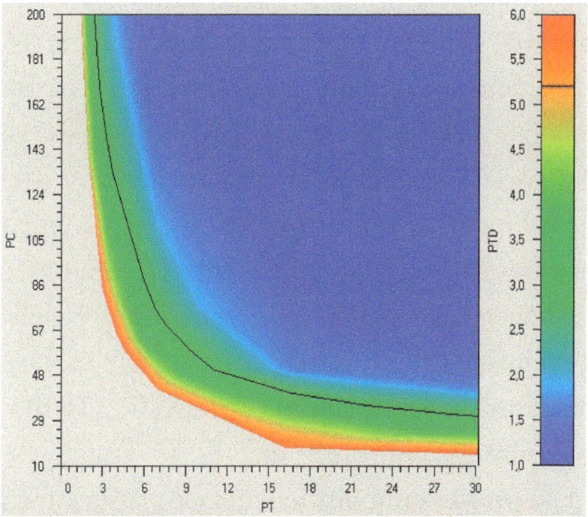

Figure 7. Marginal feasible proliferation scenario set (interactive decision map representation, PT (days): proliferation time, PC (c.u.): proliferation cost, PTD (kgSWU/yr): proliferation technical difficulty.

4.2.2. Uncertainty Treatment

An uncertainty analysis is an inevitable step providing better grounds for judgment and enables decision makers to estimate the stability of the achieved results. Studies related to the NESs and NFCs efficiency and sustainability assessment should contain uncertainty assessment calculations. Although an application of uncertainty analysis methods requires more information about system features and experts' preferences it may, at the same time, greatly enhance the quality of decision-makers' judgments since both the objective (in indicator values) and subjective (in weights) uncertainties can be taken into account [45].

Large uncertainties in initial parameters will not always lead to large uncertainties in performance indicators; thus, an accurate evaluation of uncertainties is needed for correct uncertainty treatment. In particular, a situation may arise when all alternatives due to utilized initial data will be statistically indistinguishable (for example, 90% confidence intervals of uncertainties in parameters are overlapping).

The following methods are available among those for uncertainty evaluation in performance indicators: sensitivity coefficients method, Monte-Carlo methods, quantile uncertainty estimates, GRS-methods, *etc.* The correct implementation of these methods requires however, a modification of corresponding calculation tools to overcome the major shortcomings of the conventional approach to the NESs and NFCs efficiency and sustainability assessments in which only point estimates for the technology characteristics and parameters are required and must be used.

An uncertainty of the decision-makers preferences (weights uncertainty) may be taken into account utilizing, for example, the concepts of the fuzzy numbers, probability theory or interval algebra. The applications of the MCDA methods which allow incorporating uncertainties are, for example, Fuzzy MAVT, Fuzzy AHP, *etc.*

Within a multi-criteria problem, the application of fuzzy numbers may be considered in many cases as justified and more natural than utilization of probability distributions. The use of fuzzy sets can assist with uncertainty assimilation both for criterion values and weight coefficients [46]. The Fuzzy MAVT is intended for uncertainty treatment when solving multi-criteria problems using the 'value function' concept. Within this method, criterion values, scores, and weights are considered as fuzzy numbers, single-attribute value functions are usually considered as given by experts in the forms of usual/crisp functions. Ranking alternatives in the Fuzzy MAVT is based on comparison of overall fuzzy values by means of visual analysis. Several techniques for ranking fuzzy numbers are based on different defuzzification methods and methods for comparison of fuzzy numbers.

Case studies on multi-criteria assessment of the NES and NFC options taking into consideration uncertainties may provide robust judgment aggregation and form a base for well-founded recommendation as compared to the one using deterministic MCDA methods.

For the optimization energy planning models (such as MESSAGE, FCOPT *etc.*), which can be used for studies focused on the optimization of NES and NFC structures, the development and implementation of the robust and stochastic optimizations functionalities for the uncertainty treatment seems to be very promising. Such models are usually based on various deterministic methods adapted to large-scale optimization problem solutions (usually linear programming). The formulation of deterministic models and their solution requires the use of vast amount of data which are assumed to be known exactly, which in practice is often not the case. In the optimization of NES and NFC models data uncertainty is caused by the fact that some exact data do not exist and are replaced by their evaluations, another one cannot be measured exactly, and their true values flow around the measured values. In many cases, even small data uncertainty can make the nominal solution infeasible and practically meaningless.

Stochastic and robust optimizations are complementary approaches for treatment of data uncertainty, both of them having however their own advantages and drawbacks [47]. Both robust and stochastic optimizations enable building an uncertainty-immunized solution to an optimization problem with uncertain data. In stochastic optimization the uncertain numerical data are assumed to be random. In the simplest case, these random data obey a probability distribution known in advance. In robust optimization it is not required to know probability distribution of uncertain parameters. Instead of seeking to immunize the solution, in some probabilistic sense, to stochastic uncertainty, the decision-maker constructs a solution that is optimal for any realization type of the uncertainty in a given set.

4.3. Discussion

4.3.1. General Comments

The MCDM methods can handle judgment problems and may be adapted to the assessment of NESs' and NFCs' efficiency and sustainability. Multi-criteria decision-making techniques are valuable tools in policy formulation. A multi-criteria judgment approach can deliver quantitative information about how much a certain option is more effective and sustainable then another. These methods also provide an opportunity to identify the most vulnerable areas of considered NES of NFC and the cost-effective risk mitigation measures. Moreover, MCDM application allows assessing the impact of measures implementation and optimal risk management associated with corresponding vulnerabilities.

The intrinsic difficulty in NESs' and NFCs' efficiency and sustainability assessment is that there is a large number of performance indicators describing the characteristics of different deployment options. Moreover, indicators of binary, qualitative and quantitative type have to be used. Therefore, there is a strong tendency to develop both a proper framework including the basic concept and a prototype practice-oriented software to be integrated in the decision making process. This advanced software-supported approach has the potential to replace the traditional analytical procedure associated with the application of MCDA methods. Current activities dedicated to the adaptation of advanced methods of expert judgment aggregation in the area of NESs and NFCs performance efficiency and sustainability are focused on the development of methodologies for overall, integrated assessments. As the NES and NFC potential performance must, in some cases, be assessed using a set of conflicting indicators, MCDA are the method of choice. They allow experts to recognize and implement various trade-off based judgments on conflicting alternatives to reach a compromise and to balance solutions on costs, benefits and risks, taking into consideration different decision maker groups involved in the decision process. Each group—decision makers, experts, stakeholders, environmental protection activists—brings along different criteria and point of view (preferences) which must be resolved by applying the multi-criteria decision framework to support overall judgments. Revising possible performance indicators of NESs and NFCs in each problem area, approaches and tools can be developed for performing multi-criteria analysis and on their basis and a structured and practically oriented toolbox can be implemented along with detailed recommendations on its applications.

Framework application of both software and MCDA method should be examined on case studies in order to demonstrate their potential for performing effective comparative evaluations of NFCs and ranking of NESs based on a defined set of evaluated performance indicators.

As concerns the MCDM application framework and tools it seems rational to split the problem into parts and to proceed according to the logic 'from simple to complicated'. This pathway will simplify the analysis providing a clear understanding of results which may be obtained making first the simplest assumptions and, then increase the complexity of the problem and the methods. This approach will benefit from 'learning by doing' effect and give an opportunity to analyze the use of more sophisticated MCDM methods *versus* the simpler ones.

Special attention should be given to both mathematical and application boundaries of the MCDM tools for and to a particular problem. General recommendations for the application in multi-criteria assessments without a reference to specific methods, performance indicator sets, objects of evaluation, *etc.*, should be avoided. Application framework test process should include elaboration of specific test cases, case studies, numerical examples which show the application limits of the MCDM tools in numerous problem areas preferably, on the basis of well-known studies in order to clearly demonstrate the added value resulting from application and to formulate recommendations for decision–maker groups.

It seems more appropriate to emphasize not only the specific method or software, but to consider an expanded landscape of different MCDM tools (*i.e.*, the more representative ones), which can be applied to perform a specific assessment. Particular attention should be paid to the comparative analysis of the results obtained while using the similar model assumptions but the different tools. It is important to get consistent, not conflicting results.

4.3.2. Recommendations on the MCDM Application

Despite the different theoretical frameworks, tools and calculation algorithms, various methods usually lead to a similar final ranking of alternatives. Selecting the best MCDA method suited for a given problem is a separate task. If the selection will be made based on choosing one among all the possible methods, the questions may rise as far as the completeness of the research and the confidence of the expert who proposed the method to solve multi-criteria problems is concerned.

In the general case, if two different expert groups will analyze a multi-criteria problem, the final ranking of the alternatives may not overlap even while using the same method. This is caused by different viewpoint regarding weights, types and parameters of value functions, utility and preferences functions *etc.* A similar situation is to be expected if two expert groups perform their research under the supervision of different experts skilled in using different methods. In this case, the probability that the same ranking will result, even if these groups have similar judgments about the problem, is even smaller. Moreover, the results of weight identification by means of different methods (for example, by swing and pair-wise procedures) may significantly differ, even in case of their implementation within the same MCDA framework.

Differences in rank order may occur due to different representation of a performance table, showing qualitative and quantitative assessments of alternatives on criteria, in case the experts use principally different MCDA methods. Unfortunately, there are no specific rules for conversion between different variants of the performance table into a universal form suitable to be used in different methods. This is especially true for the AHP method which is based on pair wise comparisons. Although the pair wise comparison of alternatives with regard to quantitative criteria seems useful, an automatic conversion of quantitative assessment to pair wise comparison ratios may differ from the coefficients that experts may give directly in the scale of the AHP method.

In numerous studies it was shown that application of the MCDA methods (MAVT, MAUT, TOPSIS, PROMETHEE, AHP *etc.*) for multi-criteria comparative assessments of the NESs or NFCs give, despite some differences in the ranking alternatives, well-coordinated results even while using different methods. The multi-criteria approach provides more detailed differentiation of the alternatives, for specific costs, benefits and associated risks criteria and allows different trade-off options.

For a particular problem it is recommended to make the final choice of the most appropriate method on the basis of the problem context analysis and the initial quality of information provided by experts in the subject matter. In dependence on whether the precise assessment of all indicator values without uncertainty will be done or, on the contrary, the consideration will be limited by qualitative judgments regarding indicator values, different methods should be used. MAVT, TOPSIS, PROMETHEE methods are more appropriate in the former case whereas the AHP method is more suitable in the latter. If it is necessary to consider the uncertainty in indicator values, the MAUT method is more suitable as compared to, for instance MAVT or others. Obviously, a content of relevant method set should not be limited to those considered previously. Experts may express their preference to other ranking/aggregation methods, but in this case, it is essential to make a verification of their applicability and results. Moreover, it should be checked if under comparable conditions, the results obtained using different methods are well coordinated and consistent.

It is recommended to perform an additional sensitivity analysis of the obtained ranking results which obviously will significantly increase the confidence level. MCDA methods applied to a multi-criteria assessment of the effectiveness and sustainability of NESs or NFCs deployment options should respect following recommendations. Despite the fact that the results of ranking alternatives are affected by the expert preferences with respect to certain criteria, there exist stability regions in which for a wide range of variation of the preferences, the ranking order is preserved. On the base on the sensitivity analysis results, on additional analysis of alternatives using expert judgment and on the total set of graphical and attribute information the best alternative may be chosen.

One of the important areas for further development is an extension of MCDA method implementation to problems with uncertainties. In such problems, the weights are not determined by their average values, but are distributed within certain intervals. This is relevant for and in demand of most practical multi-criteria problems. In spite of the fact that in the considered studies the uncertainty analysis was not performed at a full scale, the conditions necessary for extension with uncertainty calculations should be identified not least because it is an inevitable step providing better grounds for judgments with conclusions on the stability of results.

5. Conclusions

The successful development of the application framework of the MCDM methods and of advanced tools dedicated to a comparative assessment of NESs and NFCs achieved so far along with the experience gained to date lead to a confident conclusion that the problem understanding and conceptualization phase has come to a closure. The next phase needed in order to progress is an active and targeted application of the developed toolkits to practical problems addressing possibly a great variety of test cases.

Wide application of the MCDM (both: MCDA and MODM) techniques allows searching for compromises between the conflicting factors that determine the efficiency and sustainable deployment of nuclear reactors and related NFC, calculating corresponding trade-off rates according to expert knowledge based judgments and the decision-makers preferences. To support a decision-making a great deal of alternative comparative multi-criteria analyses should be performed for choosing, ranking, and sorting of the different options.

Growing demand for reliable assessing frameworks for NESs and NFCs potential, efficiency and sustainability requires R&D work to be actively embarked upon building integrated calculation concepts. These are expected to form the basis for dedicated software development featuring flexibility, upgradability and an availability of advanced user interface which still require implementation efforts. In order to upgrade the elaborated framework to a broad application range new case studies analyses and an intensive educational process via specialized training courses should be envisaged.

The MCDA toolkit shall be used to streamline systemic applied activities, to formulate specific guidelines with respect to approaches, in order to increase the NESs and NFCs efficiency and sustainability at technological, institutional and political levels, respectively. If successfully implemented, this bundle of efforts will lay down a firm foundation for the development of the MCDM tools application culture in nuclear engineering area which, in turn, will contribute to raising the overall nuclear energy sustainability.

Acknowledgments

The authors would like to acknowledge the IAEA INPRO Section and the IAEA Coordinated Project INPRO/KIND for an organizational support in performing this study.

Author Contributions

All authors equally contributed to the development of the concepts presented in this paper.

Abbreviations

AHP	Analytic Hierarchy Process
ADS	Accelerator Driven System
BP	Basic Principle
CR	Criteria
DOE	Department of Energy
ELECTRE	Elimination Et Choix Traduisant la Realite
EP	Evaluation Parameter
EG	Evaluation Group
GAINS	Global Architecture of Innovative Nuclear Energy Systems Based on Thermal and Fast Reactors Including a Closed Fuel Cycle
GIF-IV	Generation IV International Forum
HLW	High Level Waste
IAEA	International Atomic Energy Agency
INPRO	International Project on Innovative Nuclear Reactors and Fuel Cycles
KI	Key Indicator
KIND	Key Indicators for Innovative Nuclear Energy Systems
LWR	Light Water Reactor

MAVT	Multi-Attribute Value Theory
MAUT	Multi-Attribute Utility Theory
MCDA	Multi-Criteria Decision Analysis
MCDM	Multiple Criteria Decision Making
MIT	Massachusetts Institute of Technology
MODM	Multi-Objective Decision Making
NEA	Nuclear Energy Agency
NES	Nuclear Energy System
NESA	Nuclear Energy System Assessment
NFC	Nuclear Fuel Cycle
PROMETHEE	Preference Ranking Organization METHod for Enrichment Evaluations
P&T	Partitioning & Transmutation
R&D	Research and Development
RTA	Reactor Technology Assessment
SNF	Spent Nuclear Fuel
SSM	Simple Scoring Model
TOPSIS	Technique for Order Preference by Similarity to the Ideal Solution
TRU	Transuranic Actinides
UR	User Requirement

Conflicts of Interest

The authors declare no conflict of interest.

References

1. International Atomic Energy Agency (IAEA). *Nuclear Energy Basic Principles*; IAEA Nuclear Energy Series No. NE-BP; IAEA: Vienna, Austria, 2008. Available online: http://www-pub.iaea.org/MTCD/publications/PDF/Pub1374_web.pdf (accessed on 30 March 2015).
2. International Atomic Energy Agency (IAEA). International Project on Innovative Nuclear Reactors and Fuel Cycles (INPRO). Available online: http://www.iaea.org/INPRO (accessed on 30 March 2015).
3. International Atomic Energy Agency (IAEA). *Guidance for the Application of an Assessment Methodology for Innovative Nuclear Energy Systems*; Volume 1 of the Final Report of Phase 1 of the International Project on Innovative Nuclear Reactors and Fuel Cycles (INPRO), IAEA-TECDOC-1575; Vienna, Austria, 2007. Available online: http://www-pub.iaea.org/MTCD/publications/PDF/TE_1575_web.pdf (accessed on 30 March 2015).
4. International Atomic Energy Agency (IAEA). *Framework for Assessing Dynamic Nuclear Energy Systems for Sustainability, Final Report of the INPRO Collaborative Project on Global Architectures of Innovative Nuclear Energy Systems with Thermal and Fast Reactors and a Closed Nuclear Fuel Cycle*; IAEA Nuclear Energy Series NP-T-1.14; 2013. Available online: http://www-pub.iaea.org/books/IAEABooks/8873/Framework-for-Assessing-Dynamic-Nuclear-Energy-Systems-for-Sustainability-Final-Report-of-the-INPRO-Collaborative-Project-GAINS (accessed on 30 March 2015).

5. International Atomic Energy Agency (IAEA). *Nuclear Reactor Technology Assessment for Near Term Deployment*; IAEA Nuclear Energy Series NP-T-1.10; IAEA: Vienna, Austria, 2013. Available online: http://www-pub.iaea.org/MTCD/Publications/PDF/Pub1597_web.pdf (accessed on 30 March 2015).
6. National Environment Agency (NEA). *Trends in the Nuclear Fuel Cycle: Economic, Environmental and Social Aspects*; OECD: Paris, France, 2001. Available online: https://www.oecd-nea.org/nea-news/2001/trends19-2.pdf (accessed on 30 March 2015).
7. National Environment Agency (NEA). *Advanced Nuclear Fuel Cycles and Radioactive Waste Management*; OECD: Paris, France, 2006.
8. National Environment Agency (NEA). *Transition towards a Sustainable Nuclear Fuel Cycle*; OECD: Paris, France, 2013; No. 7133.
9. National Environment Agency (NEA). *Nuclear Fuel Cycle Synergies and Regional Scenarios for Europe*; OECD: Paris, France, 2009; No. 6857.
10. National Environment Agency (NEA). *Potential Benefits and Impacts of Advanced Nuclear Fuel Cycles with Actinide Partitioning and Transmutation*; OECD: Paris, France, 2011.
11. National Environment Agency (NEA). *Generation IV Cost Estimating Guidelines Revision 4*; GIF EMWG, OECD: Paris, France, 2007. Available online: https://www.gen-4.org/gif/upload/docs/application/pdf/2013–09/emwg_guidelines.pdf (accessed on 30 March 2015).
12. Bari, R.A.; Whitlock, J.J.; Therios, I.; Peterson, P.F. *Proliferation Resistance and Physical Protection Working Group: Methodology and Applications*; OECD NEA, Paris, France, 2013; No. 7141.
13. Massachusetts Institute of Technology (MIT). *The Future of Nuclear Power, an Interdisciplinary MIT Study*; 2003. Available online: http://web.mit.edu/nuclearpower/pdf/nuclearpower-full.pdf (accessed on 30 March 2015).
14. Massachusetts Institute of Technology (MIT). Update of the MIT 2003 Future of Nuclear Power Study. Available online: http://web.mit.edu/nuclearpower/pdf/nuclearpower-update2009.pdf (accessed on 30 March 2015).
15. An Interdisciplinary MIT study. The Future of the Nuclear Fuel Cycle. Available online: https://mitei.mit.edu/system/files/The_Nuclear_Fuel_Cycle-all.pdf (accessed on 30 March 2015).
16. Wigeland, R.; Taiwo, T.; Ludewig, H.; Halsey, W.; Gehin, J.; Buelt, J.; Stockinger, S.; Kenni, K.; Todosow, M.; Jubin, R.; *et al.* Nuclear Fuel Cycle Evaluation and Screening. Final Report "Fuel Cycle Research & Development", October 2014; U.S. Department of Energy (DOE), FCRD-FCO-2014-000106. Available online: https://inlportal.inl.gov/portal/server.pt/community/nuclear_science_and_technology/337/fuel_cycle_evaluation_and_screening_overview/11124 (accessed on 30 March 2015).
17. Keeney, R.; Nair, K. *Evaluating Potential Nuclear Power Plant Sites in the Pacific Northwest Using Decision Analysis*; International Institute for Applied System Analysis: Laxenburg, Austria, 1976; p. 46.
18. The Nirex Case study. *Generic Repository Studies*; 1988–1990. Available online: http://www.davidsmythe.org/nuclear/nirex%202005%20review%20of%201987-91_site_selection.pdf (accessed on 30 March 2015).
19. Dyer, J.; Butler, J.; Edmunds, T.; Jia, J. A Multivariate Utility Analysis of the Alternatives for the Disposition of Surplus Weapons-grade plutonium. *Oper. Res.* **1998**, *46*, 749–762.

20. Hämäläinen, R.P.; Lindstedt, M.R.K.; Sinkko, K. Multi-Attribute Risk Analysis in Nuclear Emergency Management. *Risk Anal.* **2000**, *20*, 455–467.
21. Butler, J.; Chebeskov, A.; Dyer, J.; Edmunds, T.; Jia, J.; Oussanov, V. The United States and Russia evaluate plutonium disposition options with multi-attribute utility theory. *Interfaces* **2005**, *35*, 88–101.
22. Charlton, W.S.; LeBouf, R.F.; Gariazzo, C.; Ford, D.G.; Beard, C.; Landsberger, S.; Whitaker, M. proliferation resistance assessment methodology for nuclear fuel cycles. *Nuclear Technol.* **2007**, *157*, 143–156.
23. Kim, S.K.; Park H.S.; Lee, K.W.; Jung, C.H. MAUT Approach for selecting a proper decommissioning scenario. In Proceedings of the Waste Management Symposium 2007 Conference, Tucson, AZ, USA, 25 February–1 March 2007.
24. PSI/NEEDS project. 2009. Available online: http://www.needs-project.org/ (accessed on 30 March 2015).
25. Figueira, J.; Salvatore G.; Matthias, E. *Multiple Criteria Decision Analysis: State of the Art Surveys*; Springer Science and Business Media, Inc.: Boston, MA, USA, 2005.
26. Belton, V.; Stewart, T. *Multiple Criteria Decision Analysis: An Integrated Approach*; Kluwer Academic Publishers: Dordrecht, The Netherlands, 2002.
27. Keeney, R.; Raiffa, H. *Decision with Multiple Objectives*; J. Wiley & Sons: New York, NY, USA, 1976.
28. Von Winterfeldt, D.; Edwards, W. *Decision Analysis and Behavioral Research*; University Press: Cambridge, UK, 1986.
29. Hwang, C.-L.; Yoon, K. *Multiple Attribute Decision Making: Methods and Applications*; Verlag. Springer: Berlin, Germany, 1981.
30. Brans, J.; Vincke, P. A Preference Ranking Organization Method: The PROMETHEE Method for Multiple Criteria Decision-making. *Manag. Sci.* **1985**, *31*, 647–656.
31. Saaty, T. *The Analytic Hierarchy Process*; McGraw-Hill: New York, NY, USA, 1980.
32. Bathke, C.G.; Ebbinghaus, B.B.; Sleaford, B.W.; Wallace, R.K.; Collins, B.A.; Hase, K.R.; Jarvinen, G.D.; Bradley, K.S.; Ireland, J.R.; Johnson, M.W.; *et al.* The attractiveness of materials in advanced nuclear fuel cycles for various proliferation and theft scenarios. In Proceedings of the Global 2009, Paris, France, 6–11 September 2009; Paper No. 9544.
33. Saito, M. Development of methodology for plutonium categorization. *React. Phys.* **2008**, *98*, 669.
34. Krakowski, R.A.; Bathke, C.G. Reduction of worldwide plutonium inventories using conventional reactors and advance fuels: A systems study; LA-UR-97-2809. Available online: http://fas.org/sgp/othergov/doe/lanl/lib-www/la-pubs/00412580.pdf (accessed on 30 March 2015).
35. Kowalski, K.; Stagl, S.; Madlener, R.; Omann, I. Sustainable energy futures: Methodological challenges in combining scenarios and participatory multi-criteria analysis. *Eur. J. Oper. Res.* **2009**, *197*, 1063–1074.
36. EUROPEAN COMMISSION. *Key Performance Indicators for the European Sustainable Nuclear Industrial Initiative*; 2012. Available online: https://setis.ec.europa.eu/system/files/Key_Performance_Indicators_Nuclear.pdf (accessed on 30 March 2015).

37. Ashley, S.F.; Fenner, R.A.; Nuttall, W.J.; Parks, G.T. *Sustainability Indicators for Open-Cycle. Thorium-Fuelled Nuclear Energy*; EPRG Working Paper 1217; Cambridge Working Paper in Economics 1233. Available online: http://www.econ.cam.ac.uk/dae/repec/cam/pdf/cwpe1233.pdf (accessed on 30 March 2015).
38. Andrianov, A.A.; Kanke V.A.; Kuptsov, I.S.; Murogov, V.M. Reexamining the ethics of nuclear technology. *Sci. Eng. Ethics* **2014**, 1–20, doi:10.1007/s11948-014-9578-0.
39. Kuznetsov, V.; Fesenko, G.; Andrianov, A.; Kuptsov, I. INPRO activities on development of advanced tools to support judgment aggregation for comparative evaluation of nuclear energy systems. *Sci. Technol. Nuclear Install.* **2015**, *2015*, doi:10.1155/2015/910162.
40. Andrianov, A. Approaches and software for multi-objective optimization of nuclear power structures. *Sustainability* **2012**, *4*, 721–739.
41. Andrianov, A.; Kuptsov, I.; Murogov, V. Towards sustainable nuclear power development. *Int. J. Nuclear Power* **2014**, *59*, 287–293.
42. Lotov, A.; Bushenkov, V.; Kamenev, G. *Interactive Decision Maps: Approximation and Visualization of the Pareto Frontier*; Kluwer Academic Publishers: Dordrecht, The Netherlands, 2004.
43. Andrianov, A.; Ponomarev-Stepnoy, N.; Rumyantsev; A. *Approaches to and Methods for Quantitative Assessment of Nuclear Proliferation Risk*; INPE NRNU MEPhI: Obninsk, Russia, 2011; pp. 1–132.
44. Andrianov, A.; Kuptsov, I. Approach to proliferation risk assessment based on multiple objective analysis framework. In Proceedings of the International Nuclear Fuel Cycle Conference, GLOBAL 2013: Nuclear Energy at a Crossroads, Salt Lake City, USA, 29 September–3 October 2013; Volume 1, pp. 587–593.
45. e-Learning educational module 'Requirements and criteria for nuclear technologies'. Available online: http://lannp.iate.obninsk.ru/course5 (accessed on 30 March 2015).
46. Kahraman, C., Ed. Fuzzy multi-criteria decision making. Theory and applications with recent developments. In *Optimization and Its Applications*; Springer: Istanbul, Turkey, 2008; Volume 16.
47. Ben-Tal, A.; El Ghaoui, L.; Nemirovski, A. *Robust Optimization*; Princeton University Press: Princeton, NJ, USA, 2009.

9

Theorizing for Maintenance Management Improvements: Using Case Studies from the Icelandic Geothermal Sector

Reynir Smari Atlason *, Gudmundur Valur Oddsson and Runar Unnthorsson

Department of Industrial Engineering, Mechanical Engineering and Computer Science, Centre for Productivity, Performance and Processes, University of Iceland, Hjardarhagi 6, 107 Reykjavik, Iceland; E-Mails: gvo@hi.is (G.V.O.); runson@hi.is (R.U.)

* Author to whom correspondence should be addressed; E-Mail: rsa3@hi.is

Academic Editor: Vincenzo Doví

Abstract: As renewable energy sectors evolve and grow within a country, the need for expertise to maintain its infrastructure grows. Such expertise is often provided by foreign industries. It is in the global interest to facilitate expertise to grow domestically, eventually leading to widespread clusters of industries around a renewable energy sector and a global growth of expertise. This ultimately fast tracks the development in the renewable energy sector since more players become active in developing solutions. In this article the factors influencing domestic development are identified from previous studies conducted within the Icelandic geothermal sector. The cause and effect relationships between the identified factors are then mapped. A system dynamics causal loop diagram based on Icelandic case studies is presented to visualise how the formation of industrial clusters in the renewable energy sector can be initiated. This visualisation, based on the Icelandic geothermal sector, can be of use for other industries in the renewable energy sector who are attempting to conduct their maintenance procedures domestically and increase the rate of innovation within a country.

Keywords: cluster theory; renewable energy; innovation; geothermal; maintenance

---

1. Introduction

It is estimated that easily reachable oil and gas will be depleted within the next fifty years. It is further anticipated that easily reachable coal will be depleted within the next century or so [1]. Fossil fuel energy sources have driven industrial processes and the ever increasing quality of life amongst the western nations over the past century. They have allowed for previously unknown levels of consumption and wealth relative to pre-industrial times. The depletion of these sources is of increasing concern because the effect of their depletion on human societies is virtually unknown. It is also known that on-going consumption of these sources is accompanied by severe environmental impacts [2]. One way to mitigate the effects of climate change and adapt to the ever increasing scarcity of non-renewable energy sources is to increase the use of renewable energy. The current development in the renewable energy sector is, however, not fast enough for renewable energy to compensate for a significant amount of fossil fuel use. British Petroleum (BP) expects global primary energy demand to increase by 41% before the year 2035, most of which will happen within the non-OECD (Organisation for Economic Co-operation and Development) countries [3]. By 2035, BP anticipates that renewable energy sources will grow from their current level of 2% to contribute about 8% of the world's energy supply [3]. An increase in the share of renewable energy sources can contribute to social and economic development, and can also accelerate energy access (especially in the developing world). It can increase energy supply security and some renewables can contribute to reduced greenhouse gas (GHG) emissions [4]. With this in mind, it should be of great importance to increase the rate of innovation and increase efficiency in the global renewable sector. These developments could potentially contribute to the benefits that were previously mentioned. Even though economic crisis is sometimes considered to have a negative impact on innovation, research, and development [5], crisis is also often seen to be the source of innovation. For example, the oil crisis in the 1970s pushed Icelandic society towards geothermal energy utilisation because domestic oil heating became very expensive.

During the 2008 global financial crisis, the cost of maintaining the Icelandic geothermal power plants grew immensely. This can be attributed directly to the devaluation of the Icelandic Krona, rising oil prices, and the relationship between power plant maintenance and energy prices. Even though some power plant maintenance had previously been conducted domestically, this crisis pushed many Icelandic energy companies to begin attempting to solve problems domestically that had previously been outsourced. This was mostly focused on expensive machinery such as turbines, that required costly, specialised, knowledge. Conducting this maintenance domestically required a build up of skills and knowledge within Iceland, since it had either not been present or was in shortage. This knowledge and skill transfer (KAST), originating to some extent in the 2008 global financial crisis, has resulted in a growth of expertise within the country. Icelandic industries are now almost fully capable of servicing the geothermal sector themselves. The knowledge and skills to conduct specialised, costly maintenance were previously sought internationally from a dispersed group of specialised industries. As the Icelandic geothermal companies began to seek solutions domestically, Icelandic industries, such as machine shops, began to address and solve problems that had previously been solved internationally. The KAST can also be regarded as an industrial cluster formation because the Icelandic industries who were beginning, or improving, their domestic services for the geothermal energy companies were all located within the same geographic region. It is simply a matter of time until these Icelandic industries enter global markets, providing domestically developed solutions.

During a time of crisis, problems may become to large to ignore. This was the case for the Icelandic geothermal industry, whose major maintenance activities that had previously been outsourced became simply to expensive. A method to facilitate this innovation without having an initial crisis would, however, be preferred in every case because a crisis may often lead to significant financial loss and cannot be controlled as desired. Even though various methods and tools are available to improve the rate of innovation, it is the intent of this article to show how the Icelandic geothermal sector managed to move major and expensive maintenance activities to Iceland. As a method to visualise the process of this cluster formation, a system dynamics causal loop diagram is presented, which is based on the Icelandic geothermal industry as it moved its major maintenance procedures to Iceland after the 2008 global financial crisis. The model provided was generated using a series of case studies within the geothermal industry, which were conducted by the authors and published in the scientific literature. The contents of this article are investigated in the context of cluster theory, where the theory is used to describe the concentration of industries around the renewable energy sector, in particular the Icelandic geothermal sector.

1.1. Cluster Theory

Industrial clusters can be defined as "geographic concentrations of interconnected companies, specialised suppliers, service providers, firms in related industries, and associated institutions in a particular field that compete but also cooperate [6]".

Players within a cluster include providers of specialised products and services, infrastructure providers, governmental institutions, think tanks, and trade associations who provide technical support that benefits or contribute to a specific sector. Clusters are an important competitive advantage because other factors that were previously important, such as access to non-scarce resources, are becoming less important as global logistics serve the need for resource transportation. For example, aluminium smelters are located in Iceland but the country lacks any bauxite resources. In addition, deploying sophisticated technology is not a factor because industries can freely use modern technology in their production. In pre-modern times, the technology that was available in one region was not so easily transferred or available to another region, today this is not the case. It then becomes clear that infrastructure, the legal environment, and the services that are located in geographic proximity to a particular industry have become a significant factor in how competitive the industry eventually becomes [7,8]. Being a part of a cluster increases productivity as access to inputs, information, technology and relevant institutions improves. As a cluster forms, the formation becomes self-reinforcing. This is further increased when the public sector is supportive and competition is present [7]. In cluster theory, the role of the government is to remove obstacles to industrial growth and achieve macroeconomic and political stability. It should, according to cluster theory, improve general microeconomic capacity "through improving the quality and efficiency of general-purpose inputs to business and the institutions that provide them" [6]. Regardless of the effectiveness of public policy, it has been shown that a cluster takes a decade or more to develop a competitive advantage [7].

A cluster's absorptive capacity is the "capacity of firms to establish intra- and extra- cluster knowledge linkages" [9]. This is the capacity of a cluster to gather knowledge from the outside and effectively distribute this knowledge on the inside. However, when digging deeper into cluster theory, it can be seen that the knowledge flow is not equally distributed between firms within a cluster. In fact, clustering may isolate some firms while others increase their collaboration. In addition, even though business flows are frequent between firms within a cluster, knowledge flow does not necessarily follow. This has been observed when wine clusters have been studied in Italy and Chile [10]. A sectoral system is in essence the same as a cluster. The players within such a system interact through cooperation, competition, exchange, and communication [8]. Clusters, or sectoral systems, are also a dynamic phenomena that is constantly changing [8]. This happens because firms who seek new markets tend to modify their business behavior or begin interacting with other components of a cluster in a different manner. It has furthermore been stated that a firm's value cannot only be seen from the patents issued, staff, and machines owned, but should also be seen in its participation and involvement within a cluster [11].

The literature is rich with information on the effects of cluster cooperation and how innovations are more likely to grow out of such environments [6,12,13]. Benefits have been shown to be partly due to the close geographic proximity of relevant industries, information, complementary relationships, and competitive pressure [14]. An example of this can be found in the technology industry in Silicon Valley, California. Industrial clusters are, however, not bound to be based on a single geographical location for each industry. A sector possibly benefits from having multiple clusters that are spread globally. If multiple clusters are operating around the same sector, an improved rate of innovation and development may possibly be experienced. The structure of a cluster is dependent on the "characteristics of technologies used, social norms and institutional factors that outline the rules followed" within the clusters [11]. Industrial clusters are a key factor in a nation's innovative capacity [13], it is therefore of great importance to facilitate clusters within a nation where the foundations for such a collaboration are sound. This importance is amplified when the need for faster development within a sector is desperately needed, such as within the renewable energy sector. Cluster formations, rather than the dispersion of industries, could potentially serve as the catalyst needed for the renewable sector to have the significant impact it needs to have within the global energy use portfolio. An impact that increases the access to renewable energy globally.

Even though the literature is rich with information about the benefits of cluster cooperation, it suffers from a lack of methods to initiate clusters where they do not currently operate. It has been attempted to fill this gap, with some success [14]. By conceiving a non-simulated system dynamics model, it has been shown how an industrial cluster effect may be achieved [14]. The previously proposed model is, however, based on a literature review rather than on case studies that are conducted by the authors. Others have attempted to simulate the agglomeration of industries within a region and show how knowledge and proximity effect the behaviour of a technology district [15]. These studies have, however, not shown how a cluster can be formed but rather the benefits of the proximity of players within an industrial cluster. The benefits of using a system dynamics causal loop diagram have been shown to be a convenient way of depicting a cluster behaviour [14,15]. Behaviours of industries within sectors, and the sectors themselves are not static. Industries evolve, compete, collaborate, perish and flourish. One way to describe a swift change in behaviour is to look at unforeseen, catastrophic, unwanted or any events forcing a change in behaviour for industries. Such events are often called disruptive events, leading to a so called quantum shift.

1.2. Quantum Shifts

Disruptive events often lead to a shift in behaviour, this is well known in natural evolution as well as in the corporate sector. Disruptive events are non-controllable events that force industries to modify their operational behaviour. Disruptive events can be felt in various forms for different industries, such as increased competition, modification in the legal or policy environment, or even ecological change, such as climate change. As mentioned, this process also happens in nature, but the focus of this article is on business. Disruptive events leading to adaptation are called quantum shifts or punctuated equilibria [16,17]. For industries to survive after a disruptive event, they must radically improve competitive innovation, new value creation, and create and distribute knowledge [16]. A graph depicting a business development before, during, and after a quantum shift can be seen in Figure 1. This graph, however, assumes that the industry has survived the disruptive event and continues to operate after the quantum shift. This is not necessarily correct because many businesses cease to exist after a disruptive event. A disruptive event can also be felt in the form of competition, such as an introduction of a service or a product on the marketplace by a competitor. However, a quantum shift does not necessarily lead to an elevated performance in the same business or operational direction prior to the disruptive event initiating the shift. Indeed, corporations may need, or see potentials, in shifting their direction of operation. This may include new market opportunities or introducing a new emphasis in their operations.

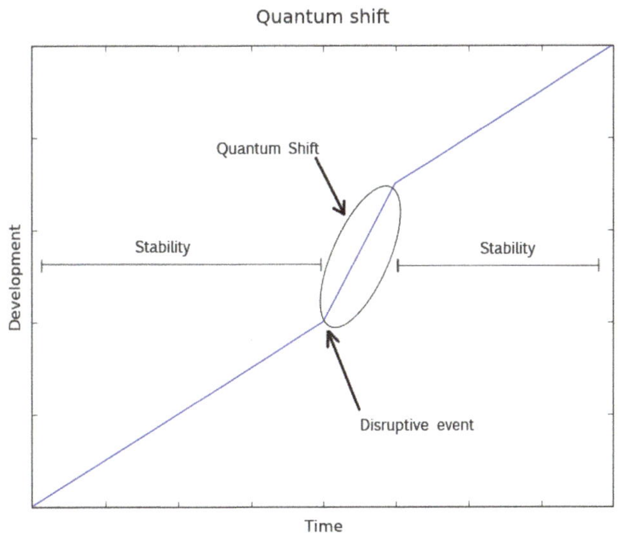

Figure 1. The effects of disruptive events on the operation and development of corporations.

Figure 2 depicts a quantum shift when the operational direction is subsequently altered. A disruptive event, such as a financial crisis, may facilitate a quantum shift for a whole industrial sector of a country rather than individual companies. In Iceland for example, the 2008 global financial crisis served as a disruptive event that forced the geothermal sector as a whole to face new challenges. As the Icelandic currency devalued dramatically, oil prices rose as did the price of spare parts for the power plants, since the price of oil and spare parts is linked. The value of Iceland's debts also rose because they were issued in foreign currency. This all lead to increased costs in the operation and maintenance of Icelandic power plants. The challenges faced were addressed effectively and this has led to further domestic collaboration that continues to thrive. To gain a good understanding of the Icelandic geothermal sector, one must understand who the main players are within the sector. The following section aims to provide such understanding.

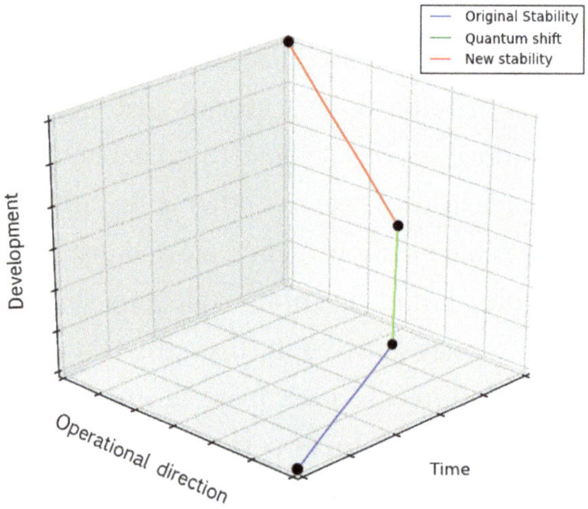

Figure 2. A quantum shift that includes a possible change in operational direction.

2. Key Facts about the Icelandic Geothermal Sector

In this chapter the Icelandic geothermal power plants and the industrial cluster environment that they engage in are described.

2.1. Icelandic Geothermal Power Plants

Six geothermal power plants operate within Iceland. In 2008, the accumulated production amounted to 4.038 GWh (Gigawatt hours) of electricity [18]. The plants are Bjarnarflag (3.2 MWe (Megawatt electrical)), Hellisheidarvirkjun (303 MWe), Krafla (60 MWe), Nesjavallavirkjun (120 MWe), Reykjanesvirkjun (100 MWe), and Svartsengi (76.4 MWe) [18]. Staff from all of the energy companies operating these power plants participated in this research. Icelandic power plants manage to keep a small amount of staff on site while keeping the plants operating without major problems. At Svartsengi and Reykjanesvirkjun (both operated by HS Energy), a staff of approximately 20 people are responsible for the plant maintenance and operations. This staff consists of mechanical and electrical engineers, and earth and environmental scientists. The employees attend to 12 turbines (five steam and seven Organic Rankine Cycle), 36 cooling fans, 17 geothermal wells, wellheads, 70 control valves, 100 pumps, 20 km of pipelines, and a vast amount of valves that all need regular maintenance [19]. A similar situation can be found in the other Icelandic plants, such as Nesjavellir and Hellisheidi, who operate with a total of only 26 employees in each plant [20]. The engineers operate the plant and conduct maintenance. The head of the plant engineering department answers directly under the plant manager. The company CEO is in all cases in direct contact with the plant managers. The plant engineers conduct a great amount of maintenance on site, in collaboration with domestic workshops. It has been stated publicly that the largest geothermal utilisers in Iceland are attempting to fully transfer all of their maintenance activities to Iceland. According to the plant manager at Nesjavellir, the efficient management of the power plants can be attributed to a great extent to the experience that their engineers acquired when working as marine engineers on fishing vessels [21]. Marine engineering education in Iceland is also more detailed than international standards require. At sea, marine engineers can only rely on themselves for repairs, regardless of the prevailing weather conditions, which requires them to adapt a certain mentality, this is especially true for the chief engineer. This mentality is much appreciated by the power plant chief engineers and it includes the notion of unity at site and resourcefulness in order to keep the vessel (in this case the plant) running at all costs. Consequently, the energy companies tend to prefer marine engineers that have worked as chief engineers.

2.2. Domestic Collaboration

Corporations in general tend to outsource more of their non-core activities than they have previously done [22]. This allows companies to focus on other important issues, such as differentiation in the market place or cost leadership [12]. However, industries may depend on non-core activities for successful operations. Non-core activities may become a financial burden that the corporation can solve with lower costs. Outsourcing can also lead to failures because the customer's expectations are not met. Outsourcing also creates dependency on the contractor. Not outsourcing allows the corporation to build knowledge and potential value; however, this is followed by an increase in other costs. Therefore, an industry must analyse the potentials fully before deciding to conduct previously outsourced procedures themselves. Furthermore, by not outsourcing. the corporation becomes independent and frees it from relying on a contractor for its operations. However, a balance is needed between outsourced services and those addressed by the organisation itself.

Various players within the Icelandic geothermal sector have engaged in a conversation on further collaboration in the operation and management of the Icelandic geothermal power plants [23]. These players include energy companies, utility companies, machine shops, banks, universities, research facilities, engineering consulting firms and others [24]. It has been shown that by domestically conducting the operational and maintenance tasks, sometimes overseen by foreign specialists, domestic know-how and experience could be increased and currency be kept within the country. Such procedures could potentially allow domestic service providers to enter foreign markets. Experiments have been conducted on geothermal steam turbine parts and the possibility for such parts to be produced domestically has been analysed. Such experiments have often proved successful. Some experiments were conducted fully by domestic machine shops but some were also conducted in collaboration with foreign specialists. Collaboration with foreign industries contributes to KAST to Iceland and can, therefore, be seen as beneficial to some extent.

Two mutual platforms have been established to smooth the communication platform between all parties: one platform is business driven, while the other is research driven. The business driven platform is called Iceland Geothermal and the research driven platform is called the Geothermal Research Group

(GEORG). The business driven platform is organised by a private entity, which facilitates lectures and provides a communication platform.

Discussion has been ongoing in both the scientific and popular literature about the possibilities of a domestic cluster that builds on Iceland's knowledge of geothermal energy utilisation. This is expected to be a good way to improve the opportunities for domestic industries to serve the geothermal sector in the coming future [23]. The knowledge created within this cluster is expected to be of value to all participating companies. The Icelandic geothermal cluster has been defined as a business driven cluster. GEORG is also a geothermal cluster but is research driven and funds various geothermal based research [25]. GEORG consists of 22 multi-national research driven partners, ranging from universities and public companies to private partners. A connection, both internally and to private industries within the cluster, can also be considered to be a factor for the efficient operation of these power plants. Given the small scale of the Icelandic economy, relatively few service projects are conducted per year and domestic service providers are forced to use their employees for servicing other industries, such as fishing and aluminium. Together, these industries work together to help to support the buildup of domestic knowledge and expertise. By servicing various industries, Iceland's service providers can keep up the number of projects required to maintain their employees. In addition, solutions developed and implemented for one industry can be adapted to another.

The collaboration described above became increasingly visible after the global financial crash. The Porter report [23] can be viewed as a landmark for the visualisation of its benefits. The cluster and the collaboration were already forming before the 2008 financial crisis. However, the 2008 events pushed the Icelandic geothermal industry towards further developments domestically, especially with regards to the most expensive maintenance procedures, namely the turbines. The benefits after this quantum shift occurred are, however, attractive for industries involved in the renewable energy sector, although the initial seed is not. Theorizing for such developments is partially the focus of this article.

2.3. Article Intent and Content

The aim of this article is to gather vital information from a series of case studies that were conducted by the authors and then present a causal loop diagram depicting how cluster formation may partially be initiated. The introduction provides background information describing the importance of domestic industrial clusters to the global renewable energy sector. Subsequently, the methodologies used in each case study are briefly described. The results from the case studies are then used to form the results chapter. A causal loop diagram is shown to demonstrate the behaviour of the Icelandic geothermal industrial cluster. The limitations of this research, suggestions for future research, and general discussions about the developed diagram are provided in the discussions chapter.

3. Methodologies

This section will discuss theory building and describe how the research was planned, eventually providing a description of how and why each case study was conducted.

3.1. Theorizing and Case Studies

The construction of theory from case studies is a strategy that includes one or several case studies that are used to conceive theories or propositions [26]. Case research has been used as a powerful research tool in the development of new theory [27]. Calls have been made for more field-based research simply to keep up with technological changes in managerial methods within operations managements [28]. By conducting field case research, theory is not merely improved or enriched but, equally or more importantly, the researcher themselves are also enriched [27]. However, case research has various challenges, such as time consumption [29], the need for skilled interviewers, and the care needed to draw conclusions [27]. This is amplified when a series of case research studies are to be conducted in theory building. Case research assists with answering how and why questions and it is also good for theory building. Conducting case based research has several benefits [30]:

- The study is conducted within the natural environment of the phenomena, which allows for an observation of real practice.
- Why, what, and how questions can be answered with a great understanding of the phenomenon.
- Early investigations where variables are unknown and the phenomenon is unclear are suitable for case studies.

Various researchers have defined the methodology for case based theory building in a similar manner [27,31]. The steps are outlined in Table 1. Interestingly, even though case based research is praised by some researchers and calls have been made for such research, only a minority of operations management articles are actually case based. The research model has been defined as having five stages, as follows: (1) define the research question; (2) instrument development; (3) data gathering; (4) data analysis; and (5) dissemination [29,31]. In fact, the case based theory building process that is presented seems to be generally accepted by researchers in operations management [31]. Furthermore, it has been stated that service design is one of the potential areas for future operations management research, which is partially the focus of this article [29].

3.2. This Research in a Theoretical Context

In the context of this research, it was decided to follow the guidelines that are shown in Table 1 [31]. Case studies were conducted to answer the questions posed in each step. The theory building process used in this article is also found to be relevant to other theory building procedures demonstrated in the literature, hence its validity [32].

The results of each step were published in the scientific literature for approval, either in journals or peer reviewed conference proceedings.

Even though the methodology to fully construct a theory has been described [31,32], because the proposed model has not been tested, it is not our intent to fully form a theory in this article but rather to theorize. This includes the first three steps in Table 1. This article is a step on the way towards a theory.

Table 1. The theory building process as presented in the literature [31].

Purpose	Research Question	Research Structure	Examples of Data Collection Techniques	Examples of Data Analysis Procedures
1a. Discovery * Uncover areas for research and theory development	* What is going on here? * Is there something interesting enough to justify research	* In-depth case studies * Unfocused, longitudinal field study	* Observation * Interviews * Documents * Elite interviewing	* Insight * Categorization * Expert opinion * Descrptions
1b. Description * Explore territory	* What is there? * What are the key issues? * What is happening?	* In-depth case studies * Unfocused, longitudinal field study	* Observation interviews * Documents * Elite interviewing * Critical incident * Technique	* Insight * Categorization * Expert opinion * Descriptions * Content analysis
2. Mapping * Identify/describe key variables * Draw maps of the territory	* What are the key variables? * What are the salient/critical themes, patterns, categories?	* Few focused case studies * In-depth field studies * Multi-site case studies * Best-in-class case studies	* Observation * In-depth interviews * Diaries survey questionnaires * History * Unobtrusive measures	* Verbal protocol * Analysis * Cognitive mapping * Repertory grid technique * Effects matrix * Content analysis
3. Relationship building * Improve maps by identifying the linkages between variables * Identify the "why" underlying these relationships	* What are the patterns or links between variables? * Can an order in the relationship be identified? * Why should these relationship exist?	* Few focused case studies * In-depth field studies * Multi-site case studies * Best-in-class case studies	* Observation * In-depth interviews * Diaries survey questionnaires * History * Unobtrusive measures	* Verbal protocol * Analysis * Cognitive mapping * Repertory grid technique * Effects matrix * Content analysis * Factor analysis * Multidimensional * Scaling * Correlation analysis * Nonparametric analysis
4. Theory validation * Test the theories developed in the previous stages * Predict future outcomes	* Are the theories we have generated able to survive the test of empirical data? * Did we get the behavior that was predicted by the theory?	* Experiment * Quasi-experiment * Large scale sample of population	* Structured interviews * Documents * Open and closed-ended questionnaires * Lab experiments * Field experiments * Quasi experiments * Surveys	* Triangulation * Analysis of variance * Regression * Analysis * Path analysis * Survival analysis * Multiple comparison procedures * Nonparametric statistics
5. Theory extension/Refinement * Expand the map of the theory * Better structure the theories in light of the observed results	* How widely applicable/ genera-lizable are the theories developed? * Where do the theories apply? * Where do they not apply?	* Experiment * Quasi experiment * Large scale sample of population	* Structured interviews * Documents * Open and closed-ended questionnaires * Lab experiments * Field experiments * Quasi experiments * Surveys * Documentation * Archival research	* Triangulation * Analysis of variance * Regression * Analysis * Path analysis * Survival analysis * Multiple comparison procedures * Nonparametric statistics * Meta analysis

3.3. Case Studies Conducted

The cluster formation will be visualised using a qualitative method. A system dynamics approach is used without simulation because it is not our intent to simulate the rate of formation of a cluster but rather to visualise which factors may initiate such formation.

To visualise the connection between different players within the Icelandic geothermal industry and to create a systems dynamics model, case studies were carried out that focused on various themes. The themes were selected after multiple meetings with chief maintenance engineers at the power plants and machine shops, and meeting with the staff at innovation and cluster centers. The themes that we studied required us to answer some questions that fitted within the theory building process as described in the literature [31].

- Domestic service providers in the geothermal industry.

  This part describes how domestic industries are currently addressing major maintenance issues with regards to geothermal steam turbines in Iceland. It will examine how they collaborate and which products are being manufactured domestically [33,34]. When critical failures occurred, the repair process was visualised and the communication chain was identified. It was seen that domestic repairs were conducted in a more economical and faster way. This resulted in shorter down-time of the turbine and a subsequent lower loss of output. It was also shown that knowledge was building up within the machine shops and the energy company. During one study, where the current turbine operations and maintenance procedures were examined, it was seen that the operations and frequency of overhauls on geothermal turbines is changing as the staff becomes more experienced. Problems are analysed in collaboration with the Icelandic innovation center and also with the machine shops. If a solution is viable when looking at performance, then it was developed further. In this case, a faulty setting on a valve lead to the breakdown of the labyrinth packing. The problem was analysed and a repair, with an improved version of the labyrinth packing, was conducted on-site in collaboration with a domestic machine shop.

- Corporate culture with regard to innovation.

  The development of a geothermal control valve was examined: first, from the corporate viewpoint of how the development process occurred within the company; and secondly, from the technical side of how the valve operates, is manufactured, and tested [35,36]. This allowed for a clear visualisation of a successful innovation process within the geothermal industry. A willingness to try the development of a solution posed by a staff member was observed because of a lack of solutions available to the problem observed and also because of the possible financial viability of the solution since it was to be produced domestically. It was shown that the CEO of the energy company allowed the staff member a certain degree of freedom to develop the proposed valve solution. This included funds for prototyping, and specialised consulting and testing. The valve became the standard for control valves within that particular company. A machine shop was included in the development process, allowing for the knowledge about the manufacturing side of the valve to be located in a geographic proximity to its final use. The valve is currently fully manufactured and used domestically.

- The effects of the operation engineer's previous experience.

  This research also outlined the Icelandic geothermal cluster, and described who the main players are and how they are interconnected. It was seen that maintenance engineers operating within the energy companies possess a certain characteristic, perhaps because of their naval experience. The mentality brought to the geothermal power plants by naval engineers was statistically examined [24]. The engineers were found to be less considerate and less likely to seek supervisory opinion than regular workers on the market. It was noted that the naval engineers need to repair any failures on board while on the ocean. The same was seen with maintenance engineers in the geothermal power plants, who are very confident and willing to try developing domestic solutions.

- The effects of operational experience

  The way that operational experience effects planning was visualised with regards to the wellheads at Hellisheidi geothermal power plants. Real data was gathered and statistically analysed [37]. This study was conducted to analyse how a maintenance pattern evolves with time as operational experience is gathered among staff. By using a Weibull survival distribution for the analysis, it was shown that the frequency of maintenance does in fact change with time, diverting from the original recommendations made by engineering consulting firms. This is also in line with the previous observations of the engineers' characteristics.

- Identification of future developments

  A quantitative Kano model was used to identify the solutions sought after by Icelandic geothermal power plant maintenance engineers [38]. This study demonstrated how a model for customer satisfaction can be utilised within the renewable energy sector. When applied to the geothermal sector, it demonstrated which maintenance management tools are wanted by maintenance engineers. The problems to be addressed include the long documentation time and the uncertainty of postponing maintenance, among others. The Kano analysis tool can be used to visualise the needs of domestic industries while the other industries within the cluster can use the results to develop these solutions.

The results from the studies were then analysed and used to form a system dynamics causal loop diagram. The causal loop diagram can be used to visualise the connection between the factors examined. The causal loop diagram was not intended to be simulated but rather to demonstrate the behaviour of the industry before and after a quantum shift has occurred. A previously proposed theory building process was followed, which eventually led to this article. Table 2 outlines the case studies conducted and where they fit in with the theory construction process. The result from this article is essentially step 3, as defined in the theory building process [31]. This step includes relationship building, identifying an order in different relationships previously observed through field and case studies, eventually resulting in cognitive mapping. In this study, case studies have been conducted and the mapping was done using causal loop diagrams.

Table 2. A list of the case studies conducted, the intention of the studies and where they fit within the theory building process.

Case Study	Intention	Stage in Theory Building according to [31]	Research Structure	Method of Data Collection according to [31]
[35]	Explore how staff has influenced innovation within geothermal firms.	1a. Discovery	- Interviews - Observation	- Insight - Expert opinion - Descriptions
[24]	Get an overview of the Icelandic geoth. sector. Who are the players and what are the characteristics of plant maintenance engineers.	1a. Discovery	- In-depth case study	- Description - Insight
[38]	Visualise how operations management can be improved using staff knowledge.	1a. Discovery	- Interviews	- Expert opinion - Insight
[33]	Identify the current benefits of the geothermal cluster collaboration within Iceland. an acute repair of a geoth. turbine was examined	1b. Description	- In-depth case study	- Critical incident - Documents
[36]	Visualise how domestic industries can influence innovation process.	1b. Description	- In-depth field study	- Observation interviews - Documents - Technical specifications
[34]	Get an overview on how maintenance on geoth. turbines is conducted. Who are the main actors and what are the challenges ahead.	2. Mapping	- In-depth field study	- Verbal protocol - Analysis
[37]	Explore how experience influences the operations management in geothermal power plants.	2. Mapping	- In-depth field study	- Analysis

4. Results

In Figure 3, one can see the established causal loop diagram, which is coded with three colours. The initial condition before the quantum shift occurs is coloured in orange. As the quantum shift occurs, a process coded in blue is initiated. Third is the green graph, which has not been visualised throughout the case studies but has been stated to be a goal suitable for the Icelandic geothermal industry [23]. The relationship between the variables is explained in the following sections. Not all of the variables are equally weighed, and some can be regarded as key variables. The variables "willingness to try" and "KAST" are shown to be key variables in the system dynamics model.

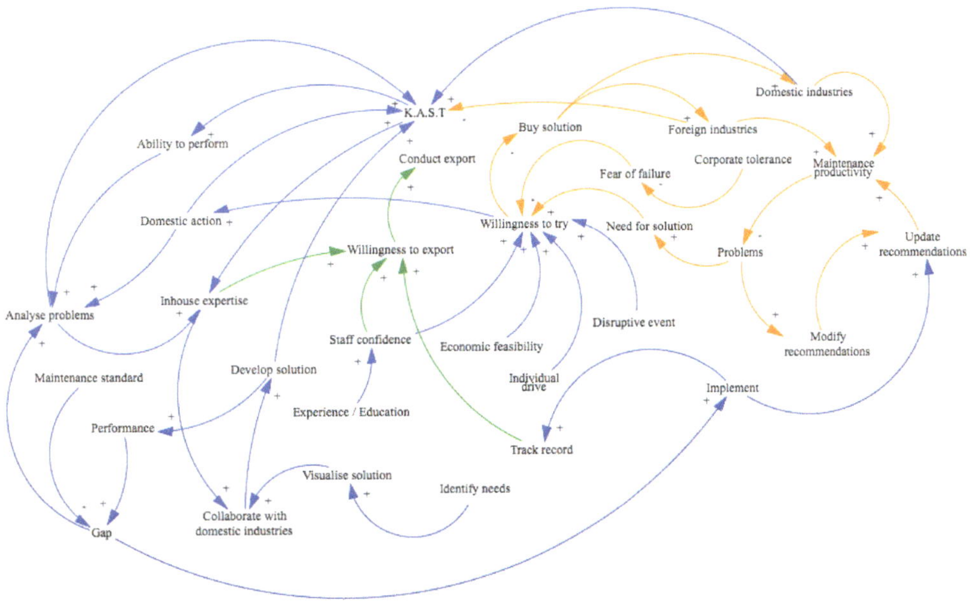

Figure 3. A causal loop diagram describing the behaviour leading to increased domestic development.

4.1. Initial Behaviour

The variable "maintenance productivity" can be seen as the efficiency (doing things the right way) or effectiveness (doing the right thing) of maintenance procedures within a given industry. The more frequent that the occurrence of an unexpected or unwanted maintenance is, the more problematic it will be. Therefore, maintenance recommendations should be reviewed and modified, and subsequently official recommendations should updated for the staff [34,37]. In many cases, a need to purchase an external service or solution (such as software) is needed. This leads to the "willingness to try" variable in Figure 3. Initially, there is not much willingness to try because the status quo has worked previously or a fear of trying a new method is prevalent. A solution is then purchased from an external company. Initially, this expertise comes from a foreign industry, leaving no build up of expertise or innovation within the country. However, in some cases domestic industries are able to provide the needed solution, especially if the risk / benefit ratio is favourable and tests can be made on site with minimal or acceptable risk. Domestic solutions are not necessarily custom made for the geothermal industry but may be a spinoff from other industries, such as the aluminium industry. When solutions are modified to serve the geothermal industry, the operations are shifted slightly for a certain industry that previously did not serve the geothermal sector. This leads to KAST domestically as previously non-related industries become familiar with problems faced in the geothermal sector and begin to focus on them. However, the fear of failure variable is very sensitive to corporate tolerance and especially top level tolerance. This initial behaviour, coloured in orange in Figure 3, can be seen as the first stability phase, as demonstrated in Figures 1 and 2. The industry has developed a certain behaviour that operates in a sufficient manner but does not initiate or support either cluster formation or knowledge buildup domestically.

4.2. Willingness to Try

When a problem occurs and the "willingness to try" variable in Figure 3 has reached a certain level within the organisation, domestic action will be initiated. The "willingness to try" variable is the willingness within an industry to seek the development or improvement of solutions domestically rather than using the solutions that were used in the past. This variable is mostly relevant among senior maintenance engineers or operations managers within the industry. There are certain ways for the willingness to try to reach high levels. This may either be a disruptive event (such as a financial crisis), individual drive within the organisation, or high confidence of the staff, in this case maintenance engineering staff. Staff confidence has been shown to be influenced by previous experience or education. In the case of the Icelandic geothermal industry, it was seen that marine engineers are preferred to engineers that have no marine experience [24] because of their high problem solving skills, ability to work under pressure, and previous nautical experience. Also, if the need for a solution is great but it is not available internationally, then the willingness to try variable increases greatly, which leads to domestic action [35,36]. There will always be a fear of failure linked to the "willingness to try" variable. "Corporate tolerance" is the only balancing variable influencing the fear of failure, underlining the importance of tolerance with regards to failure and experimentation within an organisation. As the tolerance increases within an organisation with regards to failures, the fear of failures decreases, giving in turn less impact to the willingness to try variable. The tolerance is, however, case by case dependent and is often the subject of a favourable cost / benefit ratio.

4.3. Domestic Action

As can be seen in Figure 3, only one variable ("willingness to try") leads to domestic action. When domestic action is initiated, a given problem is analysed. This increases not only in-house expertise but also KAST. In addition, KAST has been observed when foreign industries are hired to service the domestic geothermal sector but in-house staff are allowed to study the procedures. As more domestic expertise builds up, collaboration with domestic industries increases [24,33,36]. The variable "identify needs" influences the visualisation of solutions and collaboration with domestic industries to create such a solution. This variable was identified as a method to provide cascade industries with a method to visualise which solutions to develop to service the geothermal sector. Solutions are more rapidly developed when more domestic collaboration occurs. The performance of a solution that has been developed domestically is measured and a performance gap is seen between the current maintenance standard and the proposed solution that is developed domestically. If the gap is found to be favourable, then the new solution is implemented and the current recommendations are updated. This also increases the track record of domestic solutions and services. If the gap is found to be non-favourable, then this leads to the problem being analysed further and the cycle repeats itself. "Problem analysis", "domestic action" and "develop solution" are all variables leading to KAST. Furthermore, KAST leads to an increase in the in-house expertise as well as the domestic ability to perform. A need analysis serves as a good tool to take predetermined steps and service the observed sector [38]. Even though a "cluster formation" variable is not defined within the diagram, the environment for a cluster development is essentially facilitated for use when the blue portion of the model is initiated. Although the KAST is a key factor in the cluster formation, it is not the cluster formation itself. One can observe a negative effect on KAST when the cluster has been initiated but a solution is purchased from a foreign industry. This negative impact is observed because no build up of knowledge or skills occurs when foreign expertise is used to solve operational problems without the inclusion of any domestic observers, eventually domestic industries suffer.

4.4. Export

As in-house expertise increases, staff confidence increases and the track record of successful solution implementations grows. The variable "willingness to export" reaches a level where export of new solutions become viable. Although this connection has not been observed in the case studies, it is a logical continuation from previous behavior. As a final product, a cluster around the particular sector has grown domestically, serving global markets. A time delay between the start of the cluster initiative and the ability to start exporting solutions can be expected because domestic products need to be fully tested, their track record needs to be monitored, and expertise needs to built up domestically. A theoretical demonstration of such a delay is depicted in Figure 4.

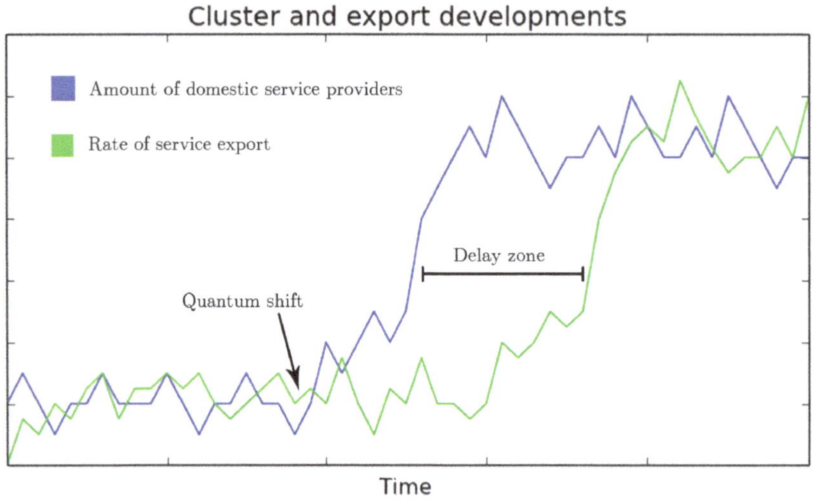

Figure 4. A theoretical relationship between a domestic cluster development and service exports. The delay zone shows a period between the cluster development and service exports.

4.5. Balancing Variables

Balancing (or negative) variables are likely to be important factors in the formation of industrial clusters. The "fear of failure" variable behaves as a balancing variable on the "willingness to try" because staff are less likely to try or develop new solutions if "corporate tolerance" is low. "Maintenance productivity" also serves as a balancing variable on the "problem" variable because fewer problems will be observed as the industry becomes more productive. The use of foreign industries influences the KAST variable in a negative way because foreign industries did not contribute towards KAST in Iceland. The variable "maintenance standard" balances the "gap" variable because a lower gap between previous solutions and developed solutions is going to be observed if the previous maintenance standard is high, thereby decreasing the possibility of new solutions being implemented.

5. Discussion

As demonstrated, three variables had a positive impact on the willingness to try variable. However, one of those variables, the disruptive event, is preferably avoided. Excluding the disruptive event puts an increased emphasis on the other two, staff confidence and individual drive. Individual drive can be impacted by an array of motivations, financial being only one of these. As has been shown, staff confidence is highly dependant on previous experience and education. It was found in Iceland that marine engineers are good candidates to oversee and conduct geothermal power plant maintenance. It is, however, not given that marine engineers possess the same qualities elsewhere as in Iceland. It is also not given that marine engineers are able to serve other renewable energy industries as efficiently. This may be contributed to by the use of steam turbines in geothermal power plants. Marine engineers might, therefore, have little knowledge about the operation and maintenance of solar or other types of power plants. Which characteristics are needed by the maintenance engineers to serve a given technology need to be identified. Although marine engineers are perhaps well suited to serve other industries, this needs to be researched in more depth. The "fear of failure" variable had a negative impact on the willingness to try. According to this model, fear of failure can only be minimised and does never increase the willingness to try. It is, therefore, in the role of corporate leaders to minimise this fear to minimise the impact on the willingness to try variable. This is especially true when the disruptive event is to be avoided. Methods to minimise the fear of failure can include improved prototyping and testing facilities, innovation seminars leading to organisational skill development, learning and growth [39].

A delay can be expected from when the cluster collaboration is initiated until exportation begins. This delay happens because products and services need to be developed and tested domestically until the track record, staff confidence and in-house expertise is sufficient and the willingness to export is enough for it to be conducted. This delay can be visualised in Figure 4. Given that the Icelandic geothermal sector has experienced a disruptive event, gone through, or is going through a quantum shift but has yet to enter global markets with products and services, it can be estimated that the Icelandic geothermal sector is currently located within the delay zone in Figure 4.

If the willingness to try variable reaches a significant level within different countries and domestic action is initiated, one can see that the resulting development increases the knowledge and skill within an industry in that particular country. However, it cannot be overlooked that by increasing the share of maintenance conducted domestically, the contractors that were previously used are no longer conducting those particular jobs. Under certain circumstances, foreign contractors keep consulting the domestic industry after the procedure has been transferred to a domestic industry. However, this consultation is merely marginal compared to previous procedures. This may have a negative effect in other regions as businesses are reduced. This could be observed as a disruptive event by international industries, pushing other industries to develop further because of the increased competition. The question also arises about how spread out clusters can actually become global before becoming a disoriented spread of small groups of companies with little connection to each other.

Building domestic clusters of adequate size, while not causing permanent harm to other clusters but rather increasing global competition, should be pursued by policymakers. These developments can potentially assist global society to develop renewable energy solutions at a faster pace, minimising the carbon footprint in energy generation and increasing the share of renewable energy technologies in the global energy portfolio.

6. Conclusions

This article uses the evidence from the case studies to demonstrate that the willingness to try domestic development within industries may be of importance when initiating industrial cluster development. If external factors will cause little willingness, then the particular industry will not participate in or initiate domestic developments but will rather continue using previous methods. The willingness to try variable needs to be triggered without a disruptive event. This may be initiated by a relevant corporate strategy (therefore minimising fear of failure within the industry), increasing the numbers of source staff with relevant experience (and therefore confidence), and facilitating individual drive. The economic benefits for the industry in question should also lead to an increased willingness to try domestic developments. Laying this groundwork should lead to domestic developments and assist with the formation of an industrial cluster around the sector in question. The model put forward, backed by case studies, shows the possible benefits of cluster collaboration for industries. Knowing the variables that influence the "willingness to try" variable may help leaders of industries to facilitate domestic development and innovation.

Future Work

It would benefit the literature if models such as the one presented in this article are quantified. This would make a practical tool available for policy makers who wish to facilitate the formation of domestic industrial clusters in the renewable energy sector. The model provided in this article demonstrates an idea, a framework, based on multiple observations, but it does not predict the rate of industrial cluster formations. Even though the role of government has been stated in the literature, governmental interventions are avoided in this article because governmental actions (such as incentives) have not been studied by the authors. It would benefit the model to include governmental actions because the importance of policy is likely to influence an industrial cluster formation.

Acknowledgments

Our gratitude goes to GEORG (Geothermal Research Group) for financial support. Saemundur Gudlaugsson, Gudmundur Hjaltalin at Reykjavik Energy and Thrandur Rognvaldsson and Steinn A. Stensson at Landsvirkjun for assisting with this research. Hreinn Halldorsson and Albert Albertsson at HS Energy for assistance. A special thanks goes to Harald Sverdrup for inspiration and assistanceion developing the causal loop diagram.

Author Contributions

Reynir S. Atlason conducted field observations, majority of analysis and the composition of the article. Gudmundur O. Valsson contributed to the design of the article, the theoretical background structure, analysis of data and model construction. Runar Unnthorsson contributed to the design of the article, oversaw the project and assisted with the data analysis and model building.

Conflicts of Interest

The authors declare no conflict of interests.

References

1. Shafiee, S.; Topal, E. When will fossil fuel reserves be diminished? *Energy Policy* **2009**, *37*, 181–189.
2. Field, C.; van Aalst, M. *Climate Change 2014: Impacts, Adaptation, and Vulnerability*; IPCC (Intergovernmental panel on climate change): Geneva, Switzerland, 2014; Volume 1.
3. *BP Energy Outlook 2035*; BP plc: London, UK, 2014.
4. Mitigation, C.C. *IPCC Special Report on Renewable Energy Sources and Climate Change Mitigation*; Technical Report; Cambridge University Press: Cambridge, UK, 2011.

5. Filippetti, A.; Archibugi, D. Innovation in times of crisis: National systems of innovation, structure, and demand. *Res. Policy* **2011**, *40*, 179–192.
6. Porter, M.E. Location, competition, and economic development: Local clusters in a global economy. *Econ. Dev. Q.* **2000**, *14*, 15–34.
7. Porter, M.E. Clusters and the New Economics of Competition. *Harvard Bus. Rev.* **1998**, *14*, 77–90.
8. Malerba, F. Sectoral systems of innovation and production. *Res. Policy* **2002**, *31*, 247–264.
9. Giuliani, E. Cluster absorptive capacity why do some clusters forge ahead and others lag behind? *Eur. Urban Regional Stud.* **2005**, *12*, 269–288.
10. Giuliani, E. The structure of cluster knowledge networks: Uneven and selective, not pervasive and collective; In Proceedings of DRUID Tenth Anniversary Summer Conference, Copenhagen, Denmark, 27–29 June 2005; pp. 27–29
11. Kogut, B. The network as knowledge: Generative rules and the emergence of structure. *Strateg. Manag. J.* **2000**, *21*, 405–425.
12. Porter, M.E. *Competitive Advantage: Creating and Sustaining Superior Performance*; Simon and Schuster: New York, NY, USA, 2008.
13. Furman, J.L.; Porter, M.E.; Stern, S. The determinants of national innovative capacity. *Res. Policy* **2002**, *31*, 899–933.
14. Lin, C.-H.; Tung, C.-M.; Huang, C.-T. Elucidating the industrial cluster effect from a system dynamics perspective. *Technovation* **2006**, *26*, 473–482.
15. Dangelico, R.M.; Garavelli, A.C.; Petruzzelli, A.M. A system dynamics model to analyze technology districts' evolution in a knowledge-based perspective. *Technovation* **2010**, *30*, 142–153.
16. Youngblood, M.D. Winning cultures for the new economy. *Strateg. Leadersh.* **2000**, *28*, 4–9.
17. Gould, S.J.; Eldredge, N. Punctuated equilibria: The tempo and mode of evolution reconsidered. *Paleobiology* **1977**, *3*, 115–151.
18. Orkustofnun. Jardvarmavirkjanir, Retrieved 09.01.2013. Available online: http://www.orkustofnun.is/jardhiti/jardhitanotkun/jardvarmavirkjanir/ (accessed on 9 January 2013).
19. Thorolfsson, G. Maintenance history of a geothermal plant: Svartsengi Iceland; In Proceedings of World Geothermal Congress 2005, Antalya, Turkey, 24–29 April 2005.
20. Geothermal Power Plant Technician. Keilir Technological Institute. Available online: http://en.keilir.net/kit/kit/education/geothermal-power-plant-technician (accessed on 10 February 2013).
21. Hagalin, G.; (Reykjavik Energy); Reynir S. Atlason (University of Iceland). Personal Communication, 2013.
22. Fuller, N. Beyond the core. *Supply Manag.* **2002**, *29*, 39.
23. Porter, M. The Icelandic geothermal cluster: Enhancing competitiveness and creating a new engine of Icelandic growth. In Proceedings of Icelandic Geothermal Conference, Reykjavik, Iceland, 1 November 2010.

24. Atlason, R.S.; Unnthorsson, R. Operation and maintenance in Icelandic geothermal power plants: Structure and Hierarchy. In Proceedings of the POWER Conference. American Society of Mechanical Engineers, San Diego, CA, USA, 15–21 November 2013.
25. Geothermal Research Group. Geothermal Research Group, Retrieved 01.08.2013. Available online: http://georg.hi.is/efni/georg_geothermal_research_group (accessed on 1 August 2013).
26. Eisenhardt, K.M.; Graebner, M.E. Theory building from cases: Opportunities and challenges. *Acad. Manag. J.* **2007**, *50*, 25–32.
27. Voss, C.; Tsikriktsis, N.; Frohlich, M. Case research in operations management. *Int. J. Oper. Prod. Manag.* **2002**, *22*, 195–219.
28. Lewis, M.W. Iterative triangulation: A theory development process using existing case studies. *J. Oper. Manag.* **1998**, *16*, 455–469.
29. Stuart, I.; McCutcheon, D.; Handfield, R.; McLachlin, R.; Samson, D. Effective case research in operations management: A process perspective. *J. Oper. Manag.* **2002**, *20*, 419–433.
30. Meredith, J.; Vineyard, M. A longitudinal study of the role of manufacturing technology in business strategy. *Int. J. Oper. Prod. Manag.* **1993**, *13*, 3–14.
31. Handfield, R.B.; Melnyk, S.A. The scientific theory-building process: A primer using the case of tqm. *J. Oper. Manag.* **1998**, *16*, 321–339.
32. Wallace, W.L.X. *The Logic of Science in Sociology*; Transaction Publishers: Piscataway, NJ, USA, 1971.
33. Atlason, R.S.; Gunnarsson, A.; Unnthorsson, R. Turbine repair at nesjavellir geothermal power plant: An icelandic case study. *Geothermics* **2015**, *53*, 166–170.
34. Atlason, R.S.; Unnthorsson, R.; Oddsson, G.V. Innovation and development in geothermal turbine maintenance based on Icelandic experience. *Geothermics* **2015**, accepted.
35. Atlason, R.S.; Unnthorsson, R. New design solves scaling problems on geothermal control valves. *Power* **2013**, *157*, 18–21.
36. Atlason, R.S.; Unnthorsson, R. *Wellhead Scaling Problems in Geothermal Power Plants Addressed Using a Needle Valve Derivative*; American Society of Mechanical Engineers: New York, NY, 2014.
37. Atlason, R.S.; Geirsson, O.; Elisson, A.; Unnthorsson, R. Geothermal wellhead maintenance: A statistical model based on documented icelandic experience. *Geothermics* **2015**, *53*, 147–153.
38. Atlason, R.S.; Oddsson, G.V.; Unnthorsson, R. Geothermal power plant maintenance: Evaluating maintenance system needs using quantitative kano analysis. *Energies* **2014**, *7*, 4169–4184.
39. Klein, K.J.; Knight, A.P. Innovation implementation overcoming the challenge. *Curr. Dir. Psychol. Sci.* **2005**, *14*, 243–246.

10

The Optimum PV Plant for a Given Solar DC/AC Converter

Roberto S. Faranda *, Hossein Hafezi, Sonia Leva, Marco Mussetta and Emanuele Ogliari

Politecnico Di Milano Department of Energy, Via la Masa 34, 20156 Milano, Italy;
E-Mails: hossein.hafezi@polimi.it (H.H.); sonia.leva@polimi.it (S.L.);
marco.mussetta@polimi.it (M.M.); emanuele.ogliari@mail.polimi.it (E.O.)

* Author to whom correspondence should be addressed; E-Mail: roberto.faranda@polimi.it;

Academic Editor: Jean-Michel Nunzi

Abstract: In recent years, energy production by renewable sources is becoming very important, and photovoltaic (PV) energy has became one of the main renewable sources that is widely available and easily exploitable. In this context, it is necessary to find correct tools to optimize the energy production by PV plants. In this paper, by analyzing available solar irradiance data, an analytical expression for annual DC power production for some selected places is introduced. A general efficiency curve is extracted for different solar inverter types, and by applying approximated function, a new analytical method is proposed to estimate the optimal size of a grid-connected PV plant linked up to a specific inverter from the energetic point of view. An exploitable energy objective function is derived, and several simulations for different locations have been provided. The derived analytical expression contains only the available data of the inverter (such as efficiency, nominal power, *etc.*) and the PV plant characteristics (such as location and PV nominal power).

Keywords: inverter size optimization; photovoltaic power system; PV inverter

---

1. Introduction

Renewable energy sources have become the most prevalent energy sources in recent years, and researchers have made considerable effort to enhance the efficiency of these systems and to optimize the functionality.

Grid-connected photovoltaic (PV) systems grew up very fast in the last few decades; however the energy production by PV systems is still too expensive when compared with other conventional technologies. Moreover, it should be noted that many countries implement policies of economic incentives to encourage the production of electricity from renewable sources, and in particular, the solar one is still financed in a consistent manner. In most cases, these incentives are designed in order to be more and more advantageous with the increasing energy production by the PV plants.

A grid-connected PV system consists of PV arrays, an inverter to convert the produced DC to AC power and, in most cases, a transformer to couple the system to the network [1].

A way to reduce the cost impact of this technology is to select each part of a PV plant with respect to achieving the maximum energy and economic performance [2–8]. For this purpose, the inverter rated capacity (P_{nINV}) must be matched with the rated PV array capacity (P_{PVpeak}) to reach the optimum PV system performance [9].

On the other hand, the optimal P_{PVpeak}/P_{nINV} sizing ratio depends on the local climate, the PV surface orientation and inclination, the inverter performance and the PV/inverter cost ratio [9].

Conventionally, the rated power of a DC/AC inverter is selected as equal to the total nominal power of the installed PV array. This solution often does not match the requirement of the optimum functioning of the systems: first, although the nominal power of the PV array is calculated under the standard test condition (STC), the statistics show that STC irradiance occurs very rarely; second, there are other factors affecting the PV system performance, such as ambient temperature, wind blowing, tilt angle, and so on. In addition, sunshine and high irradiance do not necessarily lead to a high power output. For most places, more sunlight exposure usually implies a hot climate: this high correlation between irradiance level and temperature results in a counteraction between these two factors and deteriorates the performances of both the PV array and inverter. In this case, the exact optimum inverter size is not determined; therefore, a more in-depth analysis is required [10].

Under-sizing the DC/AC inverter regarding the nominal PV array is often a suggested solution to enhance the performance and decrease the costs of the whole system; however, there is not a common definition that can be adopted for all locations in order to calculate the precise ratio between inverter and PV plant rated power.

In [11], it is reported that in Central Europe, the optimum performance of a grid-connected PV system can be achieved for an inverter size of 0.6–0.7-times the PV rated capacity. This means that the P_{PVpeak}/P_{nINV} ratio varies between 1.43 and 1.67. Kil and Van der Weiden in [12] found that PV system performance remained unaffected when the inverter/PV rated power ratio was 0.67 ($P_{PVpeak}/P_{nINV} = 1.49$) in Portugal and 0.65 ($P_{PVpeak}/P_{nINV} = 1.54$) in the Netherlands. In [13], Burger and Ruther have shown that this practice might lead to considerable energy losses, especially in the case of PV technologies with high temperature coefficients of power operating in cold climate sites and also with low temperature coefficients of power operating in warm climate sites (which means that the energy distribution of sunlight is shifted to higher irradiation levels).

The optimum inverter sizing ratio in Madrid (40.5N) and Trappes (48.7N) was reported as 1.25 ($P_{PVpeak}/P_{nINV} = 0.80$) and 1.42 ($P_{PVpeak}/P_{nINV} = 0.70$), respectively [14]. In [15] it was shown that the inverter/PV rated power ratio in Kassel, Germany, is 0.75 ($P_{PVpeak}/P_{nINV} = 1.33$) and 1.0 for Cairo, Egypt. In addition, it was found in [16] that the optimum sizing ratio (P_{PVpeak}/P_{nINV}) varied from 1–2 for different locations and conditions. Overall, both under-sizing and over-sizing are suggested for different considerations.

In [1], an analytical method is presented to find the optimized size of the inverter. It considers the maximum available DC power as a variable and tries to find the optimum inverter size for a specific PV system. Since it deals with an installed PV array as the input, the results for the inverter size are around the available DC maximum power, and it suggests an inverter size slightly higher than the maximum available DC power.

There are several other methods for sizing solar power plants in terms of the optimum ratio between the nominal PV array capacity and the rated inverter input capacity leading to discordant conclusions [8,15–20]. Even if these practices lead to PV systems that work without any problems, these solutions are not often the optimum from the energy point of view. Indeed, one key element in the energy optimization is the correct selection of the right profile concerning the energy production of the PV array with respect to the nominal power of the DC/AC inverter. It is very important, therefore, to find an adequate function that is able to consider all of the elements in order to find the optimum rated PV plant depending on the power of the inverter.

Instead, if we consider all of the different types of inverters, each with its specific performance, it is not possible to find the optimal one for a specific PV plant in a specific location. On the other side, it is possible to find an optimal PV plant for a given inverter in a specific location, since the production of the PV plant is proportional to its size (kW) only.

This paper, contradictory to [1], tries to find the optimal PV array power rating in a specific location considering a specified DC/AC inverter. Therefore contradictory to the literature, the goal is to set up an analytical method in order to define the optimal PV plant size in a specific location. The proposed method put in evidence that a very important factor for the optimization is the power duration curve. Consequently, the proposed method will be based on the hourly average irradiation curve and the approximated efficiency curve of the inverter [21].

This paper discusses the analysis and the procedure based on extracted data from PVsyst software. Hence, available data are used as a good example to illustrate the method, and the reliability of these data is not our concern.

2. Solar Irradiance and DC Power at the Output of PV Modules

The first parameters to be considered are the solar irradiance and the ambient temperature. These depend on the geographic location, on the time of the day and on the day of the year. Figure 1 shows the ideal irradiation curve (with clear sky) in comparison with the monthly and the hourly average curves in a PV system with a fixed tilted surface facing south (azimuth = 0°).

Adverse weather and other environmental conditions (such as pollution, clouds, shadows, and so on) affect the actual solar irradiance greatly at the ground level, which leads more often to a lower amplitude of the average curve compared to the ideal one, as is shown in Figure 1. Therefore, the trend of the actual solar irradiance over the daytime cannot be approximated by a Gaussian curve, as proposed in many papers, because the actual curve is very different. Besides, even the ambient temperature affects the PV production, owing to the PV panel current/voltage (I-V) curve depending on the solar irradiance and the cell temperature.

Starting with the I-V curve of the module, since the solar irradiance and the relative cell temperature values are known, the power-voltage (P-V) curve can be calculated. Moreover, the maximum DC power extraction is considered, so the DC/AC converter needs to impose on the PV panels the appropriate voltage. It can be assumed that the PV panel always operates at its maximum power point (MPP) for any given solar irradiance and cell temperature. Several average values of solar irradiance and cell temperature with different time resolutions (15 min, 1 h and one day) are available in the literature, and all of these data are excellent for making simulations for evaluating the yearly PV plant energy production, but they are useful for the optimization of the DC/AC converter sizing only if the time resolution is not too big. For instance, from Figure 1, monthly average representation is too smooth and far from reality.

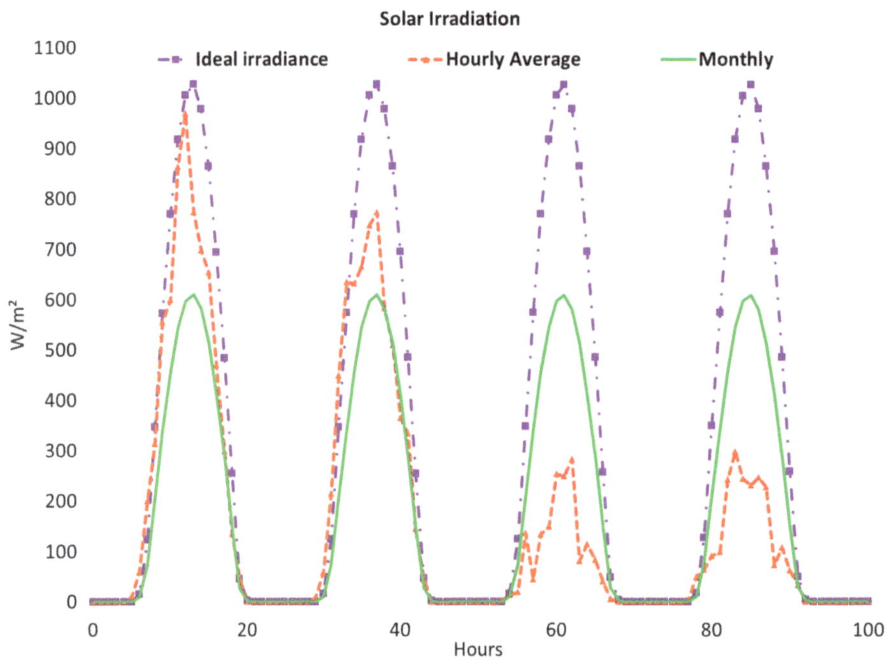

Figure 1. Example of the comparison between ideal, monthly and hourly average solar irradiance values (Paris: azimuth 0°; tilt 32°).

The electric power at the DC terminals of a PV array, $P_{DC}(t)$, is directly proportional to the solar irradiance and varies linearly with the solar cell temperature. It is expressed as:

$$P_{DC}(t) = P_{PVpeak} \cdot \frac{I(t)}{I_{STC}} \cdot [1 - DP \cdot \Delta\vartheta] \tag{1}$$

where:

- $P_{DC}(t)$ is the duration curve, defined as the curve that shows the cumulative time in a year during which the DC power is larger or equal to a certain value; therefore, it is the available DC power at time t, in kW;
- P_{PVpeak} is the installed peak power of all the PV modules under standard test conditions (STC), in kW;
- $I(t)$ is the solar irradiance at time t, in W/m^2;
- I_{STC}, equal to 1000 W/m^2, is the solar irradiance under STC;
- DP is a coefficient that expresses the power variation due to the temperature rise of the cells. A typical value of DP, for crystalline silicon cells, is −0.5%/°C;
- $\Delta\vartheta$, in °C, is the rise of the temperature of the cell above 25 °C.

Usually, the cell temperature rises 30 °C above the ambient temperature when electric current flows through the cell. Thus, $\Delta\vartheta$ can be approximately calculated from the ambient temperature, T_{amb}, using:

$$\Delta\vartheta = T_{amb} + 30\ °C - 25\ °C \tag{2}$$

The DC power, $P_{DC}(t)$, of one PV installation with P_{PVpeak} nominal power has been calculated with 1-h and with one-month resolution using the solar irradiance and ambient temperature data available in the literature. The result is shown in Figure 2.

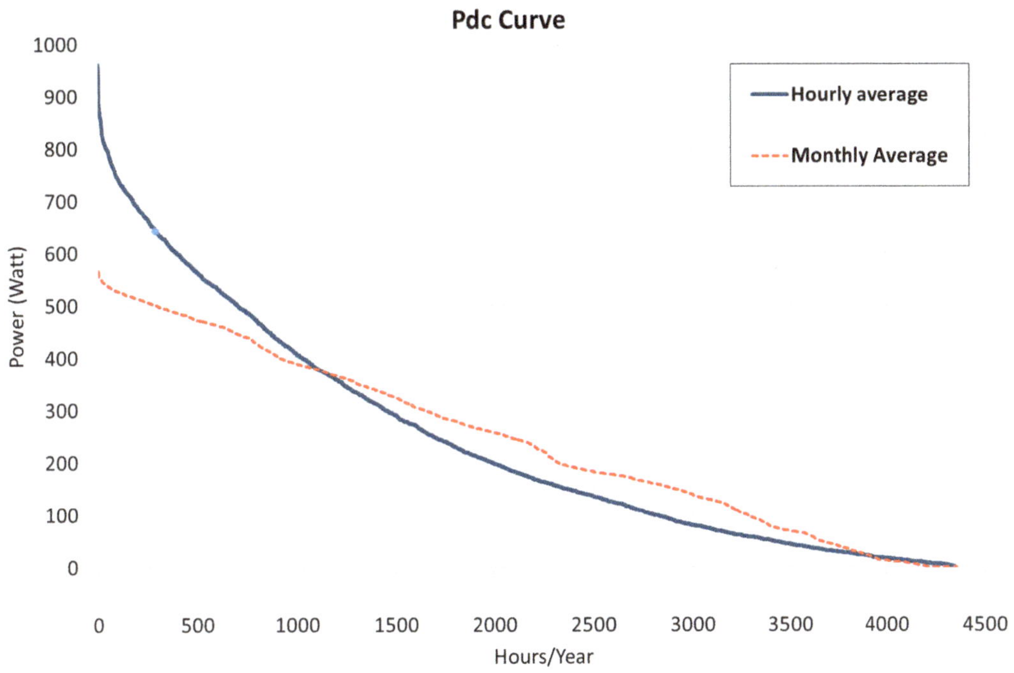

Figure 2. P_{DC} duration curves, calculated from monthly and hourly average solar irradiance values (Paris: azimuth 0°; tilt 32°).

With reference to Figure 2, there is a big difference between the two annual P_{DC} curves, coming from the hourly and the monthly average irradiation, of the same PV plant with the same nominal power and ambient temperature. The hourly average irradiation represents much better what happens in reality. In the case of the hourly curve, the maximum power of the DC power duration is higher than the monthly average curve. Besides, the shape of the two curves is very different, because the hourly average duration curve can be approximated by a parabolic function, whereas the monthly average curve by a straight line.

Even if the representation does not influence the evaluation of the yearly energy production of the PV plant, it is easy to understand that the time resolution is a very important factor in inverter sizing, as reported in [13]. Therefore, monthly, daily, hourly and shorter interval (e.g., every minute or 10 s) time representation can affect the inverter sizing strongly. For instance, monthly time resolutions, which has been widely used in the literature, are too far from reality to calculate peak spots of solar irradiance, and therefore, the results are not correct. Based on this fact, the use of the monthly average P_{DC} curve can lead to under-sizing the inverter and considerable power losses [13]. For this reason, in this paper, the authors have adopted the hourly P_{DC} curve to find the optimum solution.

3. Annual P_{DC} Analytical Representation

The P_{DC} curves can be approximated by the following analytical expression:

$$\dot{p}_{dc}(\dot{t}) = \alpha \cdot \dot{t}^2 + \beta \cdot \dot{t} + \gamma \tag{3}$$

where α, β and γ are the coefficients to be determined,

$$\dot{p}_{dc}(\dot{t}) = \frac{P_{DC}(t)}{P_{PVpeak}}$$

and:

$$\dot{t} = \frac{t}{T_{\max\_PV}}$$

where $\dot{p}_{dc}(\dot{t})$ and $\dot{t}$ stand for respective per-unit values of power and time; in particular, P_{PVpeak} is the available peak PV power and T_{max_PV} is the maximum time during one year when solar energy is available at the site; therefore:

$$P_{DC}(t) = \frac{P_{PVpeak}}{T^2_{\max\_PV}} \cdot \alpha \cdot t^2 + \frac{P_{PVpeak}}{T_{\max\_PV}} \cdot \beta \cdot t + \gamma \cdot P_{PVpeak} \tag{4}$$

It is important to underline that the maximum annual duration of the DC power, T_{max_PV}, is approximately the same for every location. This is due to the natural rotation of the Earth around its axis and to the Sun. Assuming that T_{max_PV} is constant for every location and equal to 4350 h, less than a 0.6% error is inserted. The small differences in T_{max_PV} between the locations are due to the tilt of the Earth's axis, which makes the Northern Hemisphere receive sunlight for longer periods during a year than the southern one.

Figure 3 shows the P_{DC} duration curves evaluated for some locations, at the optimum tilt angle. Table 1 shows the parabolic and correlation coefficients of the P_{DC} duration curves. In order to find the coefficients and correlations, here, "Trendline" options of Microsoft Excel software have been used.

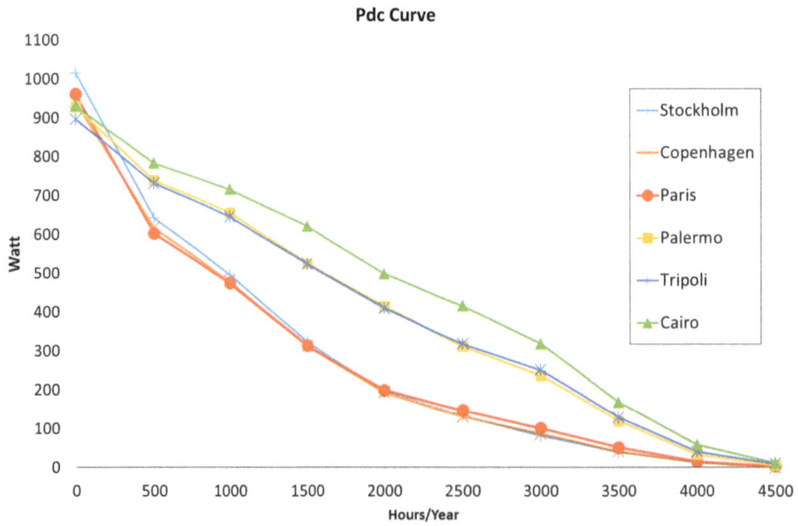

Figure 3. Duration curves of 1 kilowatts peak (kWp) PV plants located in different cities with their optimal modules' tilt.

Table 1. Coefficients of 1-kWp PV power plants in some cities by real irradiance simulated with PVsyst software. Dbase Meteonorm '97.

Location	Stockholm	Copenhagen	Paris	Palermo	Tripoli	Cairo
Latitude	59.1° N	55.4° N	49.1° N	38.0° N	32.8° N	30.1° N
Tilt (optimum) (angle°)	41	38	35	32	29	26
$P_{DC\max}$ (W)	1016	942	962	930	897	931
$T_{max\_}PV$ (h)	4323	4305	4346	4339	4358	4354
α	1.0243	0.9459	0.8829	0.1900	0.18115	−0.1433
β	−1.7946	−1.6815	−1.5979	−1.0585	−1.0261	−0.7511
γ	0.8052	0.7690	0.7487	0.8433	0.8300	0.8638
R^2 (%)	0.99610	0.9948	0.9934	0.9976	0.9989	0.9979

As regards the studied cases, the correlation coefficient range varies from 99.34% (for Paris with tilt equal to 35°) to 99.89% (for Tripoli, with tilt equal to 29°). Considering coefficients α, β and γ from Table 1, it can be noticed that the P_{DC} curves are very similar for locations with high optimum tilt values, while at the lower tilts (such as Cairo), the parabolic approximation is different. Another feature is that all of the curves begin approximately from the same point, indicating that, as expected, the peak value is almost equal in each location.

Although Equation (3) can be valid for every location and for PV panels with crystalline silicon cells, it involves some inaccuracies that should be mentioned:

1. In Equation (1), it is assumed that the P_{DC} is directly proportional to the solar irradiance, which means that the efficiency of the PV panels is constant and does not depend on the level of solar irradiance. In reality, the conversion efficiency of a PV panel drops slightly as the incident to its surface solar irradiance drops. For example, the efficiency of a typical PV panel with mono-crystalline silicon cells drops from 13.2% at 1000 W/m^2 to 12% at 200 W/m^2, so the inserted error is around 1%;
2. In Equation (2), it is assumed that the increase in the cell temperature is 30 °C independent of the power level at which it is working. The increase in temperature is in the range 22–37 °C depending on the ambient temperature, the wind speed and the level of irradiance, which varies in a stochastic way. Hence, the 30 °C increase in temperature used in (2) is a mean value. Moreover, the maximum error introduced in the $P_{DC}(t)$ calculation, when assuming a 30 °C temperature rise instead of 22 °C or 37 °C, depends on the ambient temperature. For T_{amb} = 20 °C, the respective errors are 4.4% and 4.1%. It should be mentioned, however, that this error is not due to the specific method followed in this paper, but it is added in all of the calculation methods followed in literature.

Since the respective error by both above-mentioned inaccuracies is less than 5%, these do not affect significantly the shape of the $P_{DC}(t)$ curve.

In conclusion, it is important to underline that the P_{DC} duration curve of any PV installation in any location can be represented as a parabolic curve, although in some cases, it can be a linear one.

4. The Efficiency Curve of a PV Inverter

Solar inverters are complex power electronic devices comprising often a DC/DC boost converter to boost up the PV array DC output voltage and to guarantee the MPPT working principle [20,22] and a DC/AC inverter, which is controlled as the current source to produce the output AC power in the unity power factor. This device is the interface between DC side and the AC grid. The efficiency η of a given solar inverter is defined by:

$$\eta = \frac{P_{AC}(t)}{P_{DC}(t)} = \frac{P_{DC}(t) - P_{loss}(t)}{P_{DC}(t)} = 1 - \frac{P_{loss}(t)}{P_{DC}(t)} \tag{5}$$

where $P_{DC}(t)$, $P_{AC}(t)$ and $P_{loss}(t)$ are the instantaneous DC power, AC power and power losses, respectively.

The power losses of such device consist of two distinct parts. The first is constant and involves the power to supply the control unit and the other auxiliary parts only, while the second is load dependent and consists of: switching losses on power switches, ohmic losses and the losses caused by temperature variation. Therefore, the total loss of the inverter is not constant, and the efficiency is mainly load current dependent.

Examples of variation of the efficiency of some inverters are shown in Figure 4, according to the data provided by the manufacturers in the respective technical brochures. It is evident that the conversion efficiency of a solar inverter is a load function. In PV plants, the inverter will work in every possible power level; hence, to evaluate its efficiency over the whole operating range, it is necessary to determine the working range of the PV plant.

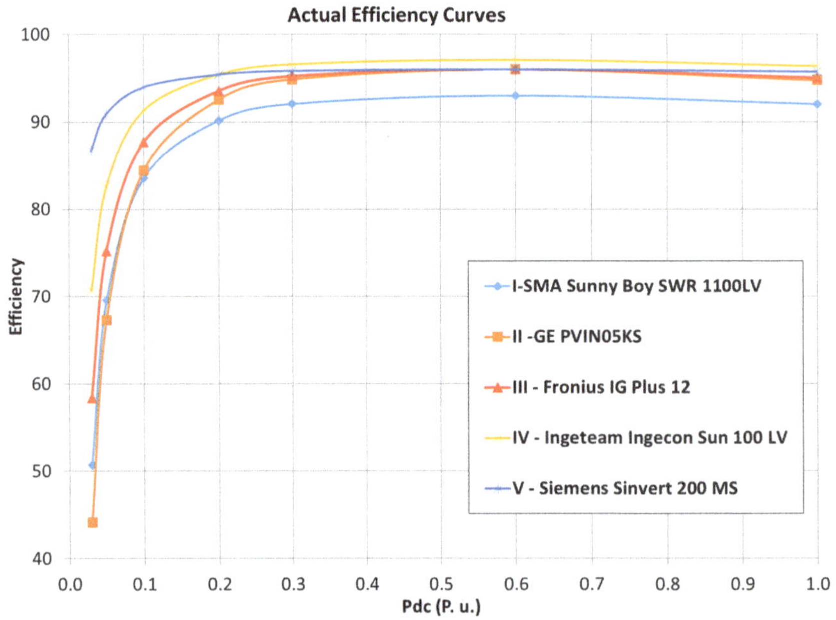

Figure 4. Typical per unit efficiency curves for grid-connected solar inverters.

The following simple mathematical function describes, with very good accuracy for loads >2%, the efficiency curve of any solar inverter [1]. By imposing:

$$\dot{p}_{INV\_DC}(\dot{t}) = \frac{P_{INV\_DC}(t)}{P_{nINV\_DC}}$$

where P_{nINV_DC} is the nominal DC power of the inverter and P_{INV_DC} is inverter DC power, we can have:

$$\eta\left(\dot{p}_{INV\_DC}\right) = A + B \cdot \dot{p}_{INV\_DC}(t) + \frac{C}{\dot{p}_{INV\_DC}(t)} \tag{6}$$

where $\eta\left(\dot{p}_{INV\_DC}\right)$ is the efficiency of the inverter as a percentage, $\dot{p}_{INV\_DC}(t) > 0$ is the per unit (p.u.). value of the DC power that the inverter can convert in AC, while A, B and C are some parameters that should be determined.

It is obvious from (6) that three pairs of $[\eta\left(\dot{p}_{INV\_DC}\right), \dot{p}_{INV\_DC}(t)]$ values are needed to determine the A, B and C parameters. These pairs are readily available from the inverter efficiency curve provided by the manufacturers. A very good choice are the pairs corresponding to:

- the central point of the curve angle $\dot{p}_{INV\_DC} = 0.1$;
- the end of the curve angle $\dot{p}_{INV\_DC} = 0.2$;
- the final point of the curve $\dot{p}_{INV\_DC} = 1.0$.

It is important to underline that these values are available in the data sheets of the manufacturers. Therefore, the efficiency curve defined by the parameters A, B and C can now be easily calculated by solving the following system:

$$\begin{cases} \eta_{10\%} = A + 0.1 \cdot B + 10 \cdot C \\ \eta_{20\%} = A + 0.2 \cdot B + 5 \cdot C \\ \eta_{100\%} = A + B + C \end{cases} \tag{7}$$

Equation (6) describes accurately the efficiency curve of any inverter, and the A, B and C parameters are estimated by the simple system of linear equations reported in Equation (7). It should also be noted that in the solar inverters considered, we have always $A > 0$, while B and C are less than zero, but those values differ. Hereunder, in Table 2, some different manufacturers' specifications, shown in Figure 4 and found in the commercial PVsyst inverter database, have been used to calculate the A, B, C and R^2 coefficients.

Table 2. Some inverter specifications found in the commercial PVsyst database and A, B, C and R^2 coefficients calculated for the same inverters.

Type	P_{nINV} (kW)		η	A	B	C	R^2
	AC	DC	MAX				
I	1	1.1	93.0	98.236	−4.786	−1.42	0.998
II	5	-	96.0	102.53	−6.019	−1.75	1.000
III	12	12.6	96.0	100.83	−4.517	−1.27	1.000
IV	100	105	97.1	100.56	−3.283	−0.89	0.998
V	200	210	96.0	97.24	−1.194	−0.31	0.995

Therefore, the efficiency curve of any inverter can be accurately described by a simple mathematical expression with three unknown parameters, which can be estimated from the data provided by the inverter manufacturer, by simply solving a system of three linear equations.

5. The Optimum PV Plant for a Specific Inverter: Parameter Evaluation

This section deals with an analytical approach to derive the parametric equation of converted energy at the AC side of a solar inverter. In order to evaluate the optimum PV plant for a specific inverter, it is useful to change (3) in the following way:

$$\underline{\dot{p}}_{dc}(t)=\frac{P_{DC}(t)}{P_{nINV\_DC}}=\frac{P_{PVpeak}}{P_{nINV\_DC}\cdot T_{\max\_PV}^{2}}\alpha\cdot t^{2}+\frac{P_{PVpeak}}{P_{nINV\_DC}\cdot T_{\max\_PV}}\beta\cdot t+\gamma\cdot\frac{P_{PVpeak}}{P_{nINV\_DC}}$$

Now, the $\underline{\dot{p}}_{dc}(t)$ is the power energy in the DC side normalized in P_{nINV_DC}. Then, it is possible to write:

$$\underline{\dot{p}}_{dc}(t)=\alpha^{1}\cdot t^{2}+\beta^{1}\cdot t+\gamma^{1} \tag{8}$$

where:

$$\alpha^{1}=\alpha\cdot\frac{P_{PVpeak}}{P_{nINV\_DC}\cdot T_{\max\_PV}^{2}},\ \beta^{1}=\beta\cdot\frac{P_{PVpeak}}{P_{nINV\_DC}\cdot T_{\max\_PV}}\ \text{and}\ \gamma^{1}=\gamma\cdot\frac{P_{PVpeak}}{P_{nINV\_DC}}$$

To find the optimum PV plant for a given inverter, it is necessary to combine Equations (6) and (8). Doing so, it is possible to study two distinctive cases:

In the first case (Case A), $P_{nINV\_DC} \geq P_{DCmax}$; therefore, it is assumed that the nominal DC power of the inverter, P_{nINV_DC}, is larger than the P_{DCmax}, which is the maximum available DC power of the PV plant (equal to P_{PVpeak} at the standard test conditions).

In the second case (Case B), $P_{nINV\_DC} < P_{DCmax}$; it is assumed that the nominal DC power of the inverter, P_{nINV_DC}, is lower than P_{DCmax}. As shown in Figure 5, in Case B, the inverter will operate for T_{nINV} h/year, deteriorating the injected power to its maximum nominal power. While for the [T_{max_PV}-T_{nINV}] h/year interval, it will operate as in Case A.

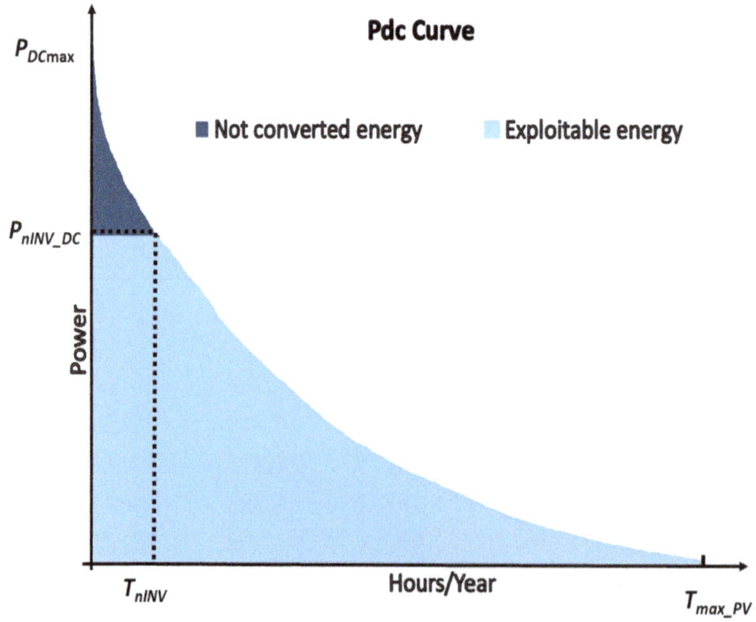

Figure 5. Illustration of the efficiency evaluation curve of Case B: the area above P_{nINV_DC} represents the energy not exploited, since the corresponding power values are greater than the inverter rated power.

The time T_{nINV} can be obtained making P_{nINV_DC} equal to the $P_{DC}(t)$. The result equation is a second order function, and the only admissible result, considering the function waveform and the minimum by the two mathematical solutions T_{nINV_1} and T_{nINV_2}, is:

$$T_{nINV} = \frac{\beta}{2 \cdot \alpha} \cdot T_{\max\_PV} \cdot \left[-1 + \sqrt{1 - \frac{4 \cdot \alpha}{\beta^2} \cdot \left(\gamma - \frac{P_{nINV\_DC}}{P_{PVpeak}} \right)} \right] \tag{9}$$

In the following, the PV plant energy production, the PV plant energy losses and the not-converted energy is evaluated for both Case A and Case B.

5.1. Case A

5.1.1. $P_{nINV\_DC} \geq P_{DCmax}$ PV Plant Energy Production

Combining Equations (6) and (8), it is possible to obtain the expression of AC power injected by the PV plant:

$$P_{ac}(t) = \eta\left(\dot{p}_{INV\_DC}\right) \cdot P_{DC}(t)$$

Additionally, considering that in this case, $\eta\left(\dot{p}_{INV\_DC}\right) = \eta\left(\underline{\dot{p}}_{dc}\right)$, it is possible to write:

$$P_{ac}(t) = \eta\left(\underline{\dot{p}}_{dc}\right) \cdot P_{DC}(t) = \eta\left(\underline{\dot{p}}_{dc}\right) \cdot \underline{\dot{p}}_{dc}(t) \cdot P_{nINV\_DC}$$

The annual energy yield, E_A, is given by:

$$E_A = \int_0^{T_{\max\_PV}} P_{ac}(t)\, dt = \int_0^{T_{\max\_PV}} \eta\left(\underline{\dot{p}}_{dc}\right) \cdot \underline{\dot{p}}_{dc}(t) \cdot P_{nINV\_DC}\, dt = E_{A1} + E_{A2} + E_{A3} \tag{10}$$

where:

$$E_{A1} = \int_0^{T_{\max\_PV}} A \cdot \underline{\dot{p}}_{dc}(t) \cdot P_{nINV\_DC}\, dt = A \cdot P_{PVpeak} \cdot T_{\max\_PV} \cdot \left(\frac{\alpha}{3} + \frac{\beta}{2} + \gamma \right) = k_1 \cdot A \cdot P_{PVpeak} \cdot T_{\max\_PV}$$

$$E_{A2} = \int_0^{T_{\max\_PV}} B \cdot \underline{\dot{p}}_{dc}^2(t) \cdot P_{nINV\_DC}\, dt = k_2 \cdot B \cdot \frac{P_{PVpeak}^2}{P_{nINV\_DC}} \cdot T_{\max\_PV}$$

$$E_{A3} = \int_0^{T_{\max\_PV}} C \cdot P_{nINV\_DC}\, dt = C \cdot P_{nINV\_DC} \cdot T_{\max\_PV}$$

The two coefficients $k_1 = \left(\frac{\alpha}{3} + \frac{\beta}{2} + \gamma \right)$ and $k_2 = \left(\frac{\alpha^2}{5} + \frac{\beta^2}{3} + \gamma^2 + \frac{\alpha \cdot \beta}{2} + \frac{2 \cdot \alpha \cdot \gamma}{3} + \beta \cdot \gamma \right)$ are linked to the installation place. They depend only on the location and the PV plant design (tilt, azimuth, PV module technology, *etc.*).

5.1.2. $P_{nINV\_DC} \geq P_{DCmax}$ PV Plant Energy Losses

The power losses can be calculated as follows:

$$P_{loss}(t) = \left[1 - \eta\left(\dot{p}_{INV\_DC}\right)\right] \cdot P_{DC}(t) \tag{11}$$

The energy losses can be calculated as follows:

$$E_{loss\_A1} = \int_0^{T_{max\_PV}} p_{loss}(t)\,dt = \int_0^{T_{max\_PV}} \underline{\dot{p}}_{dc}(t).P_{nINV\_DC}\,dt - \int_0^{T_{max\_PV}} \eta\left(\underline{\dot{p}}_{dc}\right) \cdot \underline{\dot{p}}_{dc}(t) \cdot P_{nINV\_DC}\,dt = \frac{E_{A1}}{A} - E_A \tag{12}$$

5.1.3. $P_{nINV\_DC} \geq P_{DCmax}$ PV Plant Non-Converted Energy

In this case, the non-converted solar energy is equal to zero, because the inverter rated power is greater than PV system one.

5.2. Case B

5.2.1. $P_{nINV\_DC} < P_{DCmax}$ PV Plant Energy Production

In the Case B, combining Equations (6) and (8), it is possible to obtain the expression of the AC power injected by a PV plant as:

- when $t \leq T_{nINV}$ and $P_{nINV\_DC} \leq P_{DC}(t)$,

$$P_{ac}(t) = \eta_{100\%} \cdot P_{nINV\_DC}$$

- when $t > T_{nINV}$ and $P_{nINV\_DC} > P_{DC}(t)$,

$$P_{ac}(t) = \eta\left(\dot{p}_{INV\_DC}\right) \cdot P_{DC}(t) = \eta\left(\underline{\dot{p}}_{dc}\right) \cdot \underline{\dot{p}}_{dc}(t) \cdot P_{nINV\_DC}$$

The annual energy yield, E_B, is given by:

$$E_B = \int_0^{T_{nINV}} \eta_{100\%} \cdot P_{nINV\_DC} \cdot dt + \int_{T_{nINV}}^{T_{max\_PV}} \eta\left(\dot{p}_{INV\_DC}\right) \cdot P_{DC}(t)\,dt = E_{B0} + E_{B1} + E_{B2} + E_{B3} \tag{13}$$

After this evaluation, it is possible to define the first part of the energy yield E_{B0} as:

$$E_{B0} = P_{nINV\_DC} \cdot \eta_{100\%} \cdot T_{nINV} = P_{nINV\_DC} \cdot T_{nINV} \cdot (A + B + C)$$

The second part of the energy yield E_{B1} as:

$$E_{B1} = \int_{T_{nINV}}^{T_{max\_PV}} A \cdot \underline{\dot{p}}_{dc}(t) \cdot P_{nINV\_DC}\,dt = A \cdot P_{PVpeak} \cdot \left(k_1 \cdot T_{max\_PV} - \frac{\alpha}{3} \cdot \frac{T_{nINV}^3}{T_{max\_PV}^2} - \frac{\beta}{2} \cdot \frac{T_{nINV}^2}{T_{max\_PV}} - \gamma \cdot T_{nINV}\right)$$

The third part of the energy yield E_{B2} as:

$$E_{B2} = \int_{T_{nINV}}^{T_{max\_PV}} B \cdot \underline{\dot{p}}_{dc}^2(t) \cdot P_{nINV\_DC}\,dt = B \cdot \frac{P_{PVpeak}^2}{P_{nINV\_DC}} \cdot$$

$$\cdot \left[k_2 \cdot T_{max\_PV} - \frac{\alpha^2}{5} \cdot \frac{T_{nINV}^5}{T_{max\_PV}^4} - \frac{\alpha \cdot \beta}{2} \cdot \frac{T_{nINV}^4}{T_{max\_PV}^3} - \left(\frac{\beta^2}{3} + \frac{2 \cdot \alpha \cdot \gamma}{3}\right) \cdot \frac{T_{nINV}^3}{T_{max\_PV}^2} - \beta \cdot \gamma \cdot \frac{T_{nINV}^2}{T_{max\_PV}} - \gamma^2 \cdot T_{nINV}\right]$$

The fourth part of the energy yield E_{B3} as:

$$E_{B3} = \int_{T_{nINV}}^{T_{\max\_PV}} C \cdot P_{nINV\_DC}\, dt = \int_{0}^{T_{\max\_PV}} C \cdot P_{nINV\_DC}\, dt - \int_{0}^{T_{nINV}} C \cdot P_{nINV\_DC}\, dt$$

$$= E_{A3} - \int_{0}^{T_{nINV}} C \cdot P_{nINV\_DC}\, dt = C \cdot P_{nINV\_DC} \cdot \left(T_{\max\_PV} - T_{nINV}\right)$$

5.2.2. $P_{nINV\_DC} < P_{DCmax}$ PV Plant Energy Losses

When $t \le T_{nINV}$, $P_{loss}(t) = (1 - \eta_{100\%}) \cdot P_{nINV\_DC}$, so:

$$E_{loss\_B0} = \int_{0}^{T_{nINV}} P_{loss}(t)\, dt = \int_{0}^{T_{nINV}} (1 - \eta_{100\%}) \cdot P_{nINV\_DC}\, dt = E_{B0} \cdot \left(\frac{1}{A+B+C} - 1\right)$$

When $t > T_{nINV}$, $P_{loss}(t) = \left[1 - \eta\left(\dot{p}_{INV\_DC}\right)\right] \cdot P_{DC}(t)$, so:

$$E_{loss\_B1} = \int_{T_{nINV}}^{T_{\max\_PV}} P_{loss}(t) dt = \int_{T_{nINV}}^{T_{\max\_PV}} \underline{\dot{p}}_{dc}(t).P_{nINV\_DC} dt - \int_{T_{nINV}}^{T_{\max\_PV}} \eta(\underline{\dot{p}}_{dc}).\underline{\dot{p}}_{dc}(t).P_{nINV\_DC} dt = \frac{E_{B1}}{A} - E_{B1} - E_{B2} - E_{B3}$$

In conclusion, it is possible to write the total energy loss as:

$$E_{loss\_B} = E_{loss\_B0} + E_{loss\_B1} = E_{B0} \cdot \left(\frac{1}{A+B+C}\right) + \frac{E_{B1}}{A} - E_{B} \tag{14}$$

5.2.3. $P_{nINV\_DC} < P_{DCmax}$ PV Pant Non-Converted Energy

In this case, the non-converted solar energy is positive, because the rated power of the inverter is lower than the PV plant one. This energy is equal to:

$$E_{not\_converted\_B} = \int_{0}^{T_{nINV}} \left(P_{DC}(t) - P_{nINV\_DC}\right) dt = P_{PVpeak} \cdot \left[\frac{\alpha}{3} \cdot \frac{T_{nINV}^3}{T_{\max\_PV}^2} + \frac{\beta}{2} \cdot \frac{T_{nINV}^2}{T_{\max\_PV}} + \left(\gamma - \frac{P_{nINV\_DC}}{P_{PVpeak}}\right) \cdot T_{nINV}\right] \tag{15}$$

6. The Optimum PV Plant for a Given Inverter: Objective Function

In order to find the optimum PV plant, the objective is to maximize the converted energy. From the energetic point of view, to find the optimum PV plant for a given inverter, without considering the costs of the devices, it is possible to calculate the difference between the converted energy and the sum between losses and the non-converted energy, as was evaluated in the previous section. Therefore, in Case A where $P_{nINV\_DC} \ge P_{DC}(t)$, the objective function is $E(P_{PV\_peak}) = E_A - E_{loss\_A}$, while in Case B, where $P_{nINV\_DC} < P_{DC}(t)$, the objective function has to be changed to include the non-converted energy also; thus, it becomes:

$$E(P_{DC\max}) = E_{B} - E_{loss\_B} - E_{not\_converted\_B} \tag{16}$$

As the first objective function for Case A is obviously a growing up monotone function, the optimum power can be found only in the second objective function, Case B. Therefore, with this objective function, the optimum PV plant of any given solar inverter will be always bigger than the P_{nINV_DC}.

For instance, by using simulation results of Palermo, Italy, for a 100-kW inverter, the procedure has been evaluated, and Figure 6 shows the optimum point. Only through this energetic objective function, the maximum is reached when the P_{PVpeak} is slightly higher than the inverter size.

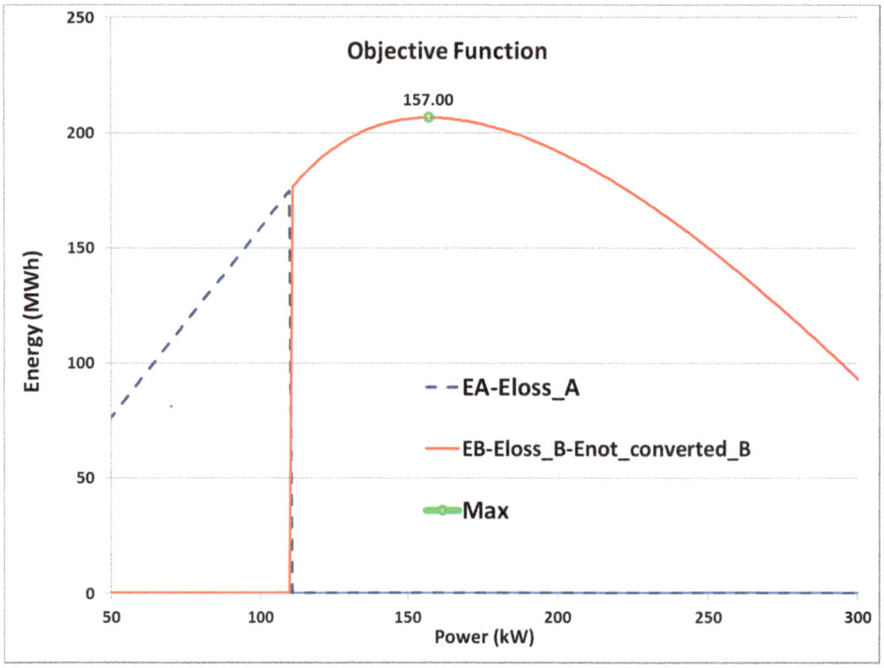

Figure 6. Objective function curve calculated for the city of Palermo, Italy.

7. The Optimum PV Plant for a Specific Inverter: Simulation Result

There exists a variety of types of solar inverters utilized in PV systems. These inverters can be categorized into three major types, which are introduced here;

- Inverter Type 1: the inverter Type 1 has peak efficiency at a very low load percentage (14.1%) and, subsequently, a very waning curve;
- Inverter Type 2: the inverter Type 2 reaches the maximum efficiency at a 31.6% load and keeps approximately constant performance up to the rated power;
- Inverter Type 3: the inverter Type 3 has a peak performance at an 81.6% load, and once, it reaches the peak value, it performs almost as a constant curve.

Figure 7 shows the performance curves of the chosen inverters, as a function of the per unit DC-side power.

All other features of the three inverters are similar, and to ensure an effective comparison of the results, it is assumed that rated power, efficiency and the overload coefficient are fixed and equal for all three types:

- P_{nom} (DC side) = 10 kW;
- η_{max} = 96%;
- k_s = 1 (overload coefficient).

According to the mathematical approximation for a solar inverter in Section 4, A, B and C coefficients are computed for the above-mentioned inverter. Results are presented in Table 3.

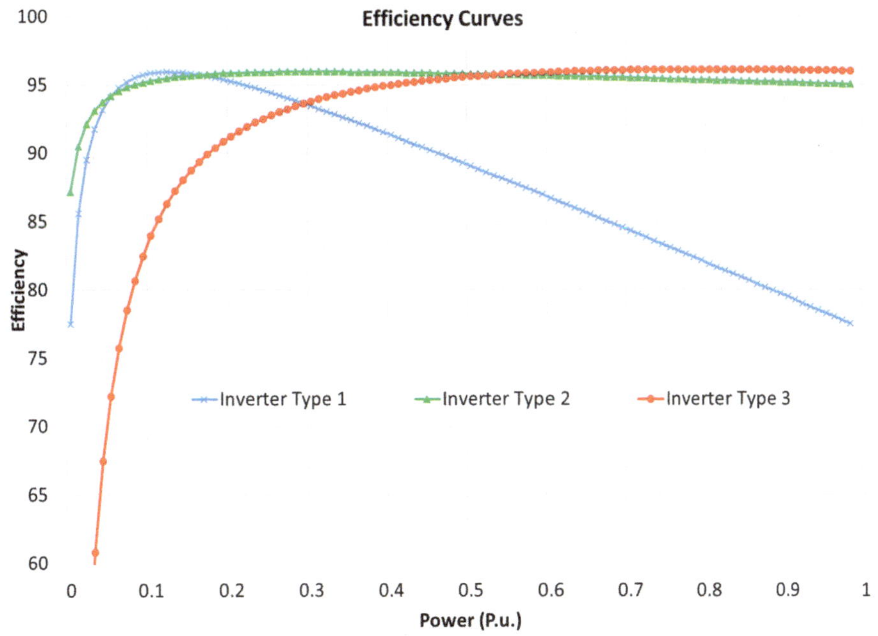

Figure 7. Example of three typical per unit efficiency curves for grid-connected solar inverters.

Table 3. Example of *A*, *B*, *C* coefficients of 3 solar inverter topologies.

Inverter type	*A*	*B*	*C*
Type 1	103.0	−25.0	−0.5
Type 2	97.2	−2.0	−0.2
Type 3	101.0	−3.0	−2.0

It should be noted that the coefficients *A*, *B* and *C* must be used in decimal form (not a percentage) for a correct application of the Equation (6). Table 4 presents simulation results of different latitudes for these inverters. It shows that inverter Type 1 has the lowest average ratio, because it works with high efficiency at low solar irradiance, which is more possible to occur for a solar inverter. Another important conclusion of Table 4 is that, for a specific inverter type, by increasing the latitude, the P_{PVpeak}/P_{nINV_DC} ratio is also increased.

Table 4. Power ratio between P_{PVpeak}/P_{nINV_DC} for different solar inverters and for different latitudes.

Type P_{PVPEAK}/P_{NINV_DC}	Stockholm	Copenhagen	Paris	Palermo	Tripoli	Cairo	Average
Latitude	59.1°	55.4°	49.1°	38.0°	32.8°	30.1°	-
1	1.63	1.71	1.76	1.44	1.46	1.34	1.556
2	1.94	2.04	2.10	1.69	1.72	1.54	1.838
3	1.98	2.08	2.14	1.71	1.74	1.56	1.868

Considering that most inverters on the market nowadays have Type 2 characteristics, by changing the latitude, the P_{PVpeak}/P_{nINV_DC} ratio is computed for the five different inverters (Type 2) represented in Section 4.

Since we considered the P_{DC} as the instantaneous DC power output from the PV modules, without any losses due to the transmission cables and without considering the inverter capability of being overloaded, these data represent the maximum power ratio available between P_{PVpeak} and P_{nINV_DC} for the latitudes shown in Table 5 at their optimum tilt angle.

Reminding that $P_{DC\max}$ is related to P_{PVpeak}, Table 5 shows clearly how the power ratio between P_{PVpeak} and P_{nINV_DC} changes by latitude. Considering η_{MAX} from Table 2 and the calculated ratios in Table 5, it can be concluded that at each location, the inverter with higher η_{MAX} has a higher P_{PVpeak}/P_{nINV_DC} ratio or, in other words, that ratio multiplied by η_{MAX} is almost constant. With this consideration for each location, it is possible to define a constant. Then, by choosing the inverter type, the ratio can be computed by the average column in Table 5.

Table 5. Power ratio between P_{PVpeak}/P_{nINV_DC} for some solar inverters and for different latitudes.

Type	Stockholm	Copenhagen	Paris	Palermo	Tripoli	Cairo	Average
Latitude	59.1°	55.4°	49.1°	38.0°	32.8°	30.1°	-
I	1.91	1.99	2.06	1.66	1.69	1.52	1.805
II	1.95	2.04	2.11	1.69	1.72	1.54	1.842
III	1.95	2.04	2.11	1.69	1.72	1.55	1.843
IV	1.97	2.05	2.13	1.71	1.73	1.56	1.858
V	1.96	2.04	2.11	1.7	1.73	1.55	1.848
Average	1.948	2.032	2.104	1.690	1.718	1.544	

In this paper, the cost effect was not considered to find the optimum solution. By introducing the cost effect in the objective function, it seems that the ratio between P_{PVpeak} and P_{nINV_DC} could be lower than the ones obtained in this study. This consideration will be addressed in future works.

8. Conclusions

The comparison between monthly average and hourly average P_{DC} curve depicts that calculations based on the monthly average can inject considerable error into the design procedure. Therefore, this paper used the hourly average curve, which, in terms of accuracy and also the number of stored data, can be considered the best choice. Then, the optimum PV size, from the energy point of view, for a given inverter can be accurately calculated using the algebraic expressions developed in this paper, avoiding in this way multiple simulation efforts. The method developed in this study can be a valuable tool for design engineers comparing different inverters, calculating the optimum PV size of a given inverter type, estimating the annual AC energy yield and the effective annual inverter efficiency without performing multiple simulations. The validity of the proposed analytical model has been tested with the results obtained by simulations and measured data.

This study also showed the importance of defining a suitable function to consider all of the elements in order to find the optimum rated PV plant based on the power of the inverter and simultaneously to meet the power consumption requirements; for this reason, further works will regard the development of analytical expressions to estimate the optimum PV size for an inverter, by considering also costs and power requirements.

Author Contributions

In this research activity, all the authors were involved in the data analysis and preprocessing phase, simulation, results analysis and discussion, and manuscript preparation. All authors have approved the submitted manuscript

Conflicts of Interest

The authors declare no conflict of interest.

References

1. Demoulias, C. A new simple analytical method for calculating the optimum inverter size in grid-connected PV plants. *Electr. Power Syst. Res.* **2010**, *80*, 1197–1204.
2. Carcangiu, G.; Dainese, C.; Faranda, R.; Leva, S.; Sardo, M. New network topologies for large scale Photovoltaic Systems. In Proceedings of the IEEE Power Tech Conference, Bucharest, Romania, 28 June–2 July 2009; pp. 1–7.
3. Grimaccia, D.F.; Leva, S.; Mussetta, M.; Faranda, R.; Gualdoni, M. Performance Analysis of a Single-Axis Tracking PV System. *IEEE J. Photovolt.* **2012**, *2*, 524–531.
4. Faranda, R.; Gualdoni, M.; Leva, S.; Monaco, M.; Timidei, A. Analysis of a PV system with single-axis tracking energy production and performances. In Proceedings of the 2011 International Conference on Clean Electrical Power (ICCEP), Ischia, Italy, 14–16 June 2011; pp. 130–136.
5. Brenna, M.; Faranda, R.; Leva, S. Dynamic analysis of a new network topology for high power grid connected PV Systems. In Proceedings of the IEEE Power and Energy Society General Meeting, Minneapolis, MN, USA, 25–29 July 2010; pp. 1–7.
6. Cucco, S.; Faranda, R.; Invernizzi F.; Leva, S. Analysis of a Fresnel lenses concentrator. In Proceedings of the IEEE 2012 Power and Energy Society General Meeting, San Diego, CA, USA, 22–26 July 2012; pp. 1–8.
7. Ogliari, E.; Grimaccia, F.; Leva, S.; Mussetta, M. Hybrid Predictive Models for Accurate Forecasting in PV Systems. *Energies* **2013**, *6*, 1918–1929.
8. Dolara, A.; Grimaccia, F.; Leva, S.; Mussetta, M.; Ogliari, E. A physical hybrid artificial neural network for short term forecasting of PV plant power output. *Energies* **2015**, *8*, 1138–1153.
9. Mondol, J.D.; Yohanis, Y.G.; Norton, B. Optimal sizing of array and inverter for grid-connected photovoltaic systems. *Sol. Energy* **2006**, *80*, 1517–1539.
10. Chen, S.; Li, P.; Brady, D.; Lehman, B. Determining the optimum grid-connected photovoltaic inverter size. *Sol. Energy* **2013**, *87*, 96–116.
11. Rieß, H.; Sprau, P. Design considerations for the PV generator/inverter matching in grid connected systems. In Proceedings of the 11th European Photovoltaic Solar Energy Conference, Montreux, Switzerland, 12–16 October 1992; pp. 1377–1378.
12. Kil, A.J.; van der Weiden, T.C.J. Performance of modular grid connected PV systems with undersized inverters in Portugal and the Netherlands. In Proceedings of the IEEE Photovoltaic Specialists Conference, Waikoloa, HI, USA, 5–9 December 1994; pp. 1028–1031.

13. Burger, B.; Ruther, R. Inverter sizing of grid-connected photovoltaic systems in the light of local solar resource distribution characteristics and temperature. *Sol. Energy* **2006**, *80*, 32–45.
14. Macagnan, M.H.; Lorenzo, E. On the optimal size of inverters for grid connected PV systems. In Proceedings of the 11th European Photovoltaic Solar Energy Conference, Montreux, Switzerland, 12–16 October 1992; pp. 1167–1170.
15. Abd El-Aal, A.E.-M.M.; Schmid, J.; Bard, J.; Caselitz, P. Modeling and optimizing the size of the power conditioning unit for photovoltaic systems. *J. Sol. Energy Eng.* **2006**, *128*, 40–44.
16. Peippo, K.; Lund, P.D. Optimal sizing of grid-connected PV-systems for different climates and array orientations: A simulation study. *Sol. Energy Mater. Sol. Cells* **1994**, *35*, 445–451.
17. Free Online Estimation of the Solar Irradiance and Ambient Temperature in Any European and African Location. Available online: http://re.jrc.ec.europa.eu/pvgis/apps4/pvest.php (accessed on 22 May 2015).
18. Krauter, S. *Solar Electric Power Generation-Photovoltaic Energy Systems*; Springer-Verlag: Berlin/Heidelberg, Germany, 2006; pp. 28–37.
19. International Electrotechnical Commission. *Photovoltaic (PV) Devices—Procedures for Temperature and Irradiance Corrections to Measured I-V Characteristics*; IEC Standard 60891; 2nd ed.; IEC: Geneva, Switzerland, 2009.
20. Berrera, M.; Dolara, A.; Faranda, R.; Leva, S. Experimental test of seven widely-adopted MPPT algorithms In Proceedings of the IEEE PowerTech: Innovative Ideas Toward the Electrical Grid of the Future, Bucarest, Romania, 28 June–2 July 2009; pp. 1–8.
21. Faranda, R.; Hafezi, H.; Leva, S.; Mussetta, M.; Ogliari, E. Enegy production estimation for suitable PV Planning. In Proceedings of the 22th International Symposium on Power Electronics, Electrical Drives, Automation and Motion (SPEEDAM), Ischia, Italy, 18–20 June 2014; pp. 248–252.
22. Faranda, R.; Leva, S.; Maugeri, V. MPPT techniques for PV Systems: Energetic and cost comparison. In Proceedings of the IEEE Power and Energy Society General Meeting, Pittsburg, PA, USA, 20–24 July 2008; pp. 1–6.

11

Hydrogen Storage in Pristine and d10-Block Metal-Anchored Activated Carbon Made from Local Wastes

Mohamed F. Aly Aboud [1,2,*], Zeid A. ALOthman [3], Mohamed A. Habila [3], Claudia Zlotea [4], Michel Latroche [4] and Fermin Cuevas [4]

[1] Sustainable Energy Technologies Center, College of Engineering, King Saud University, P.O. Box 800, Riyadh 11421, Saudi Arabia

[2] Mining, Metallurgical and Petroleum Engineering Department, Faculty of Engineering, Al-Azhar University, Nasr City, Cairo 11371, Egypt

[3] Advanced Materials Research Chair, Chemistry Department, P.O. Box 2455, College of Science, King Saud University, Riyadh 11451, Saudi Arabia; E-Mails: zaothman@ksu.edu.sa (Z.A.A.); mhabila@ksu.edu.sa (M.A.H.)

[4] ICMPE/CNRS-UPEC, UMR 7182, 2-8 rue Henri Dunant, 94320 Thiais, France; E-Mails: zlotea@icmpe.cnrs.fr (C.Z.); latroche@icmpe.cnrs.fr (M.L.); cuevas@glvt-cnrs.fr (F.C.)

* Author to whom correspondence should be addressed; E-Mail: aly@temple.edu or maboud@ksu.edu.sa

External Editor: Craig M. Jensen

Abstract: Activated carbon has been synthesized from local palm shell, cardboard and plastics municipal waste in the Kingdom of Saudi Arabia. It exhibits a surface area of 930 m^2/g and total pore volume of 0.42 cm^3/g. This pristine activated carbon has been further anchored with nickel, palladium and platinum metal particles by ultrasound-assisted impregnation. Deposition of nanosized Pt particles as small as 3 nm has been achieved, while for Ni and Pd their size reaches 100 nm. The solid-gas hydrogenation properties of the pristine and metal-anchored activated carbon have been determined. The pristine material exhibits a reversible hydrogen storage capacity of 2.3 wt% at 77 K and 3 MPa which is higher than for the doped ones. In these materials, the spillover effect due to metal doping is of minor importance in enhancing the hydrogen uptake compared with the counter-effect of the additional mass of the metal particles and pore blocking on the carbon surface.

Keywords: activated carbon; hydrogen storage; spillover effect

1. Introduction

Hydrogen has become a promising substitute for fossil fuels in automotive applications due to its abundance and its higher chemical energy. Hydrogen holds three times the specific energy of gasoline [1]. Besides, it is environmentally friendly since its only oxidation product is water [2]. Owing to its extremely low density, a lot of efforts are in progress to reversibly store hydrogen by different methods. The main avenues can be summarized in two groups. The first avenue is chemical storage where chemical bonds occur between hydrogen atoms and receptors to form hydride phases [3,4]. Some of these materials are heavy, expensive and difficult to regenerate on-board. Additionally the thermodynamic stability of these compounds is usually high and a lot of energy is needed for releasing the hydrogen back during the desorption process. The second avenue is physical storage where hydrogen is kept in its molecular state. Physical storage can be performed in three ways:

i) Liquefaction of hydrogen, which requires cryogenic temperatures and efficient insulation. This wastes around 30% of the available energy from burning the hydrogen which makes this storage method a costly process [5].

ii) Compression of hydrogen gas in high-pressure tanks raises safety concern issues mainly as results of hydrogen embrittlement effects. Additionally the low gravimetric storage should also be considered if we include the mass of the tank [6].

iii) Physisorption in porous materials, which attracts scientific interests due to its simplicity. The large surface areas of porous materials, especially of nanostructures, allow the adsorption of large quantities of hydrogen.

Activated carbon (AC) is among the potential candidates for the last route because of its abundance, low cost, non-toxicity, good chemical stability, low density, good recycling characteristics and wide range of pore structures and forms [7–12]. The main factor that limits the storage of hydrogen on the carbon surface is the low energy of adsorption, in the range of 4–8 kJ/mol H_2 for most porous carbons.

Metal atom decoration on graphitic skeletons can increase the hydrogen uptake by introducing strong hydrogen interaction sites on the carbonaceous surfaces. Besides, these metallic centers facilitate the dissociation of hydrogen molecules and eventually increase the hydrogen content over the carbon receptor by a spillover mechanism [13].

In this work, high surface area activated carbon was produced from local wastes and decorated with some d^{10}-block transition elements (Ni, Pd and Pt). Park *et al.* reported that the amount of hydrogen storage capacity for Multi Walled Carbon Nanotubes (MWCNTs) increased in proportion to the Pt content with an optimum loading value of 4 wt% [14]. Zubizarreta *et al.* reported that the optimum nickel content for improving the hydrogen storage in carbon nanospheres was 5 wt% [15], while, Kim *et al.* found that hydrogen storage increased up to 1.23 wt% after loading with 8 wt% of nickel nanoparticles due to the spillover effect [16]. For Pd doping, Lueking *et al.* reported that the optimum loading for hydrogen storage on activated carbon is 14 wt%, while it is 5.7 wt% in single wall carbon nanotubes [17]. Mu *et al.* used 20 wt% of palladium for enhancing the hydrogen storage capacity in etched carbon nanotubes [18].

After performing detailed chemical, structural and thermal characterizations, the hydrogen storage properties of the metal-anchored activated carbons were determined and compared to those of the pristine material.

2. Results and Discussion

2.1. Structural, Chemical and Thermal Characterizations

Figure 1 displays the X-Ray Diffraction (XRD) patterns of all studied samples. For the pristine activated carbon almost no diffraction peaks are detected due to its poor crystallinity. However, a tiny peak around 2θ° = 26.5° might correspond to the (002) plane of graphite [19]. After Pt-anchoring, an additional and very broad peak is detected at 2θ = 39.7°. It is assigned to the diffraction plane (111) of the *fcc* structure of Pt according to JCPDS card No. 03-065-2868. This strong peak broadening is attributed to the ultra-small size of platinum particles, which will be later supported by High-Resolution Transmission Electron Microscope (HRTEM) analysis. For the Ni-anchored sample, well-defined diffraction peaks are detected at 44.5° (111), 52° (200) and 76.5° (220) in 2θ°. They are characteristic of *fcc* nickel in agreement with the JCPDS card No. 03-065-2865. Finally, for the Pd-anchored sample sharp diffraction peaks are detected at 2θ° of 39.88° (111), 46.45° (200), 67.96° (220), 82° (311) and 86° (222) showing the occurrence of *fcc* Palladium (JCPDS card No. 03-065-2867).

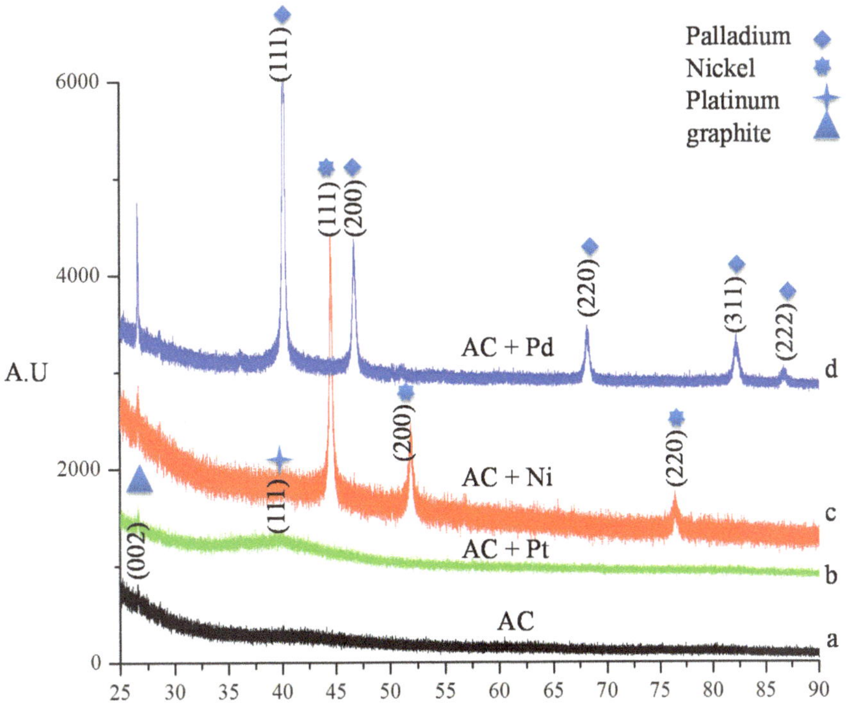

Figure 1. XRD diffraction patterns of (**a**) activated carbon (AC), (**b**) Pt-anchored AC, (**c**) Ni-anchored AC and (**d**) Pd-anchored AC.

Backscattered SEM images for pristine activated carbon AC, Pt-, Ni- and Pd-anchored AC are shown in Figure 2a to 2d, respectively. The pristine AC exhibits a fibrous layer structure with particle dimensions of ~30 μm in length and ~4 μm in thickness. Cavities and porosity are observed as result of the evaporation of the activating agent. Both Ni- and Pd-anchored particles have spherical morphology with particle size ~100 nm. Nickel particles are uniformly dispersed over the AC, while Pd particles tend to agglomerate. SEM images failed to detect the presence of platinum particles on the surface of the activated carbon (Figure 2b) due to its low spatial resolution. Therefore, HRTEM investigations were used for Pt-AC as shown in Figure 3. The platinum particles are homogeneously distributed with ultra-small particle size of ~3 nm. The formation of such ultra-small particles of platinum compared to other anchored metals is consistent with the reported observation by Okhlopkova *et al.* that referred to the aggregation effect of palladium compared to platinum [20].

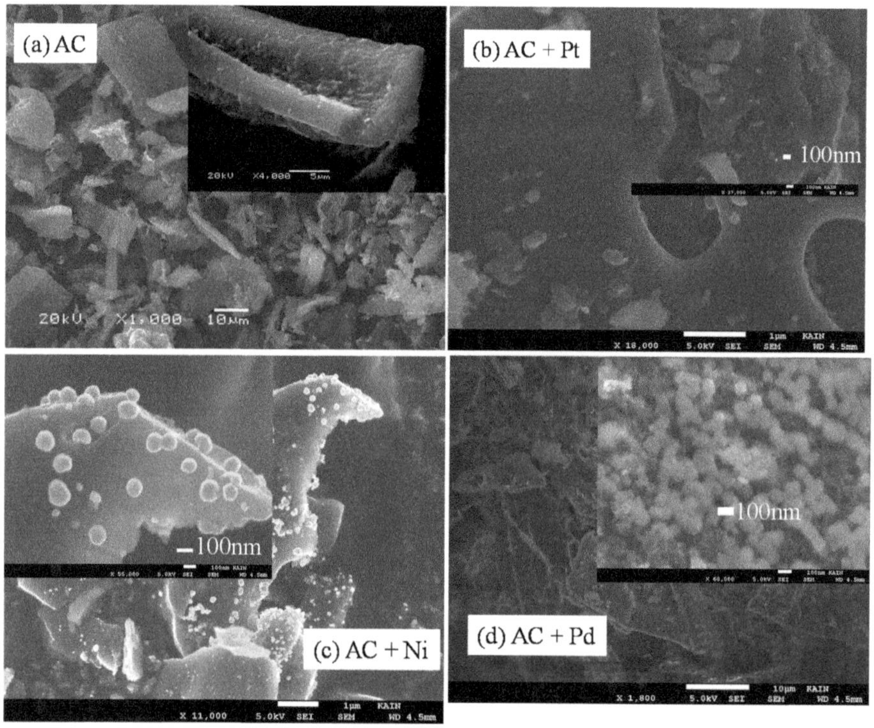

Figure 2. SEM images of (**a**) activated carbon (AC), (**b**) Pt-anchored AC, (**c**) Ni-anchored AC and (**d**) Pd-anchored AC.

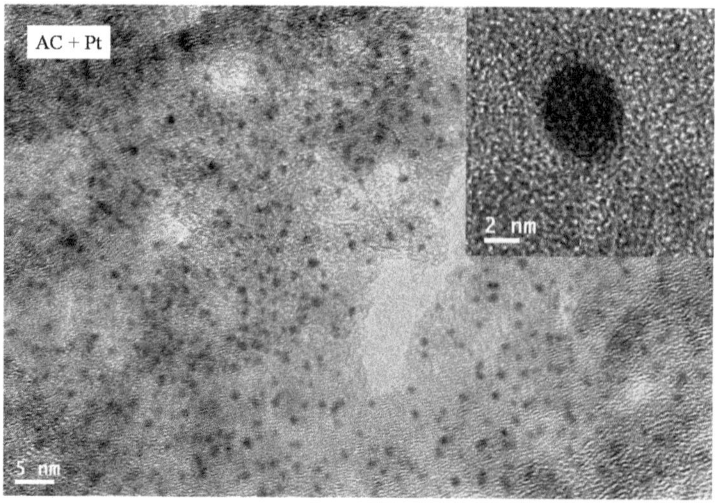

Figure 3. HRTEM images of Pt-anchored activated carbon.

The FTIR spectra recorded for all samples in the 400–4,000 cm^{-1} range are displayed in Figure 4. Almost no difference is seen between the activated carbon and the metal anchored activated carbons. The main groups are both alkene and aromatic C=C functions in addition to hydroxyl (OH^-) groups.

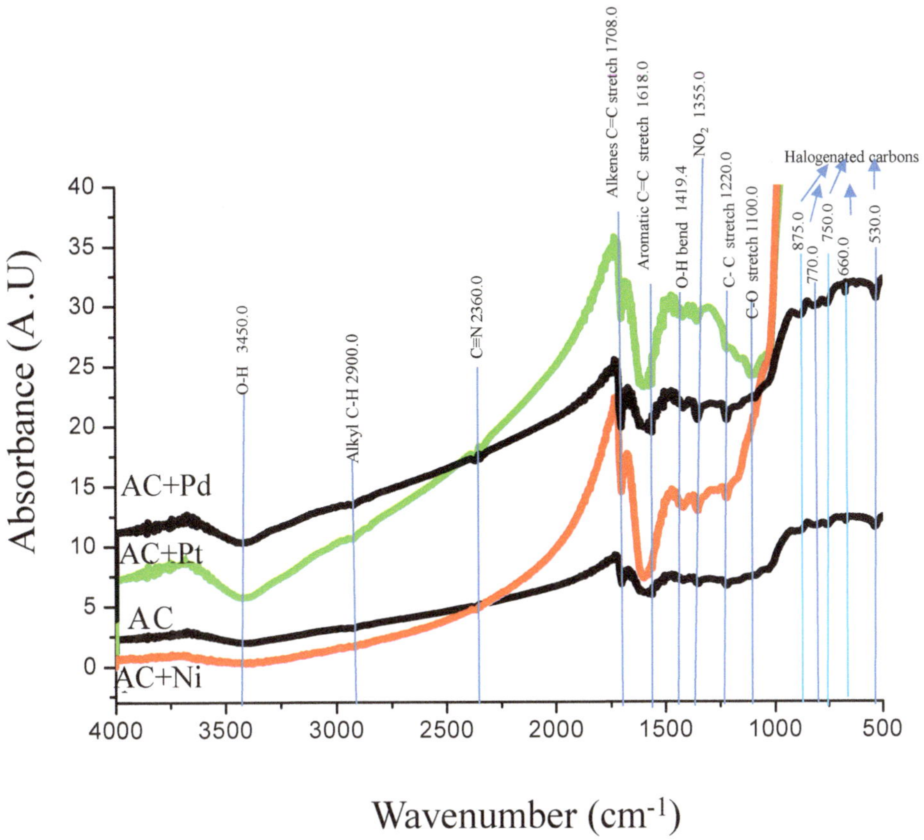

Figure 4. FTIR spectra for activated carbon (AC) and metal-anchored AC + Ni, AC + Pd and AC + Pt.

Figure 5 shows the monitored Differential Scanning Calorimetry (DSC) (Figure 5a) and Thermal Gravimetric Analysis (TGA) (Figure 5b) traces under air flow. For all samples, exothermic reactions due to carbon oxidation are observed. DSC peak temperatures are observed at ~820 K for both the AC and Ni-anchored materials, while they occur at ~770 K and 650 K for Pd- and Pt-anchored ones, respectively. The DSC onset reaction temperatures, as determined from the tangential method, are observed at ~675 K for AC, Ni- and Pd-anchored samples. Interestingly, the onset temperature for Pt-anchored AC is as low as 600 K. The lower oxidation temperature for the Pt containing sample is attributed to the ultra-fine Pt particles which enhance the oxidation reaction.

Figure 5b shows the TGA traces for all samples. Weight losses below 400 K are likely related to desorption of residual adsorbed moisture and volatile matter. Next, in agreement with the DSC analyses, release of CO_2 due to carbon oxidation takes place in the 600–873 K range leading to severe mass loss. Above this temperature range, no significant mass changes occur. The reported TGA curves allow estimating the actual metal doping of the different samples. It can be evaluated from the final weighted TGA mass considering that the amount of carbon ashes is constant for all samples (evaluated as *ca.* 9 wt% for pristine AC) and that the anchored metals form oxide species NiO, PdO, and PtO_2 by reaction with air. The thus-estimated metal doping is 6.5 wt% Ni, 16 wt% Pd, and 5 wt% Pt for nickel-, palladium- and platinum-anchored activated carbon samples, respectively.

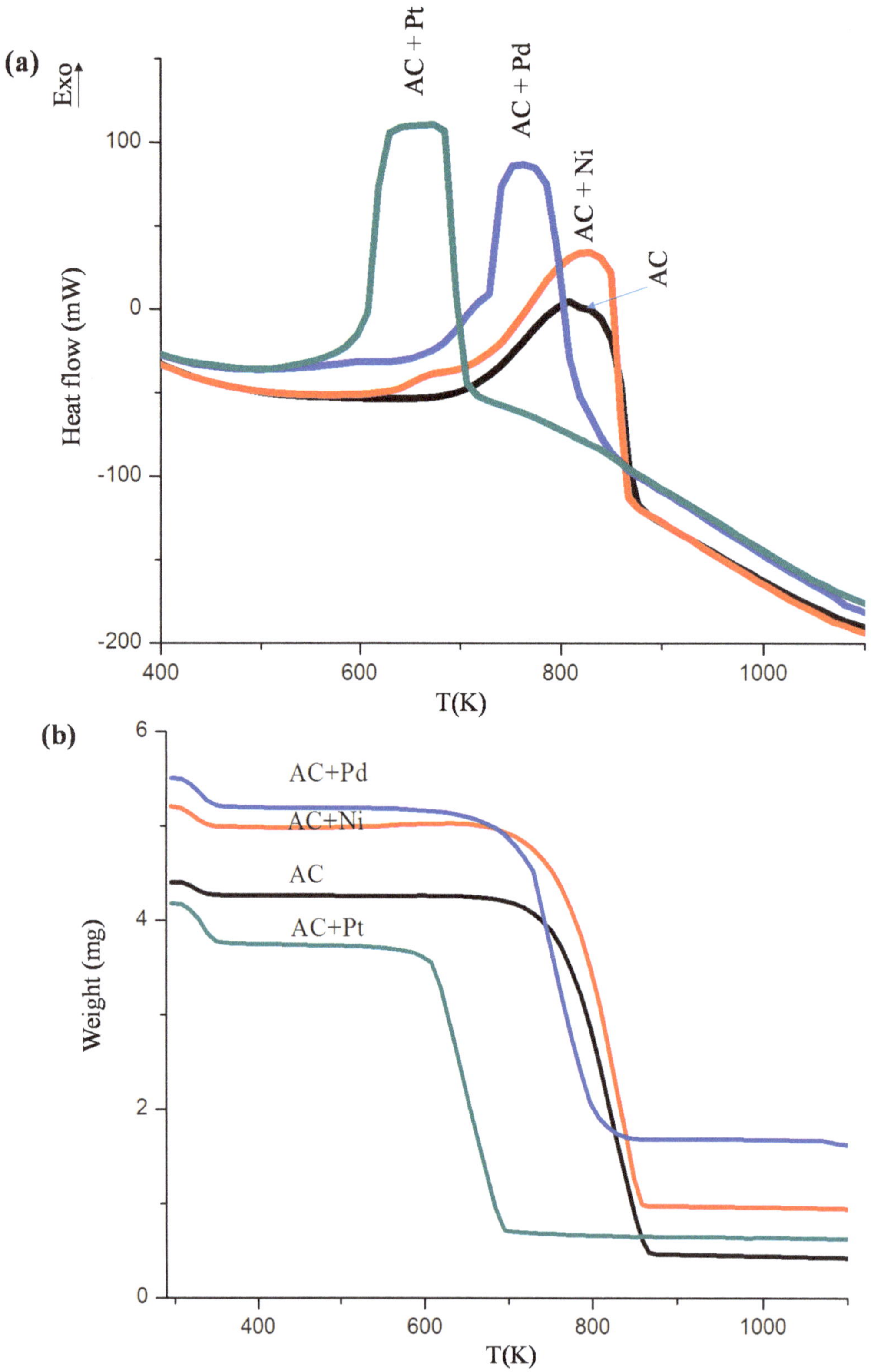

Figure 5. (**a**) DSC analysis under air flow of activated carbon (AC) and metal-anchored AC + Ni, AC + Pd and AC + Pt, and (**b**) TGA analysis under air flow of AC and metal-anchored AC + Ni, AC + Pd and AC + Pt.

2.2. Textural Characterization

Figure 6 shows the nitrogen adsorption isotherms at 77 K of pristine AC and Ni-, Pt-, Pd-anchored activated carbon. The isotherms are Type I, typical of microporous materials with a narrow pore size distribution. The textural properties are summarized in Table 1. The pristine AC as well as both Ni- and Pt-anchored samples have similar textural properties with Brunauer-Emmett-Teller (BET) surface area $A_s \sim 900$ m^2/g, total pore volume $V_{tot} = 0.42$ cm^3/g and micropore volume $V_\mu \sim 0.35$ cm^3/g. The Pd-anchored sample exhibits much lower textural properties.

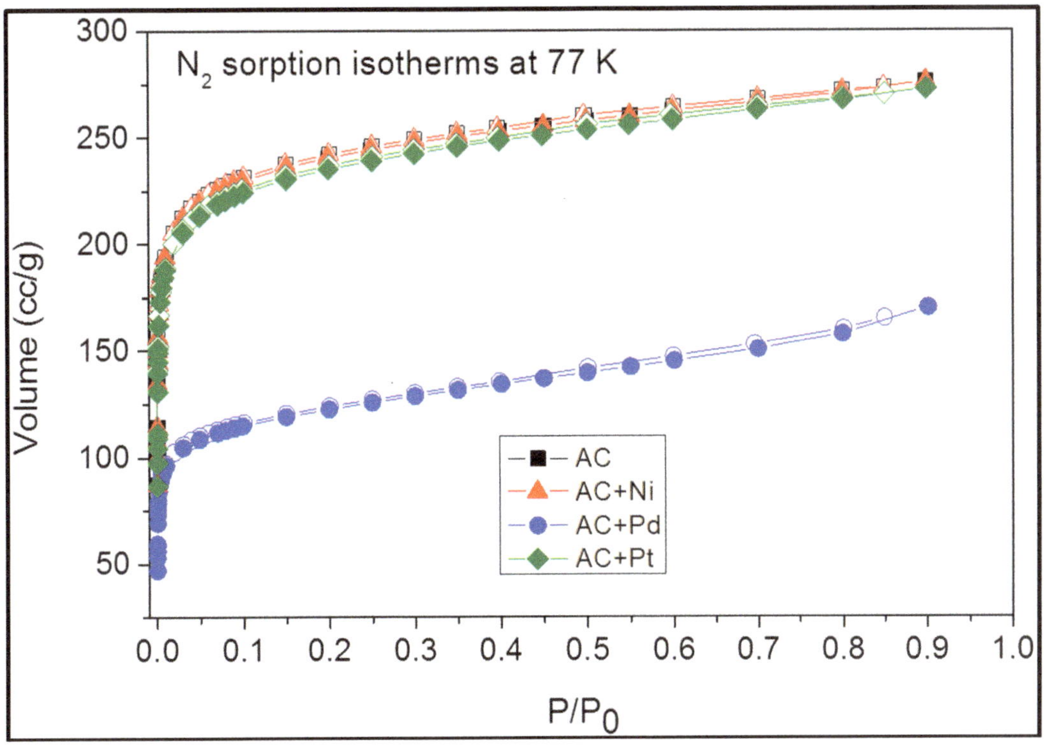

Figure 6. Nitrogen adsorption isotherms at 77 K of pristine and Ni-, Pd- and Pt-anchored activated carbon.

Table 1. Textural properties of pristine and Ni-, Pt-, Pd-anchored activated carbon.

Sample	Skeletal Density g/cm^3	A_s (BET) m^2/g	V_{tot} ($P/P_0 = 0.9$) cm^3/g	V_μ (DR) cm^3/g
Activated carbon	1.7 ± 0.1	931 ± 20	0.42 ± 0.1	0.36 ± 0.1
AC + Ni	1.9 ± 0.1	910 ± 20	0.42 ± 0.1	0.35 ± 0.1
AC + Pt	1.9 ± 0.1	889 ± 20	0.42 ± 0.1	0.34 ± 0.1
AC + Pd	2.0 ± 0.1	455 ± 20	0.26 ± 0.1	0.17 ± 0.1

Although the textural properties of activated carbons can vary depending on the processing techniques and precursors, the values reported here are almost identical to activated carbon made from Malaysian palm shells that were physically activated with steam (A_s = 941 m^2/g, V_{tot} = 0.52 cm^3/g) [21]. Moreover, they are much better than for H_3PO_4 chemically activated Malaysian palm (A_s = 615 m^2/g, V_{tot} = 0.28 cm^3/g) [22]. Also, the textural properties of our AC is superior to the activated carbon made from palm shells that underwent both chemical and physical treatments (A_s = 642 m^2/g, V_{tot} = 0.28 cm^3/g) [22].

The slight decrease for the surface area of the Ni- and Pt-anchored ACs compared with the pristine one is consistent with the reported data for Pt doped super-activated carbons [23]. This decrease was attributed to metal particles filling or blocking pores in the activated carbon. The drastic decrease for Pd-AC, with BET surface area about half that of Ni- and Pt-anchored AC, likely results from the high agglomeration of Pd-particles at the AC surface (Figure 2d).

The increase in the density of the metal anchored AC compared with the pristine one is due to the higher density of metal particles as compared to that of the adsorbent. Such an effect is consistent with the higher reported density of doped Pd-activated carbon fibers compared to non-doped ones [24]. The low surface area and pore volumes for palladium supported activated carbon due to agglomeration result in more packed and dense material compared to other doped activated carbons due its higher loading as was confirmed by the TGA measurements (Figure 5b).

2.3. Hydrogen Storage Measurements

Pressure Composition Isotherm (PCI) curves at 77 K and 298 K for all samples are shown in Figure 7 and the hydrogen uptake capacities are summarized in Table 2. In all cases, no hysteresis (below the experimental error) is observed, indicating that the adsorbed hydrogen can be desorbed reversibly when the pressure is released.

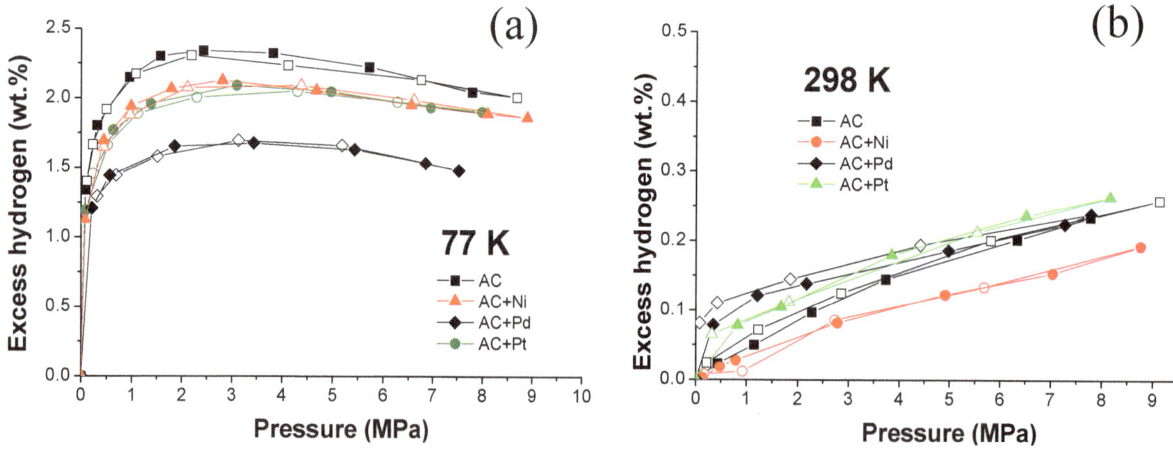

Figure 7. Pressure composition isotherm curves at (**a**) 77 K and (**b**) 298 K for the activated carbon (AC), Ni- , Pd- and Pt-anchored AC. Full and empty symbols stand for adsorption and desorption measurements, respectively.

All low temperature (77 K) the PCI curves exhibit Type I isotherms. This indicates monolayer adsorption of hydrogen molecules, which extent depends on the micropore volume and its filling. The hydrogen uptake increases up to 3 MPa and then reaches a maximum due to the completion of micropore filling. The amount of adsorbed hydrogen in the pristine activated carbon is consistent with the maximum theoretical capacity established by Ströbel *et al.* for carbonaceous materials [25]. These authors showed that the hydrogen uptake on carbon surface could be derived as:

$$H\,(\text{wt\%}) = 2.27 \times 10^{-3} \cdot A_s\,(\text{m}^2/\text{g}) \qquad (1)$$

where A_s is the surface area of the carbonaceous material. Using Equation (1) and substituting for the observed surface area A_s = 931 m^2/g, we obtain a theoretical hydrogen uptake of 2.1 wt% in close agreement with the reported value of 2.3 wt%. Moreover, these capacities are consistent with the reported data for wide range of carbon-based materials with wide textural properties [26–29]. The density of the adsorbed hydrogen at 77 K can be calculated from the hydrogen uptake (2.3 wt%) and the volume occupied by hydrogen in the micropore volume (0.36 cm^3/g). The calculated density equals 0.063 g/cm^3, lower than the reported density for liquid hydrogen at 20 K (0.071 g/cm^3) [30].

Table 2. Hydrogen uptake for pristine activated carbon and Ni-, Pt- and Pd-anchored activated carbon at 77 and 298 K.

Sample	**H_2 Excess Capacity (wt%)**	
	77 K at 3 MPa	**298 K at 8 MPa**
Activated carbon	2.3 ± 0.1	0.24 ± 0.01
AC + Ni	2.1 ± 0.1	0.17 ± 0.01
AC + Pt	2.1 ± 0.1	0.26 ± 0.01
AC + Pd	1.7 ± 0.1	0.25 ± 0.01

At 298 K and high-pressure (Figure 7b), both adsorption and desorption PCI isotherms exhibit linear behavior with low hydrogen uptake (≤0.26 wt%). Such a low capacity is attributed to the low enthalpy of hydrogen adsorption [31] and concurs with earlier reports on platinum doping in super activated carbon [18] and palladium in mesoporous carbon [32]. Similar capacities have been reported for hydrogen storage in activated carbon of even higher surface area A_s = 3,500 g/cm^3 [23]. At high pressure, pure AC, Pd- and Pt-anchored activated carbons exhibit rather similar hydrogen uptake, whereas Ni-anchored activated carbon has lower capacity. At room temperature and low pressure, Pd-anchored activated carbon has higher capacity than all other materials. This confirms the formation of palladium hydride at room temperature and low pressure, as demonstrated previously [28,29,32]. However, at high pressure, the favorable effect of palladium hydride formation is counteracted by the weight of the dopant, the filling of adsorbent micropores and pore blocking. These latter drawbacks could be counterbalanced if metal deposition would induce a significant hydrogen uptake by spillover effect. Unfortunately, the low hydrogen uptake at high pressure suggests that the spillover effect, if it exists, turns out to be of minor importance in the studied systems.

3. Experimental Section

Palm shell, cardboard and plastics collected from a municipal solid waste station in Riyadh, Saudi Arabia were used as raw material precursors in the present study. The mixing ratio was 1:1:0.33 by weight. The choice of such raw materials was given due to their abundance in the kingdom which will be in great advantage for a green environment. The elemental analysis for the raw materials and their activation process are given in [33,34]. The used metal precursors were nickel acetate tetrahydrate ($Ni(OCOCH_3)_2 \cdot 4H_2O$, 99%, E. Merck, darmstadt, Germany), palladium acetate ($Pd(OCOCH_3)_2$, 99%, Sigma Aldrich, St. Louis, MO, USA) and hexachloroplatenic acid ($H_2PtCl_6 \cdot XH_2O$, assay 40.0% Loba Chemie, Mumbai, India).

The ultrasound-assisted impregnation method was used to anchor metal particles on activated carbon surface because of its simplicity and its large scale [35–38]. Deionized water was used as dissolving media for both nickel acetate and platinum chloride while ultra-high purity benzene was used for dissolving palladium acetate. In the same dissolving media, activated carbon was dispersed using an ultrasonic bath and mixed with the dissolved metal salt to nominally form a 10 wt% of anchored metal activated carbon. The mixture was left several days under the hood with repetitive stirring to achieve complete evaporation of the solvents. The resulted impregnated powders were grounded in ceramic mortar for homogenizing the impregnated materials. Same solvents were added to the resulted impregnated powders extra two times and followed by same procedure to ensure well dispersion. The resulted metal supported activated carbons were annealed at 523 K for 2 h under a continuous flow of hydrogen atmosphere to reduce the supported metal salts.

All samples underwent structural analysis by X-ray diffraction (XRD) using a X'Pert PRO Philips diffractometer (Eindhoven, The Netherlands), operated at 40 mA and 40 kV with CuK_{α} radiation and a nickel filter. Measurements were performed over the 2-theta range from 2° to 100° by steps of 0.02°, with a sampling time of one second per step. Sample morphology was investigated using a Field-Emission Scanning Electron Microscope (FE-SEM, model FEI-200NNL, Hillsboro, OR, USA) and a High-Resolution Transmission Electron Microscope (HRTEM, model JEM-2100F, JEOL, Tokyo, Japan). Surface chemistry was studied by Fourier Transform Infrared (FT-IR) transmission spectroscopy using a Nicolet 6700 FT-IR spectrophotometer (Thermo Scientific,Waltham, MA, USA). To this aim, samples were mixed with 98% wt% of KBr and then finely ground to make pellets. Thermal stability was investigated using a Differential Scanning Calorimetry–Thermal Gravimetric Analysis (TGA/DSC1 Star^{e}) system (Mettler Toledo, Columbus, OH, USA). TGA/DSC runs were performed under air within the temperature range 298–1,100 K with a heating rate of 5 K/min.

The textural properties were determined from nitrogen adsorption/desorption isotherms using an Autosorb-iQ Automated Gas Sorption Analyzer from Quantachrome Instruments (Hartley Wintney, UK). All isotherms have been measured by volumetric means. Samples were previously degassed overnight at 473 K. The specific surface area was obtained by the Brunauer-Emmett-Teller (BET) method. The total pore volume was computed from the amount of gas adsorbed at $P/P_0 = 0.9$, and the micropore volume was calculated using the Dubinin-Radushkevich (DR) equation.

Hydrogen sorption properties of samples were determined by measuring the Pressure-Composition-Isotherms (PCI) at 77 and 298 K up to 9 MPa hydrogen pressure. The PCI curves were recorded using a manual volumetric device (Sievert's method) equipped with calibrated and thermostated volumes and pressure gauges. The samples were enclosed in a stainless steel sample holder closed with a metal seal. Before any sorption measurements, the samples were degassed under secondary vacuum at 473 K for 12 h. Weight losses were measured as 3.1, 3.5, 6.8 and 5.6 wt% for non-doped, Ni-doped, Pt-doped and Pd-doped samples, respectively. The sample holder is immersed in a liquid nitrogen Dewar at 77 K or a thermostated water bath maintained at 298 K, and high purity hydrogen (6N) is introduced step by step up to 9 MPa. The pressure variations due to both gas expansion and hydrogen sorption are measured after reaching thermodynamic equilibrium, usually in the range of minutes. Real equation of state for hydrogen gas is used from the program GASPAK V3.32. The PCI curves were measured twice (*i.e.*, two full adsorption–desorption cycles) in order to check the hysteresis effect and the measurement repeatability. Good repeatability has been obtained for all samples. All capacities reported are excess hydrogen sorption quantities and refer to the sample dry mass (*i.e.*, degassed mass). Sample volume correction is derived from skeletal density measurements and data correspond to the excess values obtained by He pycnometry (Utrapyc 1200 Quantachrome, Hartley Wintney, UK).

4. Conclusions

Activated carbon was successfully synthesized from local palm shell and other wastes. The pristine activated carbon was impregnated with nickel, palladium and platinum. Coarse metallic nanoparticles of typical size 100 nm are formed for Ni- and Pd-anchored samples, whereas ultra-small particles size of 3 nm were obtained for the Pt case.

The hydrogen adsorption mechanism for all the synthesized samples occurs via micropore filling. No significant hysteresis was observed for any of the samples, indicating that adsorbed hydrogen can be desorbed reversibly when the pressure is released. At 298 K and 8 MPa both adsorption and desorption showed linear behavior with a low amount of hydrogen uptake. Low hydrogen uptake at ambient temperature is attributed to the low enthalpy of hydrogen adsorption on carbon surface.

At 77 K and 3 MPa, the activated carbon showed hydrogen storage of 2.3 wt% which is consistent with the BET surface area of ~931 m^2/g. Under the same conditions, metal-anchored activated carbons exhibited lower hydrogen uptakes ranging between 1.7 and 2.1 wt%. For the materials used in this study, possible spillover effects induced by the presence of metallic dopants are counterbalanced in terms of hydrogen storage by the concomitant addition of metal mass and reduction of carbon porosity.

Acknowledgments

The authors gratefully thank Sustainable Energy Technologies (SET) Center at King Saud University for financing this work.

Author Contributions

Mohamed F. Aly Aboud conceived and designed the experiments, performed the doping experiments and all data analysis; Mohamed F. Aly Aboud wrote the paper; Claudia Zlotea performed the textural characterization and hydrogen storage measurements; Claudia Zlotea, Michel Latroche and Fermin Cuevas participated in data analysis, paper writing and corrections; Zeid A. ALOthman and Mohamed A. Habila provided the virgin activated carbons and participated in FTIR analysis and provided important suggestions. All authors examined and approved the final manuscript.

Conflicts of Interest

The authors declare no conflict of interest.

References

1. Schlapbach, L.; Züttel, A. Hydrogen-storage materials for mobile applications. *Nature* **2001**, *414*, 353–358.
2. Rostrup-Nielsen, J.R.; Rostrup-Nielsen, T. Large-scale hydrogen production. *Cattech* **2002**, *6*, 150–159.
3. Krasae-In, S.; Stang, J.H.; Neksa, P. Simulation on a proposed large-scale liquid hydrogen plant using a multi-component refrigerant refrigeration system. *Int. J. Hydrog. Energy* **2010**, *35*, 4524–4533.
4. Barthélémy, H. Effects of pressure and purity on the hydrogen embrittlement of steels. *Int. J. Hydrog. Energy* **2011**, *36*, 2750–2758.

5. Kleperis, J.; Wójcik, G.; Czerwinski, A.; Skowronski, J.; Kopczyk, M.; Beltowska-Brzezinska, M. Electrochemical behavior of metal hydrides. *J. Solid State Electrochem.* **2001**, *5*, 229–249.
6. Ovshinsky, S.R.; Fetcenko, M.A.; Ross, J.A. Nickel metal hydride battery for electric vehicles. *Science* **1993**, *260*, 176–181.
7. Ebbesen, T.W.; Lezec, H.J.; Hiura, H.; Bennett, J.W.; Ghaemi, H.F.; Thio, T. Electrical conductivity of individual carbon nanotubes. *Nature* **1996**, *382*, 54–56.
8. Wong, E.W.; Sheehan, P.E.; Lieber, C.M. Nanobeam mechanics: Elasticity, strength, and toughness of nanorods and nanotubes. *Science* **1997**, *277*, 1971–1975.
9. Collins, P.G.; Bradley, K.; Ishigami, M.; Zettl, A. Extreme oxygen sensitivity of electronic properties of carbon nanotubes. *Science* **2000**, *287*, 1801–1804.
10. Liang, W.; Yokojima, S.; Ng, M.F.; Chen, G.; He, G. Optical properties of single-walled 4 Å carbon nanotubes. *J. Am. Chem. Soc.* **2001**, *123*, 9830–9836.
11. Minot, E.D.; Yaish, Y.; Sazonova, V.; McEuen, P.L. Determination of electron orbital magnetic moments in carbon nanotubes. *Nature* **2004**, *428*, 536–539.
12. Motta, M.; Li, Y.L.; Kinloch, I.; Windle, A. Mechanical properties of continuously spun fibers of carbon nano tubes. *Nano Lett.* **2005**, *5*, 1529–1533.
13. Psofogiannakis, G.M.; Froudakis, G.E. Study of hydrogen spill over mechanism on Pt doped graphite. *J. Phys. Chem. C* **2009**, *113*, 14908–14915.
14. Park, S.J.; Lee, S.Y. Hydrogen storage behaviors of platinum-supported multi-walled carbon nanotubes. *Int. J. Hydrog. Energy* **2010**, *35*, 13048–13054.
15. Zubizarreta, L.; Menéndez, J.A.; Pis, J.J.; Arenillas, A. Improving hydrogen storage in Ni-doped carbon nanospheres. *Int. J. Hydrog. Energy* **2009**, *34*, 3070–3076.
16. Kim, J.H.; Han, K.S. Ni nanoparticles-hollow carbon spheres hybrids for their enhanced room temperature hydrogen storage performance. *Trans. Korean Hydrog. New Energy Soc.* **2013**, *24*, 550–557.
17. Lueking, A.D.; Yang, R.T. Hydrogen spillover to enhance hydrogen storage—Study of the effect of carbon physicochemical properties. *Appl. Catal. A Gen.* **2004**, *265*, 259–268.
18. Mu, S.C.; Tang, H.L.; Qian, S.H.; Mu, P.; Yuan, R.Z. Hydrogen storage in carbon nanotubes modified by microwave plasma etching and Pd decoration. *Carbon* **2006**, *44*, 762–767.
19. Li, Z.Q.; Lu, C.J.; Xia, Z.P.; Zhou, Y.; Luo, Z. X-ray diffraction patterns of graphite and turbostratic carbon. *Carbon* **2007**, *45*, 1686–1695.
20. Okhlopkova, L.B.; Lisitsyn, A.S.; Likholobov, V.A.; Gurrath, M.; Boehm, H.P. Properties of Pt/C and Pd/C catalysts prepared by reduction with hydrogen of adsorbed metal chlorides: Influence of pore structure of the support. *Appl. Catal. A Gen.* **2000**, *204*, 229–240.
21. Arouaa, M.K.; Daud, W.M.A.W.; Yin, C.Y.; Adinata, D. Adsorption capacities of carbon dioxide, oxygen, nitrogen and methane on carbon molecular basket derived from polyethyleneimine impregnation on microporous palm shell activated carbon. *Sep. Purif. Technol.* **2008**, *62*, 609–613.
22. Arami-Niya, A.; Daud, W.M.A.W.; Mjalli, S.F.; Abnisa, F.; Shafeeyan, M.S. Production of microporous palm shell based activated carbon for methane adsorption: Modeling and optimization using response surface methodology. *Chem. Eng. Res. Des.* **2012**, *90*, 776–784.
23. Stadie, N.P.; Purewal, J.J.; Ahn, C.C.; Fultz, B. Measurements of hydrogen spillover in platinum doped super activated carbon. *Langmuir* **2010**, *26*, 15481–15485.

24. Contescu, C.I.; Benthem, K.V.; Sa, L.; Bonifacio, C.S.; Pennycook, S.J.; Jena, P.; Gallego, N.C. Single Pd atoms in activated carbon fibers and their contribution to hydrogen storage. *Carbon* **2011**, *49*, 4050–4058.
25. Ströbel, R.; Garche, J.; Moseley, P.T.; Jörissen, L.; Wolf, G. Hydrogen storage by carbon materials. *J. Power Sources* **2006**, *159*, 781–801.
26. Zubizarreta, L.; Gomez, E.I.; Arenillas, A.; Ania, C.O.; Parra, J.B.; Pis, J.J. Hydrogen storage in carbon materials. *Adsorption* **2008**, *14*, 557–566.
27. Yürüm, Y.; Taralp, A.; Veziroglu, N.T. Storage of hydrogen in nano structured carbon materials. *Int. J. Hydrog. Energy* **2009**, *34*, 3784–3798.
28. Dibandjo, P.; Zlotea, C.; Gadiou, R.; Ghimbeu, C.; Cuevas, F.; Latroche, M.; Leroy, E.; Vix-Guterl, C. Hydrogen storage in hybrid nanostructured carbon/palladium materials: Influence of particle size and surface chemistry. *Int. J. Hydrog. Energy* **2013**, *38*, 952–965.
29. Zhao, W.; Fierro, V.; Zlotea, C.; Chevalier-César, C.; Izquierdo, M.T.; Latroche, M.; Celzard, A. Carbons doped with Pd nanoparticles for hydrogen storage. *Int. J. Hydrog. Energy* **2012**, *27*, 5072–5080.
30. Midilli, A.; Ay, M.; Dincer, I.; Rosen, M.A. On hydrogen and hydrogen energy strategies: I: Current status and needs. *Renew. Sustain. Energy Rev.* **2005**, *9*, 255–271.
31. Thomas, M.K. Hydrogen adsorption and storage on porous materials. *Catal. Today* **2007**, *120*, 389–398.
32. Zlotea, C.; Cuevas, F.; Boncour, V.P.; Leroy, E.; Dibandjo, P.; Gadiou, R.; Guterl, C.V.; Latroche, M. Size-dependent hydrogen sorption in ultrasmall Pd clusters embedded in mesoporous carbon template. *J. Am. Chem. Soc.* **2010**, *132*, 7720–7729.
33. AlOthman, Z.A.; Habila, M.A.; Ali, R. Preparation of activated carbon using the copyrolysis of agricultural and municipal solid wastes at a low carbonization temperature. *Int. Conf. Biol. Environ. Chem.* **2011**, *24*, 67–72.
34. Habila, M.; Yilmaz, E.; ALOthman, Z.A.; Soylak, M. Flame atomic absorption spectrometric determination of Cd, Pb, and Cu in food samples after pre-concentration using 4-(2-thiazolylazo) resorcinol-modified activated carbon. *J. Ind. Eng. Chem.* **2014**, *20*, 3989–3993.
35. Zha, Q.F.; Hu, X.H.; Guo, Y.S.; Wu, M.B.; Li, Z.F.; Zhang, Y.Z. Improved antioxidative ability of porous carbons by boron-doping. *New Carbon Mater.* **2008**, *23*, 356–360.
36. Li, Y.; Yang, R.T. Hydrogen storage on platinum nanoparticles doped on superactivated carbon. *J. Phys. Chem. C* **2007**, *111*, 11086–11094.
37. Matsumoto, T.; Komatsu, T.; Nakanoa, H.; Arai, K.; Nagashima, Y.; Yooa, E. Efficient usage of highly dispersed Pt on carbon nano tubes for electrode catalysts of polymer electrolyte fuel cells. *Catal. Today* **2004**, *90*, 277–281.
38. Şayan, E. Ultrasound assisted preparation of activated carbon from alkaline impregnated hazelnut shell: An optimization study on removal of copper ion from aqueous solution. *Chem. Eng. J.* **2006**, *115*, 213–218.

12

Economic Value of Li-ion Energy Storage System in Frequency Regulation Application from Utility Firm's Perspective in Korea

Wonchang Hur [1], Yongma Moon [2,*], Kwangsup Shin [3], Wooje Kim [4], Suchul Nam [5] and Kijun Park [5]

1 College of Business Administration, Inha University, Incheon 431050, Korea; E-Mail: wchur@inha.ac.kr
2 College of Business Administration, University of Seoul, Seoul 130743, Korea
3 Graduate School of Logistics, Incheon National University, Incheon 406772, Korea; E-Mail: ksshin@inu.ac.kr
4 College of Business and Technology, Seoul National University of Science and Technology, Seoul 139743, Korea; E-Mail: wjkim@seoultech.ac.kr
5 KEPCO Research Institute, Daejeon 305760, Korea; E-Mails: scnam@kepco.co.kr (S.N.); killer@kepco.co.kr (K.P.)

* Author to whom correspondence should be addressed; E-Mail: yongma@uos.ac.kr

Academic Editor: Sheng S. Zhang

Abstract: Energy Storage Systems (ESSs) have recently been highlighted because of their many benefits such as load-shifting, frequency regulation, price arbitrage, renewables, and so on. Among those benefits, we aim at evaluating their economic value in frequency regulation application. However, unlike previous literature focusing on profits obtained from participating in the ancillary service market, our approach concentrates on the cost reduction from the perspective of a utility firm that has an obligation to pay energy fees to a power exchange. More specifically, we focus on the payments between the power exchange market and the utility firm as a major source of economic benefits. The evaluation is done by cost-benefit analysis (CBA) with a dataset of the Korean market while considering operational constraint costs as well as scheduled energy payments, and a simulation algorithm for the evaluation is provided. Our results show the potential for huge profits to be made by cost reduction. We believe that this research can provide a guideline for a utility firm considering investing in ESSs for frequency regulation application as a source of cost reduction.

Keywords: energy storage system; economic value; utility firm; frequency regulation; ancillary services

1. Introduction

In a traditional generation setting, electricity, in its technical characteristics, cannot be stored, which means that if there is demand, the demand should be fulfilled right away. This implies that a power system should always secure sufficient generation capacity to meet the highest demand at peak times, even though demand stays lower for most of the time. In fact, such inefficiency is becoming more problematic as peak time demands keep soaring faster than average demands.

In this respect, Energy Storage Systems (ESS) have been paid great attention because they are believed to enhance power system efficiency due to their capability of discharging, storing, and charging electricity within a very short time interval. In fact, a variety of applications of ESS have been proposed, including peak-time shifting, load balancing, provision of ancillary services, and quality enhancement. However, despite their remarkable pace of technological advancement, adoption of ESS is still in the early stage, particularly for grid utility-related applications.

Despite the hype propagated by ESS industries, it is also true that there has been hesitation or resistance to adopting and investing in ESS. Arguably, one of the strongest obstacles to the diffusion of ESS is the absence of scientific economic validation that can be referred to or consulted by investors and policy makers [1]. Of course, some studies, such as [1–5], have discussed the economic value of ESS in the general scope, but they are still insufficient to attract investors and policy makers in particular application areas. For example, while most researchers studied the economic values of ESS obtained from arbitrage trade of energy, only a few articles investigated the economic value in terms of frequency regulation.

In addition to the frequency regulation application, a more important and specific question is who invests in and owns the ESS. In general, investors would participate in an ancillary service market to make profits by reserving capacity for regulation. However, in contrast to the general case, what happens if a utility firm that is in a position to pay regulation service fees to a power exchange market invests in ESS and utilizes it for regulation? Therefore, in this paper, we consider the case in which a utility firm invests in and operates ESS for frequency regulation.

This paper proposes an economic valuation method for a battery-based energy storage system used for frequency regulation purposes, from a utility firm's perspective. Also, unlike typical economic analyses relying on highly approximated monthly or yearly benefits, our simulation and corresponding algorithm to implement it is based on more precise estimation of hourly changes in electricity generation and on more complicated operation constraints. The method we present also considers important characteristics specific to the Korean electricity market and makes the analysis from the perspectives of both the wholesaler and generators. By providing detailed analysis of economic outcomes from both sides, the proposed method helps investors make an informed decision as well as provides meaningful implications for policy makers.

The structure of this paper is as follows: In Section 2, we investigate the literature regarding ESS investment and discuss backgrounds of frequency regulation, especially in the Korean electricity market. Then, we suggest a framework and an algorithm for ESS investment evaluation in frequency regulation application from a utility firm's perspective in Section 3. Based on the proposed framework, simulation and analysis are executed to analyze the effect of a utility adopting ESS in Section 4. Section 5 summarizes the major results, and provides implications for policy maker and utility firm investors.

2. Literature Review and Background

2.1. Energy Storage System (ESS) for Frequency Regulation (FR)

Energy Storage Systems (ESS) can charge and discharge electricity in a very short time. Since their benefits are known to be broad and include load-shift, reliability, and stabilization of the power grid, the utilization of ESS has garnered attention from practitioners and policy makers in the related fields. For example, Sandia Lab [6] provides a description and service-specific technical details for 18 services and applications associated with the use of electricity storage for electric-utility-related applications. The 18 services are categorized as follows: (1) Bulk energy service including energy time-shift and energy supply capacity; (2) Ancillary Services like regulation, spinning reserves, black start, and voltage support; (3) Transmission infrastructure like upgrade deferral; (4) Distribution infrastructure like distribution upgrade deferral and voltage support; and (5) Customer energy management services like power quality, reliability, and demand charge management.

Among the various applications of ESS, the regulation market (also referred to as frequency regulation or, in parts of the EU, called regulating power) has earned attention due to its potential for economic benefits and readiness in terms of market situation and technical issues. Regulation is used to reconcile momentary differences caused by fluctuations in generation and loads. This imbalance between generation and loads increases or decreases the frequency of electricity, which leads to degradation of power quality. The primary purpose of regulation is to maintain the grid frequency at a certain level (60 Hz).

This is accomplished by using a real-time communication signal directly controlled by the grid operator. The regulation control signal can call for either a positive or negative correction, often referred to in the industry as "regulation up" and "regulation down", respectively. If loads exceed generation, frequency and voltage drop and the grid operator relays a signal to generators requesting regulation up. Conversely, when there is a momentary excess of electric supply capacity, the grid operator requests regulation down and asks generators to reduce generation.

Regulation is typically provided by generating units that are online and ready to increase or decrease power as needed. Regulation is contracted capacity on an hourly basis and dispatched on intervals between four seconds and one minute. Neither regulation up nor regulation down is dispatched for a long duration. An important consideration for this application is that most thermal or base load generation used for regulation service is not especially well-suited or designed to provide regulation; they are not designed for operation at partial loads or to produce variable outputs, which means that they are usually most efficient when power plants operate at a specific and constant (power) output level. Similarly, air emissions and plant wear and tear are usually lowest (per kWh of output) when thermal generation operates at a full load and with constant output [6].

Therefore, storage may be an attractive alternative to most generation-based load following for at least three reasons: (1) in general, storage has superior part-load efficiency; (2) efficient storage can be used to provide up to twice its rated capacity (for regulation); and (3) storage output can be varied rapidly (e.g., output can change from none to full or from full to none within seconds rather than minutes).

2.2. Frequency Regulation in Korea

In Korea, frequency regulation is directed by KEPCO (Korea Electric Power COrporation) and KPX (Korea Power eXchange) in two forms, an Automatic Generation Control signal (AGC) and a Governor Free signal (GF). KPX and KEPCO ask generators to respond within five minutes, when the frequency is outside of the range of 0.2 Hz change. The typical amount of contracted regulation service in Korea is less than about 1,500 MW. For the past five years, governor free regulation (GFR) and automatic generation control (AGC) have been, on average, 800–1000 MW and 400 MW, respectively, as shown in Figure 1.

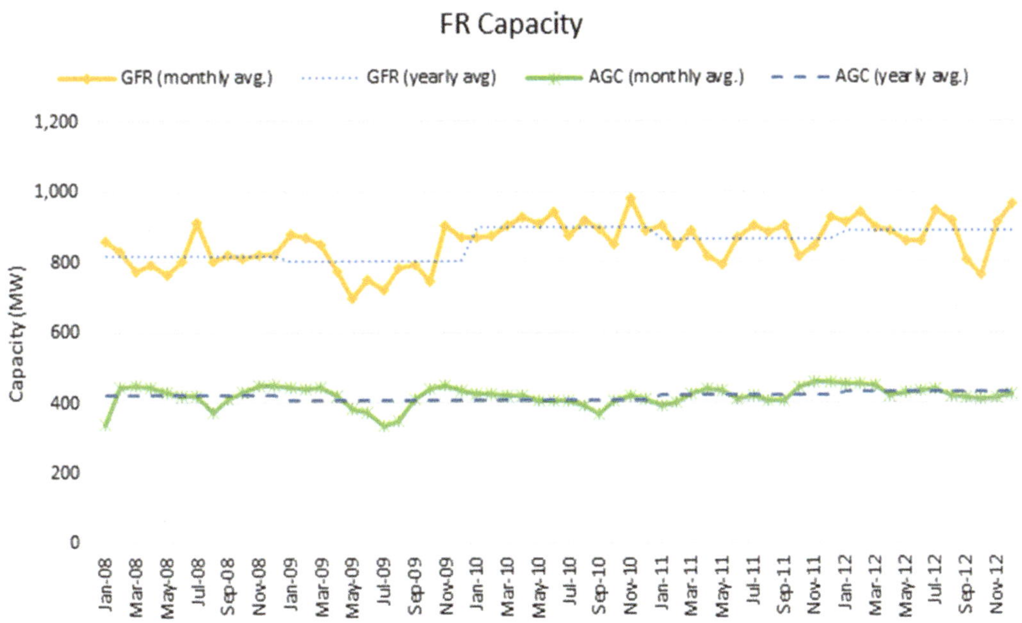

Figure 1. The capacity of frequency regulation.

In Korea, the economic validity of employing ESS for FR cannot be discussed without considering the roles that KEPCO plays in the Korean electricity market. KEPCO is a monopsony wholesale buyer and is in charge of transportation and distribution. KEPCO buys electricity in the market operated by KPX and transports/distributes electricity to customers. KEPCO is also in charge of operating ancillary services including FR. Therefore, it determines the total FR capacity and performs allocation to generators. Specifically, it secures the capacity for frequency regulation by reserving a part of the bid amount offered from generators, and generators are dispatched real time for FR based on their reserved capacity. Since KEPCO is a monopsony buyer, it has the responsibility of paying generators for their participation in ancillary services. As of now, KEPCO itself is not allowed to participate in FR service markets or power generation. In fact, since the service fee for ancillary services is being regulated, there can be no ancillary service market in the current market situation. Given this market situation, the benefits from ESS employed for frequency regulation service can only be defined in terms of KEPCO's financial benefits. More specifically, it can be expected that KEPCO can reduce energy payments made to generators, incurred as a result of managing FR-related operations.

2.3. Frequency Regulation and Constraint Cost

One important issue regarding regulation is that in the course of its operation a substantial amount of constraint costs are incurred by the grid operator. Constraint cost is the most important concept in comprehending the economic validity of applying ESS for FR service. Constraint costs arise to the extent that there are differences between the market schedule of generation and the actual dispatch by the grid operator.

Constraint costs are related with the following two types of generators that appear during the operation of frequency regulation. First, there are constrained-off generators that were scheduled to run by the market (that is, considered in determining the system market price) but which did not run in the actual dispatch or ran at a decreased level (constrained off/down). Second, there are constrained-on generators that were not scheduled to run or ran at a low level in the market, but which ran at a higher level in reality (constrained on/up). This situation happens because regulation is contracted and paid for during scheduling, but in actual operation the contracted amount was not called on to provide electricity.

In this case, in order to balance supply and demand, a generator that is constrained off/down will always result in other generators being constrained on/up and *vice versa*. That is, the generators that are constrained off/down have to pay back a constraint payment (negative) and the corresponding units that are constrained on/up receive a constraint payment (positive). As the price of the constrained on/up unit is generally greater than the constrained off/down unit, there is always a net cost associated with constraints. ESS reduces the constrained off/down capacity by replacing them. This in turns reduces the capacity of the corresponding constrained on/up generation, which results in a cost savings for the grid operators.

2.4. Economic Valuation Literature for Energy Storage System

Energy Storage Systems (ESS) are applicable in a variety of areas, as described in [6]. Among those, we focus on frequency regulation in this paper, particularly from the utility's perspective. Unlike other studies, this paper considers the viewpoint of a utility that has to pay constraint costs (constrained on/off) to a power exchange market as compensation for controlling frequency regulation. Moreover, our simulation and proposed algorithm reflects on a situation in which an introduction of ESS would change the operation and schedule of existing generators and accordingly lead to changing the payments for constraint costs as well as payments for scheduled energy.

Among the literature discussing the economic values of ESS, Alt *et al.* [7] presented a method for determining the benefits of dynamic operating cost for energy storage systems from the utility company perspective using DYNASTORE. This research showed that the operating benefits for load frequency control are fewer than for spinning reserve and load leveling. However, in this paper they drew a conclusion regarding the benefits of ESS only by comparing the production costs of energy generated by unit generators and ESS. Oudalov *et al.* [4] and Mercier *et al.* [8] suggested an optimal way of dimensioning ESS and evaluated the profitability over 20 years. They maximized the net present value of profits for frequency control reserve and excess energy sale. The main objectives of optimization for a potential ESS owner are participating in an ancillary service market and making profits from providing the service. Moreover, further studies such as Walawalkar *et al.* [5] and Bradbury *et al.* [1] focused on the benefits by participating in arbitrage price and regulation depending on energy prices and market clearing prices. Economic valuation of an energy storage system in an ancillary service market has also been studied using real data from Chile and Israel, questioning the optimal size of an energy storage system [9,10]. However, these studies focus on either the profits earned from participating in an ancillary service market or the benefits incurred through the difference in production costs between an ESS and a generator. On the other hand, our paper concentrates on the payment change, such as scheduled energy payments and constraint cost payments, in accordance with the change of operations schedule after ESSs are utilized.

Moreover, studies regarding utility side management for ESS can be found [11]. However, most papers concern different application area such as load-leveling application, vehicle-to-grid application, optimal site, or the optimal size of an ESS. Jung *et al.* [12] and Lo and Anderson [13] proposed an algorithm of economic dispatch for load-leveling and finding an optimal ESS capacity from the utility's perspective. Even though some papers have discussed the impact of vehicle-to-grid application on a utility [14,15], frequency regulation from a utility perspective is not much studied, whereas Ouldaov *et al.* [16] discussed the frequency regulation issue from the utility's perspective and derived benefits that were calculated simply through control prices and supplied power. The compensation for regulation could be made either by the market or through mandatory obligation, which has to be paid by a utility. However, in this paper, the fact that the operation schedule and payment paid by a utility can be changed by an ESS is not reflected in the model.

Therefore, in this study we consider the economic value of ESS in frequency regulation from the perspective of utility that is in a position to make payments for constraint cost as well as for scheduled energy, and also propose an algorithm to estimate the economic benefits using Korean market data.

3. Methodology

3.1. Valuation Principles

In this paper, we assume that the whole economic value of ESS can be translated into decreased energy payments by a monopsony buyer in the wholesale electricity market. The wholesale electricity market in Korea is a monopsony market in which a single buyer is responsible for buying the total electricity produced and ancillary services are also totally controlled by the monopsony buyer. Given this market institution, it can be reasonably assumed that ESS deployment for FR will result in reduction in a buyer's energy payment because ESS can reduce the amount of constrained-on generation that is usually dispatched to generators with high variable cost.

Specifically, the estimation of the economic value of ESS is conducted by measuring how much energy payments would change after an ESS is deployed. To this end we employ historical data for a benchmark scenario, which will serve as a reference for comparison. The historical data is a good representation of a current practice in which no ESS capacity is employed and the whole FR capacity is taken on by generators consuming fossil fuel. Therefore, our focus of economic valuation is on estimating how a current practice would change due to the deployment of ESS capacity and how much energy payments would be reduced due to such changes in practice.

The following variables are used to define energy payment. The notation for each variable is provided in parentheses. The variables represent the key outcomes of production scheduling, which affects the amount of energy payment by the wholesaler. All variables have an hourly record since the schedule is generated every hour.

- Variable Cost (*vc*)—Variable cost is the cost of generating electricity incurred by a generator. Variable cost of a generator is mostly dependent on the price of fuel it consumes. In Korea, variable costs cannot be determined by generators but need to be evaluated by a central committee, the Generation Cost Evaluation Committee (GCEC). The committee finally determines variable costs based on the evaluation of various elements such as energy prices, fixed costs, efficiency, and/or the generator's other technical features.
- System Marginal Price (*smp*)—*smp* is a market price of electricity determined after a dispatch schedule is established. As the name indicates, *smp* defines the system-wide cost incurred to produce a unit of electricity additional to the amount of electricity that has been generated so far. *smp* is very important in that it determines the wholesale price, which affects the energy payment given to generators involved in the production schedule.
- Price-Setting Scheduled Energy (*Q_pse*)—*Q_pse* is the scheduled amount of generation by each generator. It is determined based on offers by generators in the day-ahead electricity market. It should to be noted that *Q_pse* determines *smp* but is not equal to the actual amount of generation due to reserved capacity.
- Constrained-off capacity (*Q_coff* = Q_gfr + Q_act)—*Q_coff* represents constrained-off capacity reserved for FR service. *Q_coff* is allocated to each generator by reserving a part of the *Q_pse* it offered. *Q_coff* is the sum of Q_gfr and Q_act, which represent the two types of regulation operation: governor free and automatic generation control. If *Q_coff* is fully dispatched in real time, the actual generation of each generator becomes *Q_pse-Q_coff*.
- Constrained-on capacity (*Q_con*)—*Q_con* is the amount of constrained-on generation requested for making up the loss of scheduled supply due to *Q_coff*. Generators with available capacity are dispatched in real time. It is worth noting that *Q_con* here considers constrained-on generation only for compensating the loss due to FR service, meaning that it does not account for any other constrained-on generations that might occur in real-time situations such as forecast errors or failures in distribution/transmission.

Notice that, among the variables introduced, the only variables affected by the introduction of ESS capacity are *Q_coff* and *Q_con*. For *vc*, we can assume that it remains fixed because *vc* is mostly associated with generation efficiency and fuel costs, both of which are not related with ESS. *Q_pse* can also be assumed to remain fixed because ESS only provides capacity for FR service and is not involved in providing electricity supply to meet energy demand. Since *Q_pse* remains unchanged, the amount of scheduled supply also remains unchanged, which in turn does not change *smp*. Consequently, the availability of ESS capacity only changes *Q_coff* and *Q_con* and the payment savings are solely determined by how these two variables would change.

3.2. Payment Equations

Now we define payment equations by using the variables introduced so far. Among the various payment items made by the wholesaler, we consider the following three items that are most affected by the deployment of ESS capacity: Scheduled Energy Payment (P_sep), constrained-off payment (P_coff), and constrained-on payment (P_con). P_sep is the most basic payment item given to generators for electricity generation. Since it is payment for actual generation, the reserved capacity for FR service is not considered. Thus, P_sep is calculated as follows (subscripts i and t represent the individual generator and time (in hours), respectively):

$$P\_sep(i, t) = [Q\_pse(i,t) - Q\_coff(i,t)] \times mp(i,t) \tag{1}$$

Notice that mp (abbreviation of marginal price), instead of smp, is employed as a unit price for payment settlement; mp is an adjusted marginal price defined for individual generators. This adjustment is intended to reflect variations in variable costs among generators, preventing low cost generators (nuclear or coal) from earning relatively excessive profits compared with other generators.

Another important payment item regarding FR service is constrained payment, either constrained-off payment or constrained-on payment. Constrained-off payment can be considered as reimbursement of the opportunity cost of providing FR services; it is made to FR service providers to compensate for their lost profits due to the capacity provided for FR service. P_coff is calculated as follows:

$$P\_coff(i,t) = Q\_coff(i,t) \times [mp(i,t) - vc(i,t)] \tag{2}$$

On the other hand, constrained-on payment is incurred because of make-up generation for the capacity not dispatched due to reservation for FR generation. Constrained-on generation is usually dispatched to high-cost generators, because most low-cost generators' capacity was already brought into the production schedule. In theory, the total amount of constrained-on generation should be equal to the total reserved capacity; that is, ΣQ_coff should be equal to ΣQ_con unless reserve capacity is limited. However, in practice, there are situations in which additional generation is not possible because of insufficient reserve capacity, due to high peak demand. Constrained-on payment (P_con) is calculated as follows:

$$P\_con(i,t) = Q\_con(i,t) \times \max[smp, vc(i,t)] \tag{3}$$

Notice that the unit price for P_con should be larger than smp and vc. Because constrained-on generation is dispatched to generators whose capacity (a part or the whole) was not brought into the production schedule, they could have higher variable cost than smp. Therefore, if constrained-on generation should be dispatched to such generators, the unit price should be their variable costs instead of smp.

3.3. Q_coff and Q_con

As we have examined in the previous section, estimating the changes in Q_coff and Q_con for each generator is a crucial part of the economic valuation of ESS for FR. We use Q_coff' and Q_con' to denote the altered Q_coff and Q_con after a certain amount of ESS capacity is deployed. In order to calculate Q_coff' and Q_con', we need a rule to determine which generators become free of constrained-off/on generation when the total FR capacity charged on existing generators is reduced due to the ESS capacity deployed.

The *Q_coff'* or *Q_con'* of a certain generator could remain unchanged or become smaller depending on whether it is chosen for replacement by ESS capacity or not. The energy payment will also differ depending on the manner in which we select generators. That is, a selection rule plays a crucial role in estimating payment reduction.

Let us begin with a rule for determination of *Q_coff'*. If we reduce the *Q_coff* of a certain generator, its *P_sep* would increase but its *P_coff* would decrease. It is clear that *P_coff* reduces when *Q_coff* is reduced. The reason why *P_sep* increases is that if *Q_coff* is reduced the reduced amount will be converted into generation, which increases *Q_pse* and *P_sep*.

As a result, the following facts are obvious. If we reduce the *Q_coff* of high-cost generators first, we can expect that the decrease in *P_coff* would not be significant (because *vc* and *mp* are close), but the increase in *P_sep* would be significant due to their high *mp*. On the other hand, if we first reduce the *Q_coff* of low-cost generators, then the expected increase in *P_sep* would be much smaller but the decrease in *P_coff* must be larger compared with the case of high-cost generators. In consequence, generators with a larger gap between *mp* and *vc* and a smaller *mp* should be considered first in order to maximize payment reduction. That is, it is more advantageous if reduction in *Q_coff* starts from low-cost generators. Figure 2 describes the procedure in which *Q_coff'* is estimated.

For a better understanding of the procedure in Figure 2, consider the following simple example.

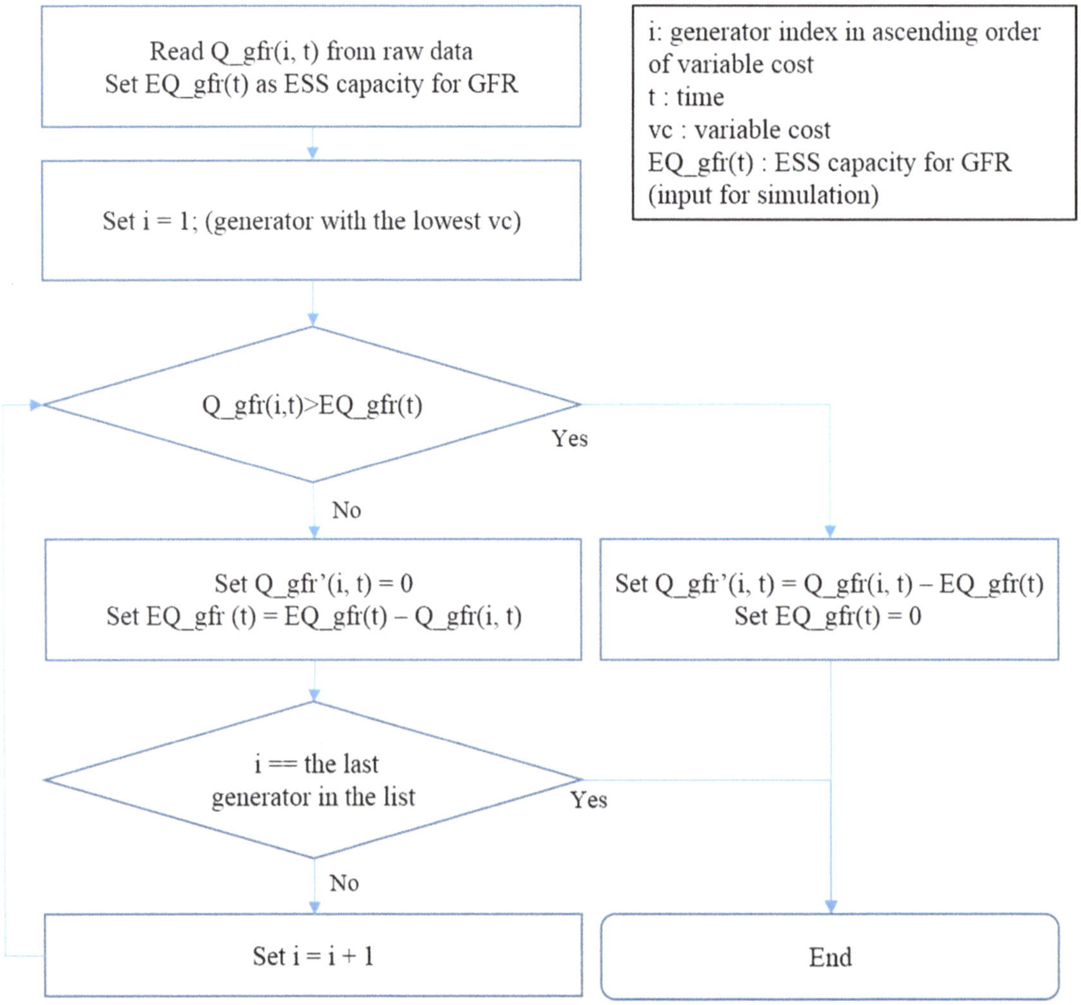

Figure 2. Calculating *Q_coff'*.

Let us assume that in this imaginary scenario *smp* is 128 \$/MWh and the ESS capacity for governor free operation is 500 MW. Table 1 lists *Q_coff* values and average variable costs for each generator type at time *t*. According to the procedure, *Q_coff* values are reduced starting from the low-cost generators as follows. At first, the coal generators' *Q_coff* is 213.8 MW; this is smaller than 500 MW, the unallocated ESS capacity. Hence the *Q_coff'* for Generator 2 now becomes 0 and the unallocated ESS capacity becomes 286.2 MW, which is the reduction from 500 to 213.8. Likewise, the *Q_coff* for Generator 3 is 13.5 MW, which is lower than the current unallocated ESS capacity of 286.2 MW. Hence *Q_coff'* becomes 0 for Generator 3 and the remaining unallocated ESS capacity becomes 272.7 MW, which represents a reduction from 286.2 to 13.5. In the case of Generator 4, *Q_coff* is 492.2 MW, which is larger than the currently remaining ESS capacity (272.7 MW). Therefore, the whole remaining capacity is used to reduce *Q_coff* for Generator 4, and thus its *Q_coff'* becomes 219.5 MW (= 492.2 − 272.7).

Table 1. Example of calculating *Q_coff* and *Q_coff'*.

Generator Index	Variable Cost (\$/MWh)	Q_ESS = 0 (Benchmark) Q_gfr (MW)	Q_ESS = 500 MW Q_gfr' (MW)	ΔQ_gfr (MW)
1	3.5	0	0	0
2	35.0	213.8	0	−213.8
3	48.0	13.5	0	−13.5
4	100.0	492.2	219.5	−272.7
5	126.6	9.9	9.9	0
6	162.3	132.0	132.0	0
		Total		−500

Meanwhile, estimating *Q_con'* is a little more complicated. It is aforementioned that in our scenario we confined *Q_con* to represent make-up generation corresponding to the reserved capacity for FR. However, in practice we cannot quantify *Q_con* in this way because FR is not the only reason that constrained-on generation is required. This means that there is no *Q_con* in the benchmark (*i.e.*, historical data), so before we estimate *Q_con'* we first need to estimate *Q_con* for each generator.

The estimation of *Q_con* is basically similar to the procedure for estimating *Q_coff* except that each generator's remaining capacity (This is the capacity they bid but was not finally brought into the generation schedule. This remaining capacity is the upper limit of *Q_con* for each generator.) is needed to determine the dispatch amount. The way to estimate *Q_con* is described in Figure 3. The available capacity of generators can be calculated by subtracting *Q_pse* from the capacity they offered in the day-ahead market. The procedure starts with Σ*Q_coff*, the total reserved capacity for FR, and assigns *Q_con* to generators in an ascending order of variable costs. Undoubtedly, it is more advantageous for low-cost generators to be dispatched first in this case.

Table 2 shows an example of allocating *Q_con* based on the procedure. Let us assume that *smp* is 128 \$/MWh and ΣQ_cof is 500 MW. Generators are listed in an ascending order of variable costs. Since *smp* is 128 \$/MWh, we start the procedure from Generator 3 with the lowest variable cost above *smp*. From Generator 3 to Generator 6, the whole remaining capacity is dispatched, and for Generator 7 48 MW is dispatched, which adds up to 500 MW of Σ*Q_con*.

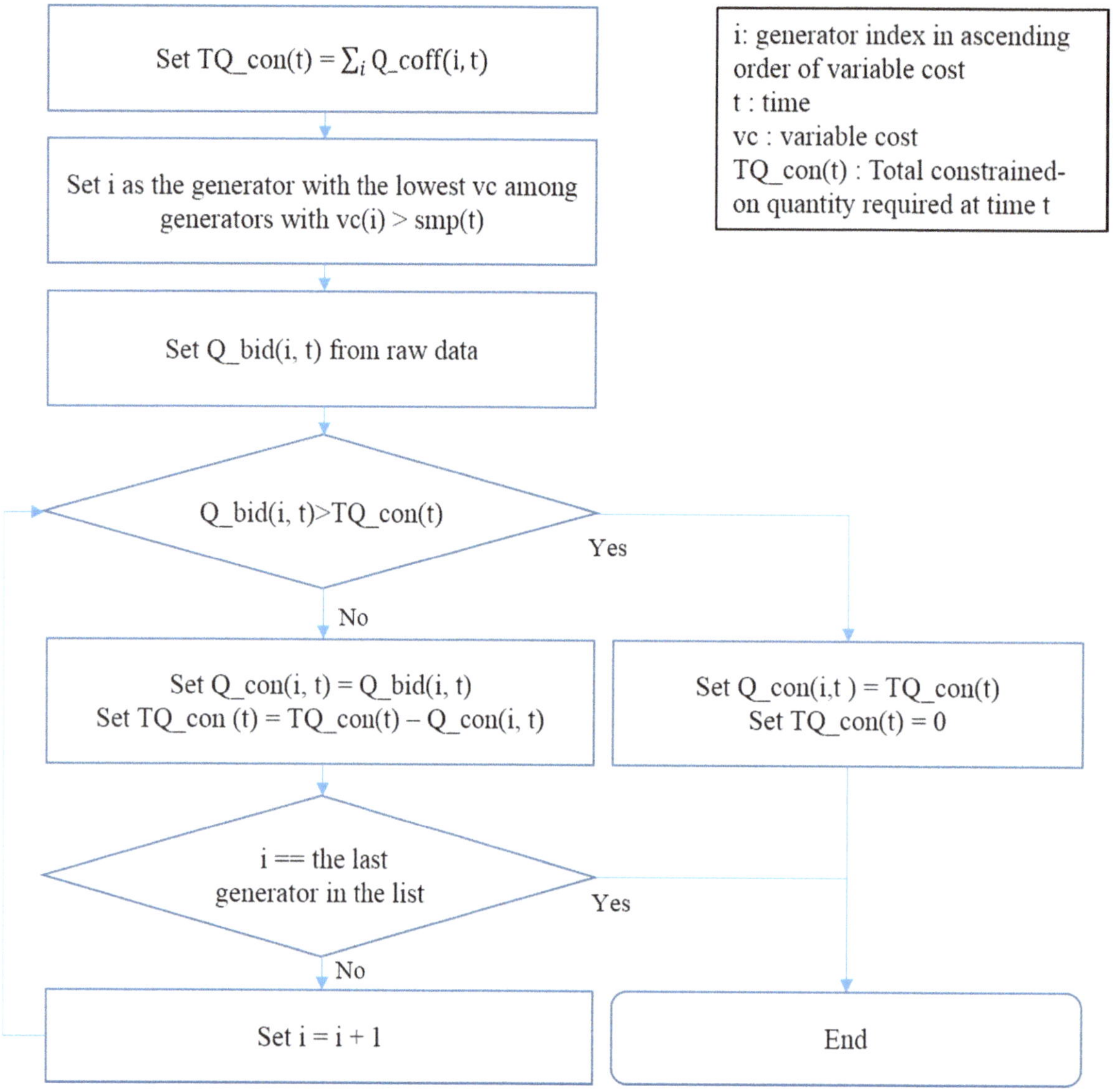

Figure 3. Calculating *Q_con*(i)'.

Table 2. Example of calculating *Q_con* and *Q_con*'.

Generator Index	Capacity (MW)	Q_bid (MW)	Variable Cost (A) ($/MWh)	Q_ESS = 0 (Benchmark)		Q_ESS = 500 MW	
				Q_con (B) (MW)	*P_con* (B × A) ($)	*Q_con*' (C) (MW)	*P_con*' (C × B) ($)
23	523.00	458.36	90.44	458.36	41,452.85	458.36	41,452.85
24	267.00	234.00	92.50	234.00	21,645.64	205.99	19,054.81
25	267.00	234.00	92.56	234.00	21,658.21		
26	267.00	234.00	93.58	234.00	21,897.04		
27	543.00	475.89	95.86	3.99	382.32		
28	708.00	620.50	96.14				
29	377.00	330.41	97.14				
Total				1,164.36	107,036.06	664.36	60,507.66

As mentioned before, *Q_con* and *Q_con*' must be determined through the same procedure described above. The only difference is that calculation of *Q_con* and *Q_con*' begins with Σ*Q_coff* and Σ*Q_coff*', respectively.

3.4. Benefit Equation

So far we have defined three payment items and estimated *Q_coff'* and *Q_con'*. Now we estimate how each payment item changes as ESS provides capacity for FR service. By using the payment equations, we can now derive a benefit equation. A benefit equation represents the estimated payment reduction after ESS capacity is introduced. Given that *vc*, *Q_pse*, and *smp* remain fixed, the amount of reduced payment for each generator can be defined as follows based on Equations (1)–(3):

$$\begin{aligned} &\Delta P\_SEP + \Delta P\_COFF + \Delta P\_CON \\ &= -\Delta Q\_coff \times \text{mp} + \Delta Q\_coff \times (\text{mp} - vc) + \Delta Q\_con \times \max(smp, vc) \\ &= \times\Delta Q\_coff \times vc + Q\_con \times \max(smp, vc) \end{aligned} \tag{4}$$

Notice that the equation represents payment reduction so profit occurs when the value of the equation becomes negative. The equation clearly shows that the amount of payment reduction depends on *smp*, variable costs, and the Δ*Q_coff* and Δ*Q_con* of each generator. Since low-cost generators are mostly constrained-off, their Δ*Q_coff* is negative but the Δ*Q_con* is almost zero. In contrast, high-cost generators are mostly constrained-on so Δ*Q_coff* becomes nearly zero and Δ*Q_con* becomes negative. Therefore, the equation shows that while there is an increase in payment for low-cost generators, payment to most high-cost generators decreases. Given that the aggregated quantities of Δ*Q_coff* and Δ*Q_con* are the same, (Of course, there are situations that this does not hold. For example, unexpected demand surges or high *SMP* (*i.e.*, insufficient capacity for constrained-on dispatch) sometimes prevent *Q_con* from matching *Q_coff*.) the decrease in payments to high-cost generators dominates the increase in payments to low-cost generators. This implies that the difference in variable costs among generators is a key factor in determining the size of benefits.

4. Results

4.1. Estimated Benefits

We implemented a simulation tool that incorporates the equations and procedures we developed by using Microsoft Excel and Visual Basic macro. The simulator has input for ESS capacity and benchmark data of *Q_pse*, *Q_coff*, *smp*, and so forth. With these input data, the simulator performs calculations for *Q_coff'*, *Q_con*, and *Q_con'* following the aforementioned procedure and then generates hourly estimates of payment reduction based on the payment equations. Figure 4 shows a portion of the results produced by the simulator.

Because benchmark data is real historical data, it can be used to test the validity of the payment equations built into the simulator. This can be done by comparing the actual payment records and the results obtained by the simulator. Figure 5 presents the yearly records of each payment item from 2008 to 2012 with the estimated payment items calculated from Equations (1)–(3). In the case of *P_sep*, both payment amounts turned out to be very close, indicating that the estimation on payment savings due to the changes in *Q_coff* (note that *P_sep* depends on *Q_coff*) can also be considered reliable as long as those changes are realistic. For constrained payments, the actual payment records turned out to be higher than the estimated results, which is natural since we exclude constrained generation due to other reasons than FR service. However, the ratio of the estimated constraint payments due to FR turns out to be about

40% of the real total constrained payments, and that ratio is fairly consistent throughout the period. In reality, it was reported that the amount of constrained generation and the FR capacity did remain constant and their ratio was about 40%. This suggests that our estimation of *P_coff* and *P_con* is likely to accurately reflect the actual payments.

Date		Q_ESS = 0			Q_ESS = 500			Payment reduction		
date	hour	P_sep	P_coff	P_con	P_sep'	P_coff'	P_con'	ΔP_sep	ΔP_coff	ΔP_con
20080101	1	1,837,419	5,065	83,849	1,858,554	1,842	44,358	-21,135	3,222	39,491
20080101	2	1,682,175	5,417	107,036	1,708,726	2,004	60,508	-26,550	3,413	46,528
20080101	3	1,619,683	6,013	116,315	1,643,845	2,052	69,371	-24,162	3,962	46,944
20080101	4	1,528,757	5,925	115,571	1,552,452	1,874	68,653	-23,695	4,050	46,918
20080101	5	1,513,396	6,449	117,737	1,535,199	2,052	70,745	-21,803	4,397	46,992
20080101	6	1,562,626	6,834	120,658	1,582,601	2,065	72,692	-19,975	4,770	47,966
20080101	7	1,516,055	6,759	115,547	1,535,213	1,874	68,630	-19,157	4,886	46,917
20080101	8	1,327,180	8,323	116,859	1,344,687	3,101	68,904	-17,508	5,223	47,955
20080101	9	831,097	4,557	76,212	845,965	1,974	37,932	-14,868	2,583	38,280
20080101	10	592,177	307	34,579	604,634	135	16,774	-12,457	172	17,806
20080101	11	787,923	3,339	66,069	802,355	1,192	34,554	-14,432	2,147	31,515
20080101	12	840,849	4,127	71,880	855,792	1,470	34,225	-14,942	2,657	37,655
20080101	13	786,813	3,355	68,293	801,252	1,201	36,194	-14,439	2,154	32,099
20080101	14	826,382	4,076	76,570	841,248	1,495	38,224	-14,866	2,581	38,345
20080101	15	681,529	1,502	63,897	694,781	535	34,784	-13,252	967	29,112

Figure 4. Simulation outputs (unit: $).

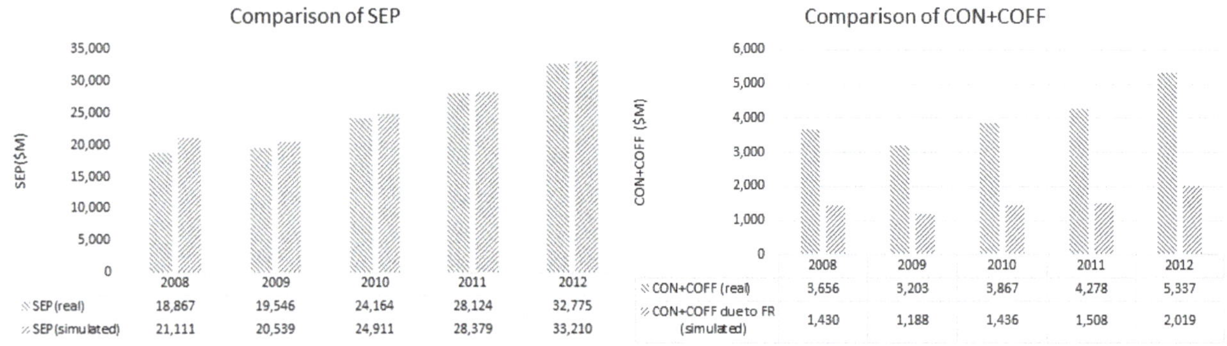

Figure 5. Comparison of simulation results with real data.

Now we consider the introduction of ESS capacity and estimate the amount of payment savings generated by using the simulator. We first assume an ESS with 500MW capacity was employed in 2008 and used for the next five years. Figure 6 presents the estimated payment savings obtained from this assumption. As expected, despite the increase of *P_sep*, the total payment turned out to be reduced thanks to more significant declines in *P_coff* and *P_con* payments. Specifically, it was estimated that the energy payment savings in 2008 were about $386.7 million and went up to about $499.4 million in 2012. This figure amounts to about 11.7% of the total real energy payments by KEPCO in 2012. According to a 2012 report by KPX, the total energy payment (including capacity payment and ancillary service fees) amounts to about $42.5 billion. In other words, it is expected that KEPCO can cut about 12% of its annual operating costs through ESS investment.

An interesting question worth asking here is where these savings come from. Since the wholesaler is a monopsony buyer in the Korean electricity market, the wholesaler's payment savings only come from declined revenues of generators. Since our simulator records estimated payment for each generator, we can compare the amount of revenue declines depending on the generator type. As shown in Figure 7, it turns out that whereas coal generators earn more revenue due to the increase in *P_sep*, liquefied natural gas (LNG) and oil generators lose significant revenue due to the marked decline in constrained-on generation. In other words, most of a wholesaler's benefits come from the revenue decline of high-priced generators like LNG or oil generators. This finding suggests that ESS provides base-unit generators like coal generators with an opportunity to make more revenue. In addition, the reduction of constrained-off generation enables them to utilize their full capacity, which also gives them a chance to save costs by enhancing generation efficiency. Consequently, the results show that ESS increases the utilization of the base-unit generators and decreases the reliance on generators consuming a high-cost energy source like LNG or oil.

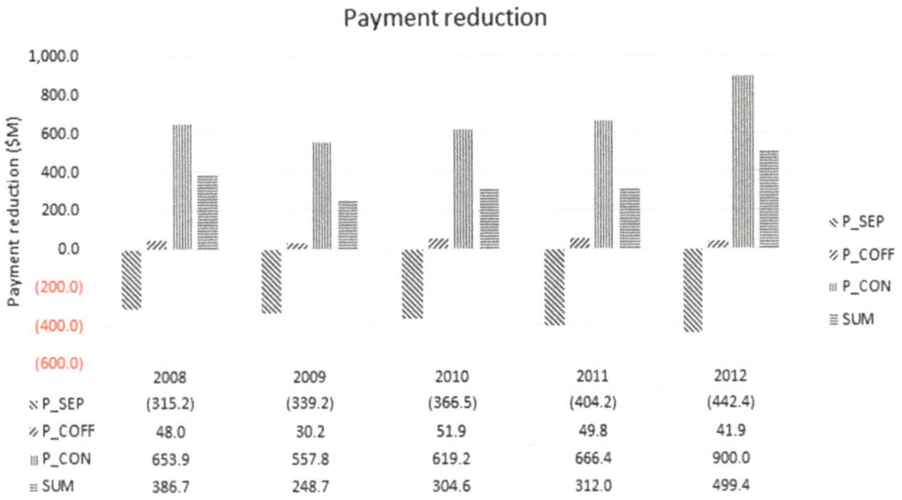

Figure 6. Estimated payment reduction (ESS capacity = 500 MW).

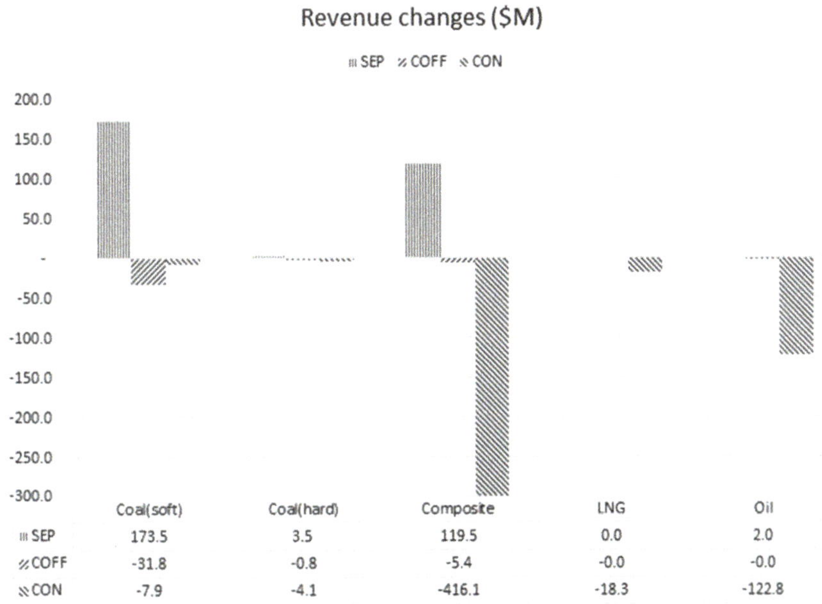

Figure 7. Estimated revenue change by fuel type.

Another important point is that the estimated benefit is highly dependent on *smp*. Since *smp* is mostly affected by the high-priced generators, we can say that the estimated benefit also relies on the variable costs of the high-priced generators. Figure 8 plots the monthly estimated payment savings with variable costs and *smp*. The figure clearly shows that the estimated payment savings are highly correlated with the energy price of LNG generators, which are responsible for determining *smp* in most cases. This suggests that the economic value of ESS gets larger as energy prices go up.

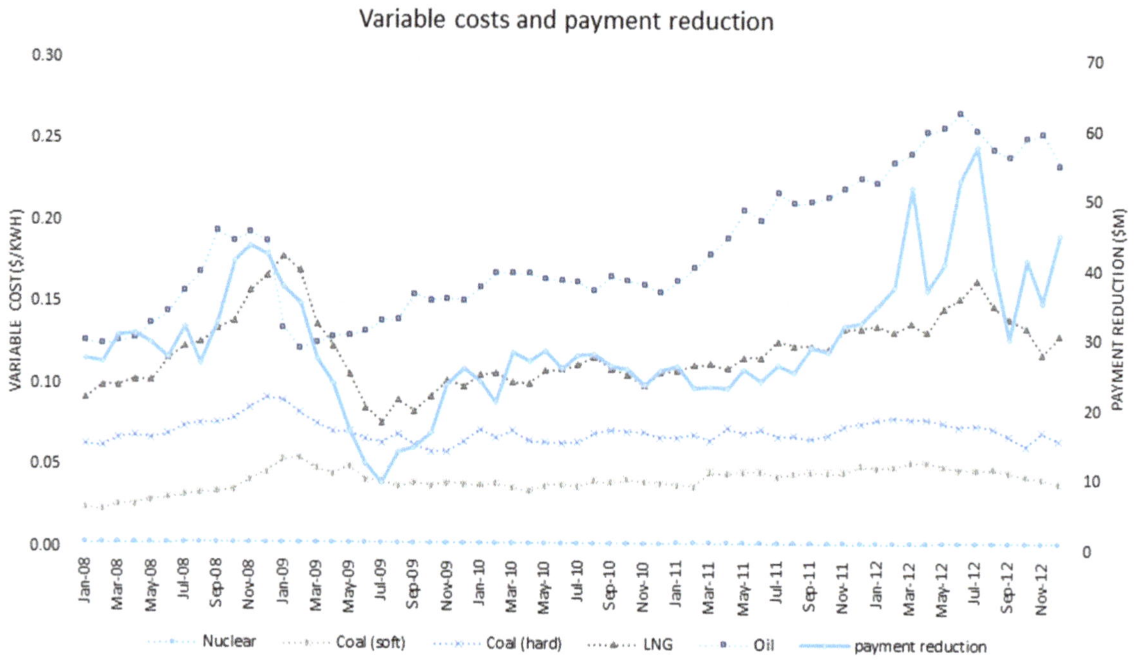

Figure 8. Estimated payment reduction and variable costs.

4.2. Estimated Benefit–Cost Ratio (BCR)

Based on these outcomes, we tried to evaluate the economic value of ESS investment from the perspective of a wholesaler in the monopsony wholesale market. To this end, we consider an investment scenario as shown in Table 3. For the first five years of the period, we considered the same data for simulation as the benchmark and during the next five years we assumed that the average data of the previous five years would repeat every year.

Given this scenario, we calculated its benefit–cost ratio (BCR) by estimating the net present values of payment savings and the ESS costs ($NPV_{Benefit}$ and NPV_{Cost}) as follows:

$$BCR = \frac{NPV_{Benefit}}{NPV_{Cost}} = \sum_{i=0}^{n} \frac{Benefit_i}{(1+r)^i} \Big/ \sum_{i=0}^{n} \frac{Cost_i}{(1+r)^i}, \tag{5}$$

where $Benefit_i$ is the benefit for the year i, $Cost_i$ is the cost for the year i, and r is a discount rate.

As for the costs of operating ESS, we referred to [17], which presents the installation costs and operational expenses of various commercial ESSs available in the market. In this handbook, various storage system costs and technical performances are provided based on surveys from multiple vendors. Among many different types of batteries such as lead-acid, sodium-nickel-chloride and CAES, we used the data on Li-ion batteries for frequency regulation support, and then calculated the annual costs. Specifically, according to their performance data, we assumed that the efficiency of ESS is 80%. Also, to estimate the operating expenses of the Li-ion battery, we utilized the information on the fixed operations and maintenance costs as well as the variable costs given in the literature [17]. Finally, we obtained the annual costs, as given in Table 3.

Based on our estimation, an investment in ESS of 500 MW provides a BCR of 2.56. Given that we did not take into account other benefits worth considering, such as energy import, CO_2 reduction, generation efficiency, and so forth, the results suggest that the investment in ESS for FR deserves our fullest consideration.

Table 3. Benefit–cost ratio of ESS investment (discount rate: 6.5%/year, 500 MW ESS, unit: $M).

Year	2008	2009	2010	2011	2012	2013	2014	2015	2016	2017	NPV
Benefit	349.6	233.4	282.3	292.5	465.6	465.6	465.6	465.6	465.6	465.6	2747.4
Cost	721.9	62.9	62.9	62.9	62.9	62.9	62.9	62.9	62.9	62.9	1071.2
BCR											2.56

5. Conclusions

Although it is well known that ESS has attractive technological features for regulation application, its economic value is relatively little known. In addition, current practice, in which thermal/base-load generators are mostly in charge of regulation service, is economically inefficient due to high constraint costs and under-utilization of low-cost generators. Thus, in this paper, we consider the frequency regulation application of ESS from a utility's perspective and investigate the economic value by providing a new simulation algorithm. Unlike the existing literature, our valuation logic includes the fact that the schedule of base-load generators would be changed when a utility operates an ESS, not to mention the change in constraint costs. Our results clearly show that an ESS provides an economically viable solution to this problem. Our method estimates that with a 500 MW ESS capacity about 11.7% of energy payments can be saved each year. This savings can in turn lead to a lower retail electricity rate. The savings in energy payment implies that a power system can be much less dependent on high-cost/pollution-causing generators.

Although we did not consider the social benefits of ESS application, a lesser dependence on high-cost generators can result in less reliance on fossil fuels, which are mostly imported from overseas. This means that ESS is also beneficial in terms of government finance. This point can be further explored by including other social benefits that we did not consider in this paper: CO_2 emissions reduction, facility upgrade deferral, or other economic impacts.

This paper also presents a challenge for policy change. According to our analysis results, deployment of ESS could raise a conflict of interest between generators and the wholesaler. In addition, most generators do not have a motive to invest in ESS due to the low rate of service fees. Sophisticated government policy measures are necessary to deal with such a subtle relationship of gain and loss among stakeholders.

As suggested by the results, if the wholesaler appropriates the excessive benefits from its market position, the benefit must be shared among other shareholders in some ways. For instance, the government can subsidize the investment of ESS and part of the benefits can be transferred to create a lower retail rate or to increase power capacity. As of 2014, the economic validity of ESS investment is under evaluation by KEPCO and the Korean government and it is expected that there will be some policy propositions regarding an energy policy including the diffusion of ESS.

Author Contributions

Wonchang Hur mainly performed the simulations in collaboration with the other authors, and Yongma Moon completed manuscript. Also, all authors contributed to the editing and reviewing of the paper, and approved the final manuscript.

Conflicts of Interest

The authors declare no conflict of interest.

References

1. Bradbury, K.; Pratson, L.; Patiño-Echeverri, D. Economic viability of energy storage systems based on price arbitrage potential in real-time U.S. electricity markets. *Appl. Energy* **2014**, *114*, 512–519.
2. Rudolf, V.; Papastergiou, K.D. Financial analysis of utility scale photovoltaic plants with battery energy storage. *Energy Policy* **2013**, *63*, 139–146.
3. Sioshansi, R.; Denholm, P.; Jenkin, T.; Weiss, J. Estimating the value of electricity storage in PJM: Arbitrage and some welfare effects. *Energy Econ.* **2009**, *31*, 269–277.
4. Oudalov, A.; Chartouni, D.; Ohler, C. Optimizing a battery energy storage system for primary frequency control. *IEEE Trans. Power Syst.* **2007**, *22*, 1259–1266.
5. Walawalkar, R.; Apt, J.; Mancini, R. Economics of electric energy storage for energy arbitrage and regulation in New York. *Energy Policy* **2007**, *35*, 2558–2568.
6. Eyer, J.; Corey, G. *Energy Storage for the Electricity Grid: Benefits and Market Potential Assessment Guide*; Sandia Report No. SAND2010-0815; Sandia National Laboratories: Albuquerque, NM, USA, 2010.
7. Alt, J.T.; Anderson, M.D.; Jungst, R.G. Assessment of utility side cost savings from battery energy storage. *IEEE Trans. Power Syst.* **1997**, *12*, 1112–1120.
8. Mercier, P.; Cherkaoui, R.; Oudalov, A. Optimizing a Battery Energy Storage System for Frequency Control Application in an Isolated Power System. *IEEE Trans. Power Syst.* **2009**, *24*, 1469–1477.
9. Suazo-Martínez, C.; Pereira-Bonvallet, E.; Palma-Behnke, R. A Simulation Framework for Optimal Energy Storage Sizing. *Energies* **2014**, *7*, 3033–3055.
10. Kottick, D.; Blau, M.; Edelstein, D. Battery energy storage for frequency regulation in an island power system. *IEEE Trans. Energy Convers.* **1993**, *8*, 455–459.
11. Divya, K.C.; Østergaard, J. Battery energy storage technology for power systems—An overview. *Electr. Power Syst. Res.* **2009**, *79*, 511–520.
12. Jung, K.-H.; Kim, H.; Rho, D. Determination of the installation site and optimal capacity of the battery energy storage system for load leveling. *IEEE Trans. Energy Convers.* **1996**, *11*, 162–167.
13. Lo, C.H.; Anderson, M.D. Economic dispatch and optimal sizing of battery energy storage systems in utility load-leveling operations. *IEEE Trans. Energy Convers.* **1999**, *14*, 824–829.
14. Yilmaz, M.; Krein, P.T. Review of the impact of vehicle-to-grid technologies on distribution systems and utility interfaces. *IEEE Trans. Power Electron.* **2013**, *28*, 5673–5689.
15. Kempton, W.; Tomić, J. Vehicle-to-grid power fundamentals: Calculating capacity and net revenue. *J. Power Sources* **2005**, *144*, 268–279.

16. Oudalov, A.; Chartouni, D.; Ohler, C.; Linhofer, G. Value analysis of battery energy storage applications in power systems. In Proceedings of the 2006 IEEE PES Power Systems Conference and Exposition (PSCE'06), Atlanta, GA, USA, 29 October–1 November 2006; pp. 2206–2211.
17. Sandia National Laboratories. *Electricity Storage Handbook in Collaboration with NRECA*; Sandia National Laboratories: Albuquerque, NM, USA, 2013.

Permissions

All chapters in this book were first published in Energies, by MDPI; hereby published with permission under the Creative Commons Attribution License or equivalent. Every chapter published in this book has been scrutinized by our experts. Their significance has been extensively debated. The topics covered herein carry significant findings which will fuel the growth of the discipline. They may even be implemented as practical applications or may be referred to as a beginning point for another development.

The contributors of this book come from diverse backgrounds, making this book a truly international effort. This book will bring forth new frontiers with its revolutionizing research information and detailed analysis of the nascent developments around the world.

We would like to thank all the contributing authors for lending their expertise to make the book truly unique. They have played a crucial role in the development of this book. Without their invaluable contributions this book wouldn't have been possible. They have made vital efforts to compile up to date information on the varied aspects of this subject to make this book a valuable addition to the collection of many professionals and students.

This book was conceptualized with the vision of imparting up-to-date information and advanced data in this field. To ensure the same, a matchless editorial board was set up. Every individual on the board went through rigorous rounds of assessment to prove their worth. After which they invested a large part of their time researching and compiling the most relevant data for our readers.

The editorial board has been involved in producing this book since its inception. They have spent rigorous hours researching and exploring the diverse topics which have resulted in the successful publishing of this book. They have passed on their knowledge of decades through this book. To expedite this challenging task, the publisher supported the team at every step. A small team of assistant editors was also appointed to further simplify the editing procedure and attain best results for the readers.

Apart from the editorial board, the designing team has also invested a significant amount of their time in understanding the subject and creating the most relevant covers. They scrutinized every image to scout for the most suitable representation of the subject and create an appropriate cover for the book.

The publishing team has been an ardent support to the editorial, designing and production team. Their endless efforts to recruit the best for this project, has resulted in the accomplishment of this book. They are a veteran in the field of academics and their pool of knowledge is as vast as their experience in printing. Their expertise and guidance has proved useful at every step. Their uncompromising quality standards have made this book an exceptional effort. Their encouragement from time to time has been an inspiration for everyone.

The publisher and the editorial board hope that this book will prove to be a valuable piece of knowledge for researchers, students, practitioners and scholars across the globe.

List of Contributors

Ana Maria Gracia Amillo, Thomas Huld, Paraskevi Vourlioti and Matthew Norton
European Commission, Joint Research Centre, Via Fermi 2749, Ispra 21027, Italy

Richard Müller
Deutscher Wetterdienst, Frankfurter Straße 135, Offenbach 63067, Germany

Yin Hua and Min Xu
National Engineering Laboratory for the Automotive Electronic Control Technology, Shanghai Jiao Tong University, 800 Dongchuan Rd., Shanghai 200240, China

Mian Li
National Engineering Laboratory for the Automotive Electronic Control Technology, Shanghai Jiao Tong University, 800 Dongchuan Rd., Shanghai 200240, China
University of Michigan-Shanghai Jiao Tong University Joint Institute, Shanghai Jiao Tong University, 800 Dongchuan Rd., Shanghai 200240, China

Chengbin Ma and Chen Zhao
University of Michigan-Shanghai Jiao Tong University Joint Institute, Shanghai Jiao Tong University, 800 Dongchuan Rd., Shanghai 200240, China

Marcin Dutka and Terese Løvås
Department of Energy and Process Engineering, Norwegian University of Science and Technology, Kolbjørn Hejes vei 1b, Trondheim 7491, Norway

Mario Ditaranto
SINTEF Energy Research, Sem Sælands vei 11, Trondheim 7034, Norway

Chun-Ching Hsiao and Jia-Wai Jhang
Department of Mechanical Design Engineering, National Formosa University, No. 64, Wunhua Rd., Huwei Township, Yunlin County 632, Taiwan

Jordi Parra, Alfredo Guardo and Eduard Egusquiza
Center for Industrial Diagnostics and Fluid Dynamics, Polytechnic University of Catalonia BARCELONATECH (UPC-CDIF), Av. Diagonal 647, ETSEIB, 08028 Barcelona, Spain

Pere Alavedra
International University of Catalonia, C\Immaculada 22, 08017 Barcelona, Spain

Jia-Shiun Chen
Department of Vehicle Engineering, National Taipei University of Technology, 1, Sec. 3, Zhongxiao E. Road, Taipei 10608, Taiwan

Yaser Khojasteh Salkuyeh and Thomas A. Adams II
Department of Chemical Engineering, McMaster University, 1280 Main St W, Hamilton, ON L8S 4L7, Canada

Vladimir Kuznetsov and Galina Fesenko
International Atomic Energy Agency, Wagramer Strasse 5, P.O. Box 100, Vienna A-1400, Austria

Aleksandra Schwenk-Ferrero
Karlsruher Institut für Technologie (KIT) Institut für Kern- und Energietechnik, Hermann-von-Helmholtz-Platz 1, Eggenstein-Leopoldshafen 76344, Germany

Andrei Andrianov and Ilya Kuptsov
National Research Nuclear University MEPhI (Moscow Engineering Physics Institute), Kashirskoe sh. 31, Moscow 115409, Russia

Reynir Smari Atlason, Gudmundur Valur Oddsson and Runar Unnthorsson
Department of Industrial Engineering, Mechanical Engineering and Computer Science, Centre for Productivity, Performance and Processes, University of Iceland, Hjardarhagi 6, 107 Reykjavik, Iceland

Roberto S. Faranda, Hossein Hafezi, Sonia Leva, Marco Mussetta and Emanuele Ogliari
Politecnico Di Milano Department of Energy, Via la Masa 34, 20156 Milano, Italy

Mohamed F. Aly Aboud
Sustainable Energy Technologies Center, College of Engineering, King Saud University, P.O. Box 800, Riyadh 11421, Saudi Arabia
Mining, Metallurgical and Petroleum Engineering Department, Faculty of Engineering, Al-Azhar University, Nasr City, Cairo 11371, Egypt

Zeid A. ALOthman and Mohamed A. Habila
Advanced Materials Research Chair, Chemistry Department, P.O. Box 2455, College of Science, King Saud University, Riyadh 11451, Saudi Arabia

Claudia Zlotea, Michel Latroche and Fermin Cuevas
ICMPE/CNRS-UPEC, UMR 7182, 2-8 rue Henri Dunant, 94320 Thiais, France

Wonchang Hur
College of Business Administration, Inha University, Incheon 431050, Korea

Yongma Moon
College of Business Administration, University of Seoul, Seoul 130743, Korea

Kwangsup Shin
Graduate School of Logistics, Incheon National University, Incheon 406772, Korea

Wooje Kim
College of Business and Technology, Seoul National University of Science and Technology, Seoul 139743, Korea

Suchul Nam and Kijun Park
KEPCO Research Institute, Daejeon 305760, Korea